Methods in Molecular Biology

Series Editor
John M. Walker
School of Life and Medical Sciences
University of Hertfordshire
Hatfield, Hertfordshire, UK

For further volumes:
http://www.springer.com/series/7651

For over 35 years, biological scientists have come to rely on the research protocols and methodologies in the critically acclaimed Methods in Molecular Biology series. The series was the first to introduce the step-by-step protocols approach that has become the standard in all biomedical protocol publishing. Each protocol is provided in readily-reproducible step-by step fashion, opening with an introductory overview, a list of the materials and reagents needed to complete the experiment, and followed by a detailed procedure that is supported with a helpful notes section offering tips and tricks of the trade as well as troubleshooting advice. These hallmark features were introduced by series editor Dr. John Walker and constitute the key ingredient in each and every volume of the Methods in Molecular Biology series. Tested and trusted, comprehensive and reliable, all protocols from the series are indexed in PubMed.

Manuscript Instructions https://drive.google.com/file/d/1K4Az5vpAFVzYadJ3jKx2-QSo7Dqa_SNA/view?usp=sharing

Conservation of Plant Genetic Resources

Edited by

Marco A. Ramírez-Mosqueda

Centro Nacional de Recursos Genéticos-INIFAP, Tepatitlán de Morelos, Jalisco, Mexico

Jorge David Cadena-Zamudio

Centro Nacional de Recursos Genéticos-INIFAP, Tepatitlán de Morelos, Jalisco, Mexico

José Luis Aguirre-Noyola

Centro Nacional de Recursos Genéticos-INIFAP, Tepatitlán de Morelos, Jalisco, Mexico

Humana Press

Editors
Marco A. Ramírez-Mosqueda
Centro Nacional de Recursos Genéticos-INIFAP
Tepatitlán de Morelos, Jalisco, Mexico

Jorge David Cadena-Zamudio
Centro Nacional de Recursos Genéticos-INIFAP
Tepatitlán de Morelos, Jalisco, Mexico

José Luis Aguirre-Noyola
Centro Nacional de Recursos Genéticos-INIFAP
Tepatitlán de Morelos, Jalisco, Mexico

ISSN 1064-3745 ISSN 1940-6029 (electronic)
Methods in Molecular Biology
ISBN 978-1-0716-5129-2 ISBN 978-1-0716-5130-8 (eBook)
https://doi.org/10.1007/978-1-0716-5130-8

This Humana imprint is published by the registered company Springer Science+Business Media, LLC, part of Springer Nature
The registered company address is: 1 New York Plaza, New York, NY 10004, U.S.A.

Preface

Conservation of Plant Genetic Resources

The purpose of this book is to recapitulate techniques, methods, and information relevant to the conservation of plant germplasm. Currently, talking about conservation techniques is a priority issue for the sustainable development of each country. In this context, the protocols for the conservation of plant genetic resources guarantee food security, rescue of endangered species and environmental sustainability. However, each plant species requires different and detailed conservation processes, in order to be established in collections or biobanks. This book contributes to the establishment of biobanks through conservation protocols, in vitro conservation, cryopreservation, and seed conservation. Likewise, it recapitulates the basic concepts and innovative technologies in the conservation of plant genetic resources. The main objective of this book is to present efficient protocols for plant conservation, from seed processing and storage, short-term conservation through plant tissue culture, medium-term conservation through in vitro conservation for minimal long-term growth, and conservation through cryopreservation.

Tepatitlán de Morelos, Mexico — *Marco A. Ramírez-Mosqueda*
Tepatitlán de Morelos, Mexico — *Jorge David Cadena-Zamudio*
Tepatitlán de Morelos, Mexico — *José Luis Aguirre-Noyola*

Contents

Contributors

ALMA ARMENTA-MEDINA • *Centro Nacional de Recursos Genéticos-INIFAP, Tepatitlán de Morelos, Jalisco, Mexico*

ÁNGELES SÁNCHEZ-CONTRERAS • *Centro de investigación y asistencia en tecnología y diseño del estado de Jalisco Subsede Sureste, Mérida, Mexico*

ANTONIA GUTIÉRREZ-MORA • *Unidad de Biotecnología Vegetal, Centro de Investigación y Asistencia en Tecnología y Diseño del Estado de Jalisco, A.C., Guadalajara, Jalisco, Mexico*

ARIADNA LINARES-GABRIEL • *Facultad de Ciencias Agrícolas, Universidad Veracruzana, Xalapa, Veracruz, Mexico*

ARTURO HERNÁNDEZ-GARCÍA • *Laboratório de Cultivo de Tejidos Vegetales, Colegio de Postgraduados Campus Cordoba, Amatlán de los Reyes, Veracruz, Mexico*

BENITO REYES-TREJO • *Natural Products Laboratory, Agricultural Preparatory Department, Autonomous University of Chapingo, Chapingo, Mexico State, Mexico*

BLANCA ELVIRA LÓPEZ-VALENZUELA • *Universidad Autónoma de Sinaloa, Facultad de Agricultura del Valle del Fuerte (UAS-FAVF), Sinaloa, Mexico*

BRENDA I. TREJO-TÉLLEZ • *Postgrado en Innovación en Manejo de Recursos Naturales, Campus San Luis Potosí, Colegio de Postgraduados, Salinas de Hidalgo, San Luis Potosí, Mexico*

CARLO MASCARELLO • *CREA Research Centre for Vegetable and Ornamental Crops, Sanremo, IM, Italy*

CARLOS ALBERTO CRUZ-CRUZ • *Facultad de Ciencias Químicas, Universidad Veracruzana, Orizaba, Veracruz, Mexico*

CARLOS HUGO AVENDAÑO-ARRAZATE • *Centro Nacional de Recursos Genéticos-INIFAP, Tepatitlán de Morelos, Mexico*

CARLOS I. CRUZ-CÁRDENAS • *Centro Nacional de Recursos Genéticos-INIFAP, Tepatitlán de Morelos, Jalisco, Mexico*

CLAUDIA BERENICE ESPITIA-FLORES • *Centro Nacional de Recursos Genéticos-INIFAP, Tepatitlán de Morelos, Jalisco, Mexico*

CYNTHIA PAOLA RANGEL-CHAVEZ • *Instituto Tecnológico Superior de Irapuato (ITESI), Irapuato, Guanajuato, Mexico*

DAFNE V. BACAB-CAAMAL • *Unidad de Biología Integrativa, Centro de Investigación Científica de Yucatán A.C., Mérida, Mexico*

DANIEL AGUILAR-JIMÉNEZ • *Agrobiotechnology Educational Program, Technological University of Izúcar de Matamoros, Izúcar de Matamoros, Puebla, Mexico*

DANIEL ALEJANDRO CADENA-ZAMUDIO • *Grupo Interdisciplinario de Investigación en Sechium edule en México, Texcoco, México*

DANIEL PADILLA-CHACÓN • *Secretaría de Ciencia, Humanidades, Tecnología e Innovación (SECIHTI)-Facultad de Ciencias Químicas, Universidad Veracruzana, Orizaba, Veracruz, Mexico*

DIEGO ALFREDO OLVERA-CASANOVA • *Centro de investigación y asistencia en tecnología y diseño del estado de Jalisco Subsede Sureste, Mérida, Mexico*

EDGAR TORRES-GARCÍA • *Centro Nacional de Recursos Genéticos-INIFAP, Tepatitlán de Morelos, Jalisco, Mexico*

Elizabeta Hernández-Domínguez • *Tecnológico Nacional de México/Instituto Tecnológico Superior de Acayucan, Acayucan, Veracruz, Mexico*
Ernestina Valadez-Moctezuma • *Department of Phytotechnics, Institute of Horticulture, Autonomous University of Chapingo, Chapingo, Mexico State, Mexico*
Esmeralda J. Cruz-Gutiérrez • *Centro Nacional de Recursos Genéticos-INIFAP, Tepatitlán de Morelos, Jalisco, Mexico*
Evelia Guadalupe Mora-González • *Universidad Autónoma Metropolitana, Unidad Iztapalapa, Iztapalapa, Ciudad de México, Mexico*
Fabiola Hernández-Ramírez • *Facultad de Ciencias Químicas, Universidad Veracruzana, Orizaba, Veracruz, Mexico*
Fernanda Díaz-Sánchez • *Postgrado en Innovación en Manejo de Recursos Naturales, Campus San Luis Potosí, Colegio de Postgraduados, Salinas de Hidalgo, San Luis Potosí, Mexico*
Fernando Santacruz-Ruvalcaba • *Departamento de Producción Agrícola, Centro Universitario de Ciencias Biológicas y Agropecuarias, Universidad de Guadalajara, Zapopan, Jalisco, Mexico*
Flor Josefina Ramos-Cruz • *Facultad de Agricultura Sustentable y Protegida, Universidad Tecnológica de Tecamachalco, Estado de México, Mexico*
Francisco F. Calvillo-Aguilar • *Centro Nacional de Recursos Genéticos-INIFAP, Tepatitlán de Morelos, Jalisco, Mexico*
Gabriela Sánchez-Viveros • *Facultad de Ciencias Agrícolas, Universidad Veracruzana, Xalapa, Veracruz, Mexico*
Gabriela Sandoval-Cancino • *Centro Nacional de Recursos Genéticos-INIFAP, Tepatitlán de Morelos, Jalisco, Mexico*
Germán Nic-Matos • *Tecnológico Nacional de México-Campus Tizimín, Tizimín, Mexico*
Gilbert José Herrera-Cool • *Campo Experimental Chetumal-INIFAP, Quintana Roo, México*
Gina Kelly Mendez-Castillo • *Department of Phytotechnics, Genetics Area, Autonomous University of Chapingo, Chapingo, Mexico State, Mexico*
Guadalupe López-Puc • *Centro de investigación y asistencia en tecnología y diseño del estado de Jalisco Subsede Sureste, Mérida, Mexico*
Gustavo Cuaxinque-Flores • *Facultad de Ecología Marina, Universidad Autónoma de Guerrero, Acapulco de Juárez, Guerrero, Mexico*
Humberto Antonio López-Delgado • *Conjunto SEDAGRO-INIFAP, Metepec, Estado de México, Mexico*
Iván Córdova-Lara • *Centro de Investigación Científica de Yucatán, Unidad de Biotecnología, Mérida, Mexico*
Ivonne Naomi Bravo-Ruíz • *Facultad de Ciencias Químicas, Universidad Veracruzana, Orizaba, Veracruz, Mexico*
Javier Mora-Macias • *Centro de Investigación y Asistencia en Tecnología y Diseño del estado de Jalisco A.C. (CIATEJ), Subsede Sureste, Merida, Yucatan, Mexico*
Jorge Cadena-Iñiguez • *Postgrado en Innovación en Manejo de Recursos Naturales, Campus San Luis Potosí, Colegio de Postgraduados, Salinas de Hidalgo, San Luis Potosí, Mexico; Grupo Interdisciplinario de Investigación en Sechium edule en Mexico, Texcoco, Estado de México, Mexico*
Jorge David Cadena-Zamudio • *Grupo Interdisciplinario de Investigación en Sechium edule en Mexico, Texcoco, Estado de México, Mexico; Centro Nacional de Recursos Genéticos-INIFAP, Tepatitlán de Morelos, Jalisco, Mexico*

José Daniel Chávez-González • *Escuela superior de Ciencias de la Tierra, Doctorado en Recursos Naturales y Ecología. Universidad Autónoma de Guerrero, Taxco el Viejo, Mexico*

José Humberto Caamal-Velázquez • *Colegio de Postgraduados campus Campeche, Campeche, Mexico*

José Juvencio Castañeda-Nava • *Unidad de Biotecnología Vegetal, Centro de Investigación y Asistencia en Tecnología y Diseño del Estado de Jalisco, A.C., Guadalajara, Jalisco, Mexico*

José Luis Aguirre-Noyola • *Centro Nacional de Recursos Genéticos-INIFAP, Tepatitlán de Morelos, Jalisco, Mexico*

José Luis Rodríguez-De la O • *Department of Phytotechnics, Genetics Area, Autonomous University of Chapingo, Chapingo, Mexico State, Mexico*

José Manuel Rodríguez-Domínguez • *Unidad de Biotecnología Vegetal, Centro de Investigación y Asistencia en Tecnología y Diseño del Estado de Jalisco, A.C., Guadalajara, Jalisco, Mexico*

José Roberto Bautista-Aguilar • *Facultad de Ciencias Agrícolas, Universidad Veracruzana, Circuito Gonzalo Aguirre Beltrán, Zona Universitaria, Xalapa, Veracruz, Mexico*

Juan M. Pichardo-González • *Centro Nacional de Recursos Genéticos-INIFAP, Tepatitlán de Morelos, Jalisco, Mexico*

Juan Ramos-Garza • *Escuela de Ciencias de la Salud, Campus Coyoacán, Universidad del Valle de México, Coyoacán, Ciudad de México, Mexico*

Julio Armando Massange-Sánchez • *Unidad de Biotecnología Vegetal, Centro de Investigación y Asistencia en Tecnología y Diseño del Estado de Jalisco A.C. (CIATEJ), Guadalajara, Jalisco, Mexico*

Kazuo N. Watanabe • *Gene Research Center, University of Tsukuba, Tsukubu, Japan*

Keila Anel Pool-Chab • *Centro de investigación y asistencia en tecnología y diseño del estado de Jalisco Subsede Sureste, Mérida, Mexico*

Liberato Portillo • *Departamento de Botánica y Zoología, Centro Universitario de Ciencias Biológicas y Agropecuarias, Universidad de Guadalajara, Zapopan, Jalisco, México*

Liliana Muñoz-Gutiérrez • *Centro Nacional de Investigación Disciplinaria en Conservación y Mejoramiento de Ecosistemas Forestales-INIFAP, Coyoacán, Ciudad de México, Mexico*

Lizandro Ramírez-Trejo • *Facultad de Ciencias Químicas, Universidad Veracruzana, Orizaba, Veracruz, Mexico*

Lorena Jacqueline Gómez-Godínez • *Centro Nacional de Recursos Genéticos-INIFAP, Tepatitlán de Morelos, Jalisco, Mexico*

Lourdes Delgado-Aceves • *Centro de Investigación y Asistencia en Tecnología y Diseño del Estado de Jalisco, Unidad de Biotecnología Vegetal, Zapopan, Jalisco, Mexico*

Lourdes Georgina Iglesias-Andreu • *Instituto de Biotecnología y Ecología Aplicada, Universidad Veracruzana, Xalapa, Veracruz, Mexico*

Luis Alberto Gómez-Reyes • *Centro Nacional de Recursos Genéticos-INIFAP, Tepatitlán de Morelos, Jalisco, Mexico*

Luis A. Barrera-Guzmán • *Universidad Autónoma Chapingo, Centro Regional Universitario Oriente, Huatusco, Veracruz, Mexico*

Luis Sáenz-Carbonell • *Centro de Investigación Científica de Yucatán, Unidad de Biotecnología, Mérida, Mexico*

MANUEL DE JESÚS BERMÚDEZ-GUZMÁN • *Campo Experimental Tecomán-INIFAP, Tecomán, Colima, México*
MARCELINA VÉLEZ-TORRES • *Colegio de Postgraduados, Posgrado en Recursos Genéticos y Productividad, Texcoco, Mexico*
MARCO A. RAMÍREZ-MOSQUEDA • *Centro Nacional de Recursos Genéticos-INIFAP, Tepatitlán de Morelos, Jalisco, Mexico*
MARCO SAVONA • *CREA Research Centre for Vegetable and Ornamental Crops, Sanremo, IM, Italy*
MARÍA DE LOURDES DELGADO-ACEVES • *Unidad de Biotecnología Vegetal, Centro de Investigación y Asistencia en Tecnología y Diseño del Estado de Jalisco, A.C., Guadalajara, Jalisco, Mexico*
MARÍA ESTHER DIAZ-MARTÍNEZ • *Facultad de Ciencias Agrícolas, Universidad Veracruzana, Xalapa, Veracruz, Mexico*
MARÍA G. PÉREZ-REYNOZO • *Centro Nacional de Recursos Genéticos-INIFAP, Tepatitlán de Morelos, Jalisco, Mexico*
MARÍA ISABEL IÑIGUEZ-LUNA • *Grupo Interdisciplinario de Investigación en Sechium edule en México, Texcoco, Mexico*
MARÍA TERESA GONZÁLEZ-ARNAO • *Facultad de Ciencias Químicas, Universidad Veracruzana, Orizaba, Veracruz, Mexico*
MARIO ALEJANDRO HERNÁNDEZ-CHONTAL • *Facultad de Ciencias Agrícolas, Universidad Veracruzana, Xalapa, Veracruz, Mexico*
MARTÍN QUINTANA-CAMARGO • *Centro Nacional de Recursos Genéticos-INIFAP, Tepatitlán de Morelos, Jalisco, Mexico*
OBDULIA BALTAZAR-BERNAL • *Laboratório de Cultivo de Tejidos Vegetales, Colegio de Postgraduados Campus Cordoba, Amatlán de los Reyes, Veracruz, Mexico*
PAULO CÉSAR PARADA-MOLINA • *Facultad de Ciencias Agrícolas, Universidad Veracruzana, Xalapa, Veracruz, Mexico*
PEDRO OSORIO-MONTALVO • *Tecnológico Nacional de México-Campus Tizimín, Tizimín, Mexico*
RAMON SOUZA-PERERA • *Laboratorio Regional para el Estudio y Conservación de Germoplasma, Centro de Investigación Científica de Yucatán A.C., Mérida, Mexico*
RICARDO ALVAREZ-ESPINO • *Secihti-Laboratorio Regional para el Estudio y Conservación de Germoplasma, Centro de Investigación Científica de Yucatán A.C., Mérida, Mexico*
ROSALINDA GONZÁLEZ-SANTOS • *Facultad de Ciencias Naturales, Universidad Autónoma de Querétaro, Santiago de Querétaro, México*
RUBÉN H. ANDUEZA-NOH • *SECIHTI-Instituto Tecnológico de Conkal, Conkal, Mexico*
VERÓNICA GARCÍA-IGLESIAS • *Facultad de Ecología Marina, Maestría en Recursos Naturales y Ecología, Acapulco de Juárez, Mexico*
VÍCTOR M. RUIZ-VERA • *Postgrado en Innovación en Manejo de Recursos Naturales, Campus San Luis Potosí, Colegio de Postgraduados, Salinas de Hidalgo, San Luis Potosí, Mexico*
ZELZIN ERENDIRA FERNÁNDEZ-VILLA • *Instituto de Biotecnología y Ecología Aplicada, Universidad Veracruzana, Xalapa, Veracruz, Mexico*

Check for updates

Chapter 1

Overview of Plant Genetic Resources Conservation

Marco A. Ramírez-Mosqueda, Carlos Hugo Avendaño-Arrazate and Rubén H. Andueza-Noh

Abstract

Biodiversity on a global scale comprises natural resources that are often exploited for the benefit of humanity. However, these natural resources are finite, making it necessary to implement preservation strategies. Genetic resources are biological species primarily characterized by their utility in human nutrition. Phytogenetic or plant genetic resources belong to the Plantae kingdom and serve a similar purpose. Each country has public policies aimed at the conservation and sustainable use of these resources, using several strategies, including in situ conservation (preservation within the natural habitat of each plant species) and ex situ conservation (preservation outside the natural habitat of each plant species). Certain phytogenetic resources have their centers of origin and diversification in specific countries or geographic areas, leading conservation efforts to focus on these specific biological forms. In particular cases, the significance of a genetic resource is so pronounced that it is safeguarded in national collections or scientific conservation centers known as biobanks. These centers aim to conserve the largest possible germplasm quantity from wild and cultivated phytogenetic resources. This chapter provides a general overview of plant genetic resource conservation, discussing its significance, strategies, and how these actions are implemented worldwide.

Keywords Genetic resources, Phytogenetic resources, Biobanks, Center of origin, Diversification

1 Introduction

The use of natural resources for human nutrition has been a practice established since the beginning of agriculture and animal domestication [1, 2]. Later, these natural resources came to be known as genetic resources, their importance was recognized, and various strategies for their preservation were developed [3]. Each country determined the priority and methods for conserving the germplasm of its genetic resources, as well as the public policies associated with this goal [4, 5]. However, on a global scale, there is concern about effectively conserving these resources, as many of them are fundamental to the nutrition of the world's population [6, 7].

Marco A. Ramírez-Mosqueda et al. (eds.), *Conservation of Plant Genetic Resources*, Methods in Molecular Biology, vol. 3011,
https://doi.org/10.1007/978-1-0716-5130-8_1,

Plant genetic resources embrace the plants that contribute to the sustenance of humanity [3, 8]. In this regard, staple crops such as maize, rice, wheat, beans, soybeans, and others are prioritized for conservation, as they are critical for the food security of many countries [9, 10]. However, agro-industrial importance crops, which produce many products used in daily life, should not be overlooked [11]. The diversity of plants used to supply food to the global population ensures the large-scale availability of germplasm. However, factors such as overexploitation, monoculture, climate change, and intensive agriculture jeopardize the current diversity of phytogenetic resources, highlighting the urgent need to develop and implement strategies to safeguard and better operate these resources [2, 12].

Moreover, it is essential to conserve wild plant genetic resources to ensure the availability of germplasm for crop improvement programs [3, 13]. Harvesting and preserving wild relatives will enable the development of future crops [14]. The ongoing threat to the preservation of these phytogenetic resources is the continuous germplasm loss, which results in the disappearance of unique and irreplaceable genetic information that serves as the foundation for plant species improvement programs [15, 16]. Recently, this loss has been exacerbated by anthropogenic activities such as land-use changes and habitat fragmentation [17, 18]. Consequently, it is essential to establish conservation strategies for plant genetic resources to prevent potential scenarios of species extinction and risks to global food security [18].

2 Global Biodiversity

The various biological forms that exist worldwide constitute global biodiversity, with the majority found in so-called megadiverse countries [19, 20]. Within this diversity, some biological species have their centers of origin in specific countries or geographic regions. A species' center of origin refers to the region of the world that contains the highest concentration of its germplasm (genetic reservoir) [21, 22]. Additionally, certain regions serve as centers of diversification for genetic resources, referring to the geographic areas where domestication and the development of subspecies, breeds, varieties, strains, or cultivars of these biological forms took place [23, 24].

Biodiversity used for human nutrition is termed genetic resources, and its sustainable use and conservation have become a priority [25, 26]. Safeguarding genetic resources is a scientific, political, and academic activity aimed at ensuring food security for the growing global population in the future [27]. The implementation of strategies that promote the sustainable use and conservation of genetic resources has become an international concern, as many of these resources form the foundation of food systems in various countries [27, 28]. If any of these genetic resources were to be compromised, national food security would be at risk [29].

The use of genetic resources must be sustainable, ensuring their utilization without compromising their availability for future generations [28]. Furthermore, the safeguarding of this germplasm must be guaranteed to ensure its availability for future human generations [18]. Numerous conservation efforts are underway, with one of the most notable being cold storage in the Svalbard Global Seed Vault [30].

3 Genetic Resources

Biological material from various plant, animal, and microorganism species is considered part of genetic resources, provided it is involved in human nutrition [31]. All genetic resources are of utmost importance, and their conservation must be addressed from the perspective of various scientific disciplines.

The germplasm of these biological forms must be safeguarded and preserved to prevent the loss of hereditary material that contains valuable genetic information [32, 33]. This valuable genetic information is primarily found in wild specimens, wild relatives, and common ancestors of current genetic resources. However, valuable genetic information is also present in already domesticated specimens, offering insight into the changes undergone by these biological forms to become what is currently cultivated or reproduced [34, 35].

Genetic resources can be categorized into plant, microbial, zoogenetic, and genomic resources [36, 37]. Each of these resources plays a critical role in food security, requiring dedicated research, improvement, and conservation efforts. They are essential links in ensuring global food security, making it necessary to establish the most suitable conservation strategy based on the characteristics of each genetic resource [38].

4 Phytogenetic Resources

The importance of global biodiversity in plant genetic resources lies in the centers of origin and diversification of the main plants used in food supply [39, 40]. Examples include the center of origin and diversification of maize in Mexico, the center of diversification of beans in Mesoamerica, the center of origin and diversification of rice in Asia, and the center of origin and diversification of wheat in the Middle East [41–43]. These geographic distributions of phytogenetic resources should influence the prioritization of conservation policies in each country.

Safeguarding the largest possible quantity of germplasm from a plant genetic resource is essential for the future development of crop improvement programs [44]. In this context, the center of origin of a phytogenetic resource contains the most extensive genetic reservoir of that crop. Therefore, it is imperative to develop conservation strategies in these geographic regions.

5 Conservation of Plant Genetic Resources

There are various techniques for conserving plant germplasm, ranging from the conservation of species in their natural habitat (in situ) to the preservation of species outside their habitat (ex situ) [45]. Both conservation types are employed worldwide, depending on the amount of germplasm to be preserved, the resources available, and the short-, medium-, or long-term conservation purposes [3, 46].

Approaches to conserving plant genetic resources are carried out across various scientific disciplines [3]. For instance, in situ conservation of plants adopts an ecological approach, aiming to preserve the interactions between biotic and abiotic factors that make up the species' habitat. Meanwhile, from a biotechnological perspective, plant species are frozen to halt metabolism, as seen in cryopreservation [47]. From a genomic standpoint, the DNA (deoxyribonucleic acid) of phytogenetic resources is extracted from cells and frozen to prevent degradation [48]. In plant physiology, germplasm is preserved through plant shoots subjected to minimal growth using tissue culture techniques [49]. Orthodox seeds are conserved at low temperatures, maintaining their dormancy [50]. Any conservation strategy is commendable and accepted, with the ultimate goal being to preserve the genetic reservoir of the biological forms that constitute these genetic resources (Fig. 1).

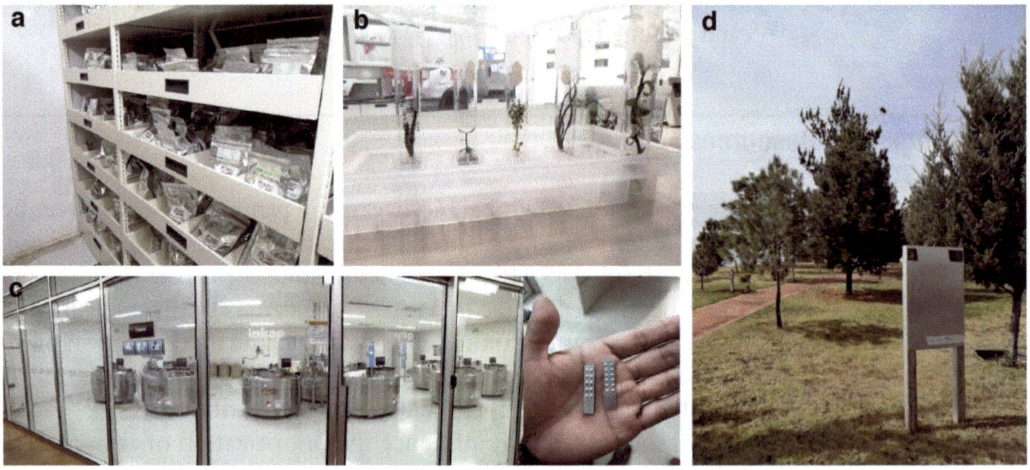

Fig. 1 Ex situ conservation techniques for plant genetic resources. (**a**) Conservation of orthodox seeds at low temperatures, (**b**) in vitro conservation by minimum growth, (**c**) cryoconservation of plant genetic resources, and (**d**) field conservation using the CNRG-INIFAP forest Arboretum

6 Global Actions for the Conservation of Plant Genetic Resources

Global efforts have focused on four main strategies: (1) in situ conservation; (2) ex situ conservation; (3) sustainable use; and (4) capacity-building for conservation [32]. Of these strategies, in situ conservation is widely utilized worldwide due to the following advantages: (a) It conserves both the genetic material and the processes that generate diversity; (b) It ensures the continuous availability of genetic variation that can be maintained and developed in farmers' fields, supporting the sustainability of breeding programs; (c) It allows for the conservation of a large number of species in a single location; and (d) Under certain conditions, depending on the crop or type of genetic resource to be conserved, it can be more cost-effective and efficient than ex situ conservation.

The main approach for ex situ conservation involves the use of germplasm banks, which preserve seeds from plants, fungi, and microorganisms [46]. Seed conservation primarily focuses on orthodox seeds, which are stored in airtight containers at low temperatures. Other methods include tissue cryopreservation, collections of vegetative propagules, and in vitro tissue culture [47]. To a lesser extent, ex situ conservation is achieved through field germplasm banks, particularly for species producing recalcitrant seeds, perennial species, or those for which other conservation techniques are challenging. Additionally, botanical gardens serve as an ex situ conservation strategy, featuring collections of plants organized by various scientific criteria, such as kinship and other parameters [45].

The sustainable use of phytogenetic resources is critical, as their value lies in meeting humans' primary needs, such as food, clothing, and shelter. All aspects related to phytogenetic resources would lose significance without considering their practical use [32].

To build sustainable institutional and human capacities, priority activities and a series of indicators are implemented to facilitate the conservation and sustainable use of phytogenetic resources. Ensuring food and nutritional security for all requires better use of and greater diversity in phytogenetic resources for global agriculture and food production. However, global climate change presents an additional challenge to the world's food supply, highlighting the need to implement conservation strategies for plant genetic resources

A notable conservation strategy example is Mexico's National Genetic Resource Center of the National Forestry, Agriculture, and Livestock Research Center. Its mission is to safeguard genetic resources of agro-food significance in Mexico. Its objective is to ensure the conservation of the country's Genetic Resources, which generate economic, social, and environmental value. This is achieved by contributing to the equitable, integral, and sustainable use of these resources for the benefit of present and future generations, always supported by scientific knowledge and cutting-edge technology.

References

1. Ulian T, Diazgranados M, Pironon S, Padulosi S, Liu U, Davies L et al (2020) Unlocking plant resources to support food security and promote sustainable agriculture. Plants, People, Planet 2(5):421–445. https://doi.org/10.1002/ppp3.10145
2. Maja MM, Ayano SF (2021) The impact of population growth on natural resources and farmers' capacity to adapt to climate change in low-income countries. Earth Syst Environ 5(2):271–283. https://doi.org/10.1007/s41748-021-00209-6
3. Salgotra RK, Chauhan BS (2023) Genetic diversity, conservation, and utilization of plant genetic resources. Genes 14(1):174. https://doi.org/10.3390/genes14010174
4. Smith S, Nickson TE, Challender M (2021) Germplasm exchange is critical to conservation of biodiversity and global food security. Agron J 113(4):2969–2979. https://doi.org/10.1002/agj2.20761
5. Engels JM, Ebert AW, van Hintum T (2024) Collaboration between private and public genebanks in conserving and using plant genetic resources. Plants 13(2):247. https://doi.org/10.3390/plants13020247
6. Engels, J, Ebert, A (2021) History of global germplasm conservation system. Scholarly Community Encyclopedia, https://doi.org/10.3390/plants10091904
7. Jain S, Singh H, Rathod M, Meena R, Deshmukh RN, Mohapatra A, Sharma R (2023) Preserving for the future: the critical role of germplasm conservation in fruit crop resilience. Int J Environ Clim Change 13(11):4651–4661. https://doi.org/10.9734/ijecc/2023/v13i113644
8. Gore PG, Gowthmi R, Tripathi K, Malav PK, Tyagi V, Singh N, Gupta V (2022) Genetic resources. Environment and climate-smart food production. Springer, Cham, 109–149 https://doi.org/10.1007/978-3-030-71571-7_4
9. Guzzon F, Arandia Rios LW, Caviedes Cepeda GM, Céspedes Polo M, Chavez Cabrera A, Muriel Figueroa J et al (2021) Conservation and use of Latin American maize diversity: pillar of nutrition security and cultural heritage of humanity. Agronomy 11(1):172. https://doi.org/10.3390/agronomy11010172
10. Khoury CK, Brush S, Costich DE, Curry HA, De Haan S, Engels JM et al (2022) Crop genetic erosion: understanding and responding to loss of crop diversity. N Phytologist 233(1):84–118. https://doi.org/10.1111/nph.17733
11. Rahman W, Cahyaningsih R, Herawati H, Aminah A, Risliawati A, Diantina S, Surya MI et al (2024) Harnessing plant genetic diversity in research on industrial crop plants for environmental conservation concerns. In: Kumar N (ed) Industrial crop plants. Springer Nature Singapore, Singapore, https://doi.org/10.1007/978-981-97-1003-4_11
12. Salem KF, Ibrahim AA (2024) Plant biodiversity in the context of food security under climate change. Sustainable utilization and conservation of plant genetic diversity. Springer Nature Singapore, Singapore, 3–30. https://doi.org/10.1007/978-981-99-5245-8_1
13. Kashyap A, Garg P, Tanwar K, Sharma J, Gupta NC, Ha PTT et al (2022) Strategies for utilization of crop wild relatives in plant breeding programs. Theor Appl Genet 135(12):4151–4167. https://doi.org/10.1007/s00122-022-04220-x
14. Bohra A, Kilian B, Sivasankar S, Caccamo M, Mba C, McCouch SR, Varshney RK (2022) Reap the crop wild relatives for breeding future crops. Trends Biotechnol 40(4):412–431. https://doi.org/10.1016/j.tibtech.2021.08.009
15. Galluzzi G, Seyoum A, Halewood M, Lopez Noriega I, Welch EW (2020) The role of genetic resources in breeding for climate change: the case of public breeding programmes in eighteen developing countries. Plants 9(9):1129. https://doi.org/10.3390/plants9091129
16. Jordanovska S, Jovovic Z, Andjelkovic V (2020) Potential of wild species in the scenario of climate change. Rediscovery of genetic and genomic resources for future food security. Springer, Singapore, 263–301 https://doi.org/10.1007/978-981-15-0156-2_10
17. Lázaro A, Fuster F, Alomar D, Totland Ø (2020) Disentangling direct and indirect effects of habitat fragmentation on wild plants' pollinator visits and seed production. Ecol Appl 30(5):e02099. https://doi.org/10.1002/eap.2099
18. Goettsch B, Urquiza-Haas T, Koleff P, Acevedo Gasman F, Aguilar-Meléndez A, Alavez V, Jenkins RK (2021) Extinction risk of Mesoamerican crop wild relatives. Plants People Planet 3(6):775–795. https://doi.org/10.1002/ppp3.10225
19. Dunning K (2022) Biodiversity conservation policy in megadiverse countries: comparing policy systems for 2020 targets to inform management in the coming decades. J Environ Manag 302:113815. https://doi.org/10.1016/j.jenvman.2021.113815

20. Villaseñor JL, Meave JA (2022) Floristics in Mexico today: insights into a better understanding of biodiversity in a megadiverse country. Botanical Sci 100(SPE):14–33. https://doi.org/10.17129/botsci.3050
21. Dong Y, Duan S, Xia Q, Liang Z, Dong X, Margaryan K et al (2023) Dual domestications and origin of traits in grapevine evolution. Science 379(6635):892–901. https://doi.org/10.1126/science.add8655
22. Harlan JR (2023) Centers of origin. CRC handbook of plant science in agriculture. CRC Press, Boca Raton, FL, 15–21
23. Larson G, Piperno DR, Allaby RG, Purugganan MD, Andersson L, Arroyo-Kalin M et al (2014) Current perspectives and the future of domestication studies. Proc Natl Acad Sci 111(17):6139–6146. https://doi.org/10.1073/pnas.1323964111
24. Gutaker RM, Purugganan MD (2023) Adaptation and the geographic spread of crop species. Annu Rev Plant Biol 75(1):679–706. https://doi.org/10.1146/annurev-arplant-060223-030954
25. Fowler C, Hodgkin T (2004) Plant genetic resources for food and agriculture: assessing global availability. Annu Rev Env Resour 29(1):143–179. https://doi.org/10.3390/genes14010174
26. Boller B, Greene SL (2010) Genetic resources. Fodder crops and amenity grasses. Springer, 13–37 https://doi.org/10.1007/978-1-4419-0760-8_2
27. Sonnino A (2017) International instruments for conservation and sustainable use of plant genetic resources for food and agriculture: an historical appraisal. Diversity 9(4):50. https://doi.org/10.3390/d9040050
28. Pathirana R, Carimi F (2022) Management and utilization of plant genetic resources for a sustainable agriculture. Plants 11(15):2038. https://doi.org/10.3390/plants11152038
29. Ebert AW (2020) The role of vegetable genetic resources in nutrition security and vegetable breeding. Plants 9(6):736. https://doi.org/10.3390/plants9060736
30. Westengen OT, Lusty C, Yazbek M, Amri A, Asdal Å (2020) Safeguarding a global seed heritage from Syria to Svalbard. Nat Plants 6(11):1311–1317. https://doi.org/10.1038/s41477-020-00802-z
31. Painting, KA, Perry, MC, Denning, RA, Ayad, WG (1995) Guidebook for genetic resources documentation. International Plant Genetic Resources Institute, Rome, p. 296
32. FAO, Organización de las Naciones Unidas para la Alimentación y la Agricultura y (FAO) (2014) Plan de acción mundial para la conservación, la utilización sostenible y el desarrollo de los recursos genéticos forestales. Comisión de recursos genéticos para la alimentación y la agricultura FAO, Roma, Italia, https://openknowledge.fao.org/server/api/core/bitstreams/010cad07-9293-45a1-8020-5353177f00c2/content, p. 37
33. Molina-Pertíñez A, Torres E, Rubio-Teso ML, Álvarez Muñiz C, De la Rosa L, Rincón V et al (2022) Estrategia Nacional de Conservación y Utilización de Parientes Silvestres de los Cultivos (PSC) y Plantas Silvestres de Uso Alimentario (PSUA). Centro de Recursos Fitogenéticos y Agricultura Sostenible (CRF), España, p. 132
34. Andersson L, Purugganan M (2022) Molecular genetic variation of animals and plants under domestication. Proc Natl Acad Sci 119(30):e2122150119. https://doi.org/10.1073/pnas.2122150119
35. Huang X, Huang S, Han B, Li J (2022) The integrated genomics of crop domestication and breeding. Cell 185(15):2828–2839. https://doi.org/10.1016/j.cell.2022.04.036
36. Gautam PL, Singh BB, Saxena S, Sharma RK (2004) Collection, conservation and utilization of plant genetic resources. Plant breeding: mendelian to molecular approaches. Springer, Dordrecht, 657–690 https://doi.org/10.1007/978-94-007-1040-5_28
37. Ude GN, Igwe DO, Onyia CO, Obih CE (2022) Biodiversity and genetic resources conservation in Nigeria. Agricultural biotechnology, biodiversity and bioresources conservation and utilization. CRC Press, 405–425
38. Anju TR, Theresa M, Mohanlal J (2024) Conservation approaches of plant genetic resources. Sustainable utilization and conservation of plant genetic diversity. Springer Nature Singapore, Singapore, 643–681 https://doi.org/10.3390/genes14010174
39. Rafiqpoor D, Kier G, Kreft H (2005) Global centers of vascular plant diversity. Nova Acta Leopoldina NF 92(342):61–83
40. Lyzenga WJ, Pozniak CJ, Kagale S (2021) Advanced domestication: harnessing the precision of gene editing in crop breeding. Plant Biotechnol J 19(4):660–670. https://doi.org/10.1111/pbi.13576
41. Singh SK, Kumar S, Kashyap PL, Sendhil R, Gupta OP (2023) Wheat. Trajectory of 75 years of Indian agriculture after independence. Springer Nature Singapore, Singapore, 137–162 https://doi.org/10.1007/978-981-19-7997-2_7
42. Acosta-Bayona J, León-Martínez G, Vielle-Calzada JP (2024) Origin and diversification of maize: two teosintes but different contributions. Mol Plant 17(2):233–235. https://doi.org/10.1016/j.molp.2023.12.024

43. Vazeux-Blumental N, Manicacci D, Tenaillon M (2024) The milpa, from Mesoamerica to present days, a multicropping traditional agricultural system serving agroecology. Comptes Rendus Biologies 347(G1):159–173. https://doi.org/10.5802/crbiol.164
44. Azhar MT, Wani SH (2021) Wild germplasm for genetic improvement in crop plants. Academic Press, p. 383
45. Faraji L, Karimi M (2022) Botanical gardens as valuable resources in plant sciences. Biodivers Conserv 31(12):2905–2926. https://doi.org/10.1007/s10531-019-01926-1
46. Engels JM, Ebert AW (2021) A critical review of the current global ex situ conservation system for plant agrobiodiversity. I. History of the development of the global system in the ontext of the political/legal framework and its major conservation components. Plants 0(8):1557. https://doi.org/10.3390/plants10091904
47. González AMT, Engelmann F (2013) Crioconservación de plantas en América Latina y el Caribe. Instituto Interamericano de Cooperación para la Agricultura, p. 35
48. Holt WV, Comizzoli P (2021) Perspective: genome resource banking for wildlife conservation: promises and caveats. CryoLetters 42(6):309–320
49. Kulak V, Longboat S, Brunet ND, Shukla M, Saxena P (2022) In vitro technology in plant conservation: relevance to biocultural diversity. Plants 11(4):503. https://doi.org/10.3390/plants11040503
50. Walters C, Pence VC (2021) The unique role of seed banking and cryobiotechnologies in plant conservation. Plants People Planet 3(1):83–91. https://doi.org/10.1002/ppp3.10121

Check for updates

Chapter 2

Strategies for the Conservation of Plant Genetic Resources

Ivonne Naomi Bravo-Ruíz, Lizandro Ramírez-Trejo, María Teresa González-Arnao, Fabiola Hernández-Ramírez, Daniel Padilla-Chacón and Carlos Alberto Cruz-Cruz

Abstract

The high vulnerability of plant species caused by alterations in their natural habitats, environmental contamination, and climate change has led to the progressive loss of genetic diversity. In situ and ex situ conservation are strategies to preserve plant genetic resources in the short, medium, and long term. This would be beneficial through approaches to enable the creation of genetic reservoirs of populations within a specific geographic region, which are helpful for ecosystem restoration, germplasm exchange, and genetic improvement. From a biotechnological perspective, plant conservation methods contribute to safeguarding sources of genotypes with potential benefits that impact food production, synthesis of secondary metabolites, and resistance to biotic and abiotic stress. Furthermore, the selection of representative accessions from diverse populations, regeneration, scalability, and genetic stability testing support the production of elite genotypes. This chapter reviews current plant conservation technologies, highlighting the advantages and limitations of each technique, as well as their contribution to plant biotechnology and ecological restoration for the future of sustainable agriculture.

Keywords Biotechnology, Plant biodiversity, Endangered plants, In situ conservation, In vitro storage, Sustainable agriculture

1 Introduction

In agriculture, the genetic diversity of cultivated plant species and their genetic components underpin the productivity, sustainability, resilience, and adaptability of agricultural systems. Nature's rich diversity provides an opportunity to integrate the beneficial characteristics of natural diversity, enhancing the evolutionary development of cultivars and facilitating the implementation of solutions to current and emerging challenges, such as soil salinity, drought, and pests [1, 2].

Plant genetic resources, which encompass this diversity, have played a pivotal role in the development of agriculture, from early domestication to the spread of domesticated species and the continuous improvement and adaptation of wild relatives from domestication

Marco A. Ramírez-Mosqueda et al. (eds.), *Conservation of Plant Genetic Resources*, Methods in Molecular Biology, vol. 3011, https://doi.org/10.1007/978-1-0716-5130-8_2,

centers subjected to dynamic environmental changes [3]. The survival of species largely depends on the maintenance of genetic diversity [2]. This heritable variation among individuals and populations—encoded in DNA as intraspecific diversity—is fundamental for the ability of species to adapt to the extent and pace of modern environmental changes, including climate change and pressures from phytopathogens.

Currently, society faces a critical juncture that demands transformative action to conserve biodiversity and support resilient ecosystems [4]. In this context, germplasm conservation strategies have emerged, aiming not only to rescue endangered plant species to preserve essential plants that underpin the survival of ecosystems [5]. These plant resources are increasingly threatened by anthropogenic factors such as habitat degradation, the spread of invasive species, overexploitation of natural resources, and climate change [6, 7].

It is estimated that approximately 75% of the Earth's land surface is under human pressure, with about 40% used for agricultural purposes. Given this scenario, the need for novel techniques to conserve biodiversity has become increasingly urgent. Various institutions have responded by implementing new strategies designed to provide viable and secure solutions for the long-term preservation of plant genetic resources (PGRs) [8].

Conservation of plant species can be approached through ex situ and in situ strategies. Ex situ conservation involves a range of techniques applied outside their natural habitats, focusing on the sampling, transfer, and storage of germplasm in controlled environments, such as botanical gardens, seed banks, or ovule storage facilities. In contrast, in situ conservation aims to protect, manage, and systematically observe species within their natural habitats, including protected wildlife reserves and community-managed areas [9]. These approaches are complementary rather than mutually exclusive, offering different conservation pathways based on a set of criteria, such as the biological characteristics of the species and the feasibility of method implementation [10].

This review presents an overview of key plant conservation techniques that can be effectively utilized to enhance biodiversity conservation efforts. Furthermore, a bibliometric analysis of scientific literature is conducted to examine thematic trends in the field, highlighting emerging areas of research and identifying gaps that warrant further investigation.

2 Conservation of Plant Genetic Resources

The conservation of plant species involves the management and protection of natural resources to prevent their exploitation, degradation, or neglect. Its primary goal is to ensure that these resources remain available for future human generations while simultaneously addressing the needs of the present population [1, 11]. Conservation involves safeguarding species, their habitats, and ecosystems against threats such as overexploitation, habitat destruction, pollution, and climate change. Strategies include both the sustainable use of resources—defined as utilizing natural resources at a rate that allows for their replenishment without causing ecological harm—and the protection of natural areas [11].

The conservation of PGRs focuses on methods designed to capture the widest possible range of genotypic diversity. It also involves the use of conservation techniques and subsequent multiplication to mitigate genetic erosion over time under controlled conditions, maximizing the diversity of samples collected from field populations or centers of origin [12]. Gene banks, as a cornerstone of ex situ conservation strategies, have successfully stored and managed a broad range of PGRs, ensuring their availability for research and breeding programs. Table 1 shows examples of gene banks with the largest collections of plant species from different regions.

The International Union for Conservation of Nature (IUCN) Red List of Threatened Species has assessed 112,432 species, concluding that approximately 25% face a high and increasing risk of extinction [32]. Complementary efforts, such as the Red List of Ecosystems, will soon enhance these assessments by evaluating the risks of ecosystem collapse [33]. A comparison between the 238,563 sites documented in the World Database on Protected Areas and the 16,366 sites listed in the World Database of Key Biodiversity Areas reveals that, on average, only 43% of key biodiversity sites are covered by protected areas [32].

Despite its fundamental importance, plant conservation has historically received considerably less attention than animal conservation, likely because plants lack the public appeal of many animal groups [34]. Consequently, plant conservation efforts are underfunded and under-resourced compared to those targeting animal species [35].

According to the national environmental policies, the National System of Science, Technology, Innovation, and Ancestral Knowledge promotes and regulates both in situ and ex situ scientific research progress, including the extraction, collection, mobilization, transport, export, and management of wildlife species, with the overarching aim of safeguarding genetic diversity. Effective biodiversity conservation thus requires the implementation of monitoring and tracking mechanisms, as stipulated by the relevant authorities [9].

The future of life on Earth critically depends on the decisions and actions undertaken in the coming years. Immediate and coordinated global measures are urgently needed, with the urgency intensifying over time [32]. Although conserving global plant diversity faces a monumental challenge, the establishment of a well-designed and effectively managed system of protected areas, supplemented by targeted ex situ conservation efforts, offers a viable path forward.

Table 1
GenBanks with the most species by region: conservation and sustainable use [13]

GenBanks	Description	Species/ accessions	Region
Millennium Seed Bank—Royal Botanic Gardens Kew	Ex situ conservation of orthodox seeds from wild plants, with a focus on ecological restoration [14] and international collaboration [15, 16].	15777/88249	Europe
Genetic Resources Research Institute—Kenya Agricultural & Livestock Research Organization	Their efforts contribute to food security, climate change adaptation, and biodiversity preservation in Kenya [17].	2525/51405	Sub-Saharan Africa
National Gene Bank of Tunisia	Focused on food security, climate change adaptation, and biodiversity preservation in Tunisia [18, 19].	849/24485	Northern Africa
Western Regional Plant Introduction Station, USDA-ARS, Washington State University	Plant germplasm conservation, management, and research, including active studies on phytopathogens [20] and pollinator efficiency [21].	2956/114442	Northern America
Embrapa Genetic Resources and Biotechnology—Brazil	Preservation and sustainable utilization of plant genetic resources [22, 23]	968/124591	Latin America and the Caribbean
National Bureau of Plant Genetic Resources—India	Contributions include both genotype characterization [24] and genomic research [25] targeting sustainable food security [26].	1580/534592	Asia
Australian Pastures Genebank	Focused on the conservation and genetic improvement of grasses and forage crops, this serves as a genetic resource bank for international research partnerships [27, 28].	1811/81133	Oceania
NordGen	Preservation, characterization, and rescue of plant genetic resources [29, 30].	432/33306	Regional
International Livestock Research Institute	Conservation, breeding, and adaptation of forage plants, with a focus on livestock production sustainability [31].	1599/18639	International

The most pressing priorities include completing a comprehensive global botanical inventory and assessing the conservation status of approximately 94% of plant species that have yet to be evaluated. Achieving these objectives is crucial to strategically direct both in situ and ex situ conservation initiatives [36].

2.1 In Situ Method for Conservation Plant Biodiversity

In situ conservation refers to the preservation of genetic resources within their natural ecosystems through the management of protected areas and the implementation of effective conservation strategies [37]. Its primary objective is to protect and restore species populations and their habitats [38], to harness genetic diversity in breeding and conservation programs to maintain genetic variability and promote adaptation to changing environmental conditions [39].

The "in situ" definition is used traditionally in naturalist contexts to denote a species' original habitat. Its formal application in conservation dates from 1967 according to The Technical Conference on the Exploration, Utilization, and Conservation of Plant Genetic Resources [40]. It is often associated with wild populations in protected areas; also in situ conservation applies to managed forests and human-altered ecosystems, providing germplasm preserved within its original environment or in locations with distinctive characteristics through cultivation [41].

In situ conservation involves the identification and protection of habitats and species through various management and monitoring strategies. For threatened species, it includes the removal or mitigation of threats as well as the reinforcement or reintroduction of populations into areas from which they have disappeared [39]. In critical cases, detailed recovery plans have been developed, incorporating considerations of taxonomy, distribution, ecology, genetic diversity, and the overall conservation status of target species. This process, known as ecogeographic surveying, is critical for the implementation of effective conservation measures [39].

A bibliometric analysis conducted over the past 25 years, using the search string "Conservation" AND "plant" AND "in situ" on the Scopus platform, identified a total of 3137 documents, comprising scientific articles (2514), book chapters (230), reviews (181), conference papers (177) , and books (10). Figure 1 illustrates the evolution of the number of publications indexed in Scopus, providing insights into research trends related to in situ plant conservation.

During the early 2000s, a gradual increase in the number of publications was observed, rising from 26 documents in 2000 to nearly 80 by 2009. From 2010 onward, annual publications consistently exceeded 120, peaking between 2017 and 2021, with the highest output recorded in 2021 (246 documents). Although a slight decrease was noted in 2025, this is likely due to the data being collected before the end of the year. Overall, the data reflects a progressive growth in scientific output, consolidating in situ conservation as a key area of interest within the academic community.

The sustained increase in research activity between 2000 and 2021 highlights the growing recognition of the crucial role that in situ conservation plays in safeguarding plant biodiversity. This trend also reflects a response to the escalating threats faced by natural ecosystems, including climate change, habitat fragmentation, and resource overexploitation [42, 43].

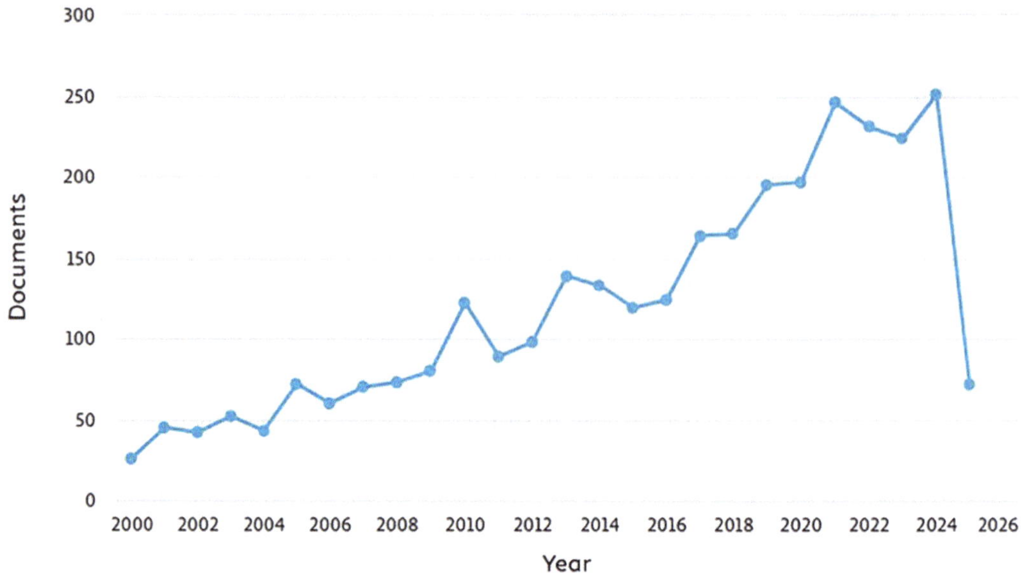

Fig. 1 Total number of publications indexed in Scopus during the period 2000–2025, using the keywords "Conservation" AND "plant" AND "in situ"

To further explore research trends, a keyword co-occurrence analysis was conducted using VOSviewer, based on Scopus publications from 2000 to 2025. Figure 2 presents a co-occurrence network grouping relevant terms into two thematic clusters, differentiated by color.

The first cluster (red color) includes concepts such as biodiversity, genetics, conservation of natural resources, and environmental protection, underscoring the emphasis on biological diversity and the safeguarding of natural resources. The second cluster (green color) is centered around the term RAD-seq (Restriction-site Associated DNA Sequencing), a genetic analysis technique, which appears interconnected with both ecological and genetic topics. This suggests an emerging trend toward integrating advanced genomic tools in the conservation of wild plant populations.

This analysis indicates that investigation on in situ plant conservation has evolved a multidisciplinary approach, incorporating ecological, genetic, and technological dimensions to tackle contemporary challenges in biodiversity conservation.

In situ conservation focuses on protecting plant species within their natural habitats, promoting both evolutionary adaptation and ecological interactions. These approaches have been fundamental in programs focused on the conservation of plant genetic resources.

The growing incorporation of in situ conservation into scientific and political agendas reflects a more holistic understanding of conservation—not solely as the preservation of genetic material, but as the conservation is complex, dynamic socio-ecological systems. Consequently, in situ strategies encompass the management

Fig. 2 Term map of publications from the Scopus database during the period 2000–2025, using the keywords "Conservation" AND "plant" AND "in situ"

of wildlife reserves, protected areas, national parks, sanctuaries, and biosphere reserves [37]. These efforts ensure the sustainability of ecosystems and the preservation of biodiversity within their natural contexts, supporting the continuity of essential ecological processes critical for life on Earth [9].

2.1.1 Protected Areas (PAs)

Protected areas (PAs) are a fundamental strategy for the conservation of biodiversity and ecosystem services, particularly in the context of increasing human pressure on natural resources. Most countries regard PAs as the cornerstone of their conservation policies, serving as a primary defense against biodiversity loss, provided they are properly managed [44].

Over the past 25 years, the number of protected areas has increased substantially, reaching approximately 202,000 sites that encompass 16.64% of terrestrial and freshwater ecosystems and 7.74% of marine environments. This expansion represents a significant achievement in global conservation efforts [45].

Beyond their ecological importance, PAs also offer utilitarian benefits, including the supply of reproductive material for commercial production and the promotion of activities such as ecotourism, which can support local economies while fostering conservation goals [46].

Nevertheless, a considerable proportion of biodiversity persists outside protected areas, rendering current coverage insufficient in terms of spatial extent, ecological representation, and the protection of key biodiversity areas [47]. Furthermore, approximately half of all existing PAs are inadequately managed, limiting their effectiveness as refuges for species [44].

Planning and management inconsistencies at the national level present additional challenges, particularly between developed and developing countries. The absence of coherent legislation and the fragmentation of responsibilities could result in suboptimal conservation outcomes. For instance, in China, the administration of PAs is distributed among different sectors and levels of government, complicating coordination efforts [45, 48].

Despite these limitations, protected areas remain a critical tool for biodiversity conservation. Their success depends on efficient planning and policies that come out of the integrated regional planning process as the integration of complementary strategies that extend conservation efforts beyond the boundaries of formally designated protected spaces.

2.1.2 National Parks

The term "national park," representing the largest and most comprehensive form of nature protection in human civilization history, was introduced in the second half of the 19th century. Since then, national parks have grown significantly, expanding and proliferating globally, becoming not only key instruments for environmental conservation but also symbols of national pride and prestige within the international community. These expansions reflect the growing awareness of the need to protect specific natural areas, and the establishment of tangible actions aimed at preserving regions in their natural state [49].

The original conservation model, based on the strict preservation of wilderness, has largely evolved into a multifunctional approach that integrates a wide range of management objectives, including conservation, recreation, and sustainable development. In some cases, the designation of "national park" has also been utilized as a branding and marketing tool, promoting the commodification of natural landscapes and rural spaces [50].

Nevertheless, the International Union for Conservation of Nature (IUCN) defines national parks as "large natural or near-natural areas set aside to protect large-scale ecological processes, along with the complement of species and ecosystems characteristic of the area, providing a foundation for spiritual and creative opportunities, scientific research, education, as well as environmentally and culturally appropriate visitation" [51].

To prevent environmental deterioration, activities such as grazing, hunting, and logging are typically prohibited within national parks. The management of these areas often emphasizes the conservation of species while simultaneously preserving broader ecosystem functions [11]. For these reasons, national parks are considered among the most valuable components of natural heritage, representing essential reservoirs of biodiversity and natural wealth [49].

2.1.3 Wildlife Sanctuaries

Wildlife sanctuaries are protected areas where permission to carry out activities such as hunting, shooting, and wildlife capture is prohibited without special permits. Unlike national parks, wildlife sanctuaries may allow certain human activities, such as forestry operations, provided that these do not negatively impact wildlife. Sanctuaries serve as essential refuges for endangered species and contribute to maintaining ecological balance [11].

Given the fragile nature of native wildlife, and in accordance with the spatial management of land cover patterns at the class level, the establishment of several pest-free urban wildlife sanctuaries is necessary to ensure the survival of vulnerable fauna among increasing pressures from predators and the spread of diseases driven by various factors [52].

In conservation biology, safe spaces for species and ecosystems have traditionally been referred to as "reserves" or "protected areas." However, the term "sanctuaries" has emerged to describe those reserves that afford an additional level of protection. For instance, under the 1987 Conservation Act that established New Zealand's Department of Conservation (DOC), sanctuaries are recognized as one of five types of specially protected conservation areas, characterized by stringent conservation requirements. More recently, the term "sanctuary" has also been applied to areas where there is intensive and sustained control of a full range of exotic pests [53].

2.1.4 Biosphere Reserves

According to the United Nations Educational, Scientific, and Cultural Organization (UNESCO) (2024), biosphere reserves are areas designed to foster learning toward sustainable development [54]. These sites employ interdisciplinary approaches that facilitate the understanding and management of changes and interactions between social and ecological systems, including conflict prevention and biodiversity administration. They serve as spaces providing local solutions to global challenges.

Biosphere reserves, established under UNESCO's Man and the Biosphere (MAB) Program, promote sustainable development through the harmonious management of interactions between communities and the natural environment. Unlike traditional protected areas, biosphere reserves included geographically delineated landscapes, typically incorporating a protected core area. Their ecosystems may be terrestrial, marine, or coastal, with each reserve aiming to balance biodiversity conservation with the sustainable use of natural resources [55, 56].

Through their global network, biosphere reserves act as platforms for scientific research and community-based initiatives, promoting innovative solutions without compromising conservation objectives. The conceptual framework, developed in the 1970s, is based on a zoning model that integrates core, buffer, and transition areas to reconcile conservation and development goals. Nevertheless, a significant challenge remains the gap between theoretical frameworks and practical implementation, as the effective management of biosphere reserves demands a deeper understanding of operational practices necessary to achieve sustainability objectives. Advances in biotechnology are increasingly providing new methods for the conservation and assessment of plant genetic resources, supporting these efforts [56].

3 Ex Situ Methods for Plant Biodiversity Conservation

Ex situ conservation of plant genetic resources aims to capture and preserve the greatest possible genetic diversity of plant species. Genetic diversity is critical for species' adaptation and resilience, particularly in the context of climate change and habitat loss [57]. The management, maintenance, and types of conservation strategies (Fig. 3) depend on the specimen's nature and current risk status. Exceptional species that cannot be preserved in conventional seed banks are maintained in living collections or cryobanks [58]. Additionally, research activities, the reintroduction of endangered plants, and genetic improvement initiatives form part of the sustainable use of PGRs [59, 60].

Storage can be achieved through the maintenance of whole plant collections and germplasm repositories. Complete plants are preserved in field collections and botanical gardens; however, these methods are susceptible to losses caused by adverse climatic conditions, pathogen attack, high agronomic management costs, land preparation requirements, and challenges in material management and exchange. Conversely, germplasm conservation involves the use of seed banks, pollen banks, in vitro banks, gene banks, and DNA banks [61, 62].

According to the bibliometric analysis, Fig. 4 illustrates the temporal evolution of the number of scientific documents indexed in Scopus between 2000 and 2025 containing the keywords "Conservation" AND "plant" AND "ex situ," highlighting the dynamics of scientific production over a 25-year period.

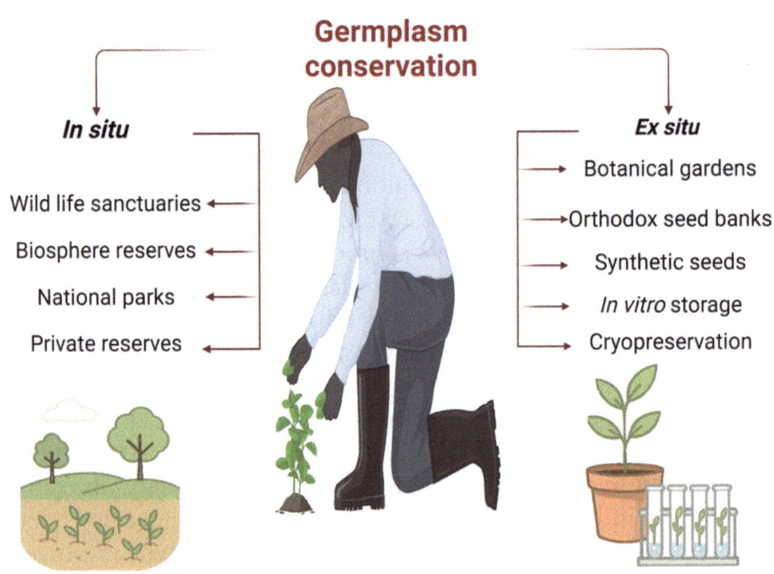

Fig. 3 Plant conservation strategies (Created with BioRender)

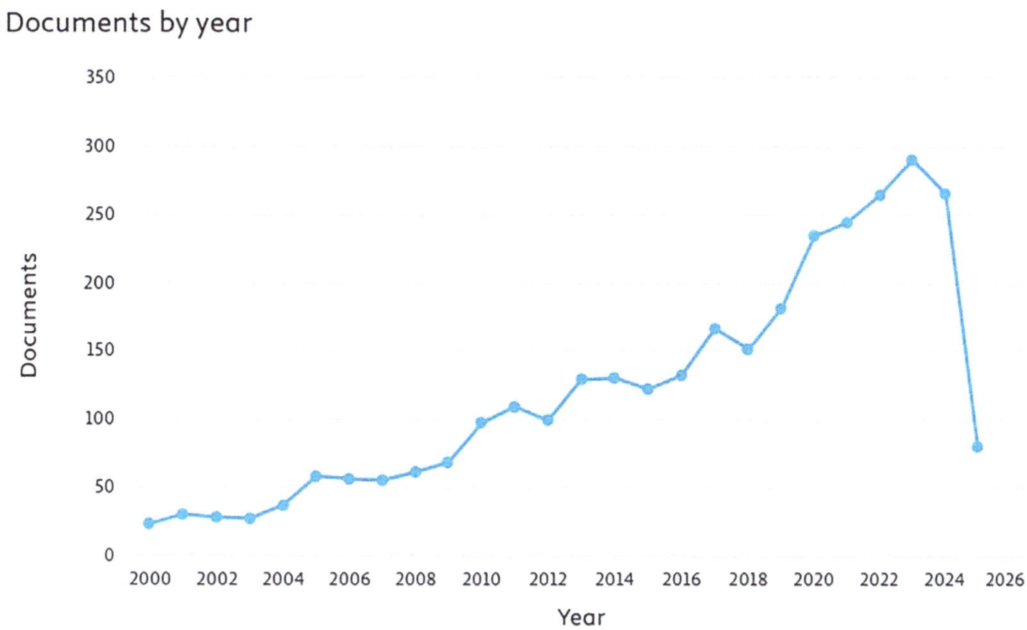

Fig. 4 Total number of publications indexed in Scopus (2000–2025) using the keywords "Conservation" AND "plant" AND "ex situ"

During the first decade of the 2000s, annual publication numbers were relatively low, increasing from 23 to 68 between the period 2000 and 2009. A marked acceleration after 2010, with the first significant peak in 2011 (109 documents). During the period of 2012 and 2017, production remained relatively stable; however, from 2018 to 2023, a sustained increase has been observed, culminating in over 290 publications. This trend suggests a growing interest in ex situ plant conservation, likely influenced by advances in biotechnology, the strengthening of germplasm banks, and the global recognition of ex situ conservation as a critical complement to in situ strategies. In 2024, publication numbers exceeded 250, although data for 2025 remain preliminary.

Overall, this trend demonstrates the strengthening of scientific research in ex situ plant conservation and underscores the need to continue fostering integrated conservation strategies to address biodiversity loss.

The visualization of keyword co-occurrence through VOSviewer analysis of Scopus-indexed publications ("Conservation" AND "plant" AND "ex situ") revealed five prominent terms: "endangered species," "genetics," "phylogeny," "procedures," and "genetics" (Fig. 5). These terms are interconnected by lines whose thickness indicates the frequency of co-occurrence.

"Genetics" has a central position within the network, significantly linking with both "phylogeny" and "endangered species," suggesting

Fig. 5 Term co-occurrence network for Scopus-indexed publications (2000–2025) using the keywords "Conservation" AND "plant" AND "ex situ"

a molecular-evolutionary focus on the conservation of threatened species. The connection to "procedures" highlights that technical and methodological protocols for ex situ conservation are heavily influenced by genetic considerations. The prominence of "endangered species" reflects the evolutionary and molecular emphasis in research aimed at species at risk of extinction.

The keyword co-occurrence analysis reveals that ex situ plant conservation methods have primarily evolved through a genetics- and phylogeny-centered approach. Endangered species constitute the thematic core driving research in this field, while technical procedures adopted reflect advances in genetic and evolutionary methodologies. This pattern suggests a trend toward a scientific conservation framework rooted in molecular data, with significant implications for biodiversity policymaking and germplasm bank management.

Globally, there are 871 germplasm banks distributed across 116 countries, employing various conservation methods, including short-, medium-, and long-term storage, as well as preservation in botanical gardens (Fig. 6).

Together, these institutions safeguard 5,941,616 accessions of plant genetic resources, corresponding to 7320 genera and 51,513 species (Fig. 7). Europe hosts the largest network of gene banks (445) (Fig. 7A), reflecting substantial investment in PGRs conservation. Europe also leads in the number of conserved accessions (2,071,866), genera (6308), and species (40,498) (Fig. 7B).

Asia ranks second in terms of accession numbers (Fig. 7C), with countries such as India and Japan making significant contributions, although species diversity is comparatively lower than in the European continent. In contrast, North America, with relatively few gene banks [30], stands out for its large storage capacity (705,699 accessions). An example is the Western Regional Plant Introduction Station of

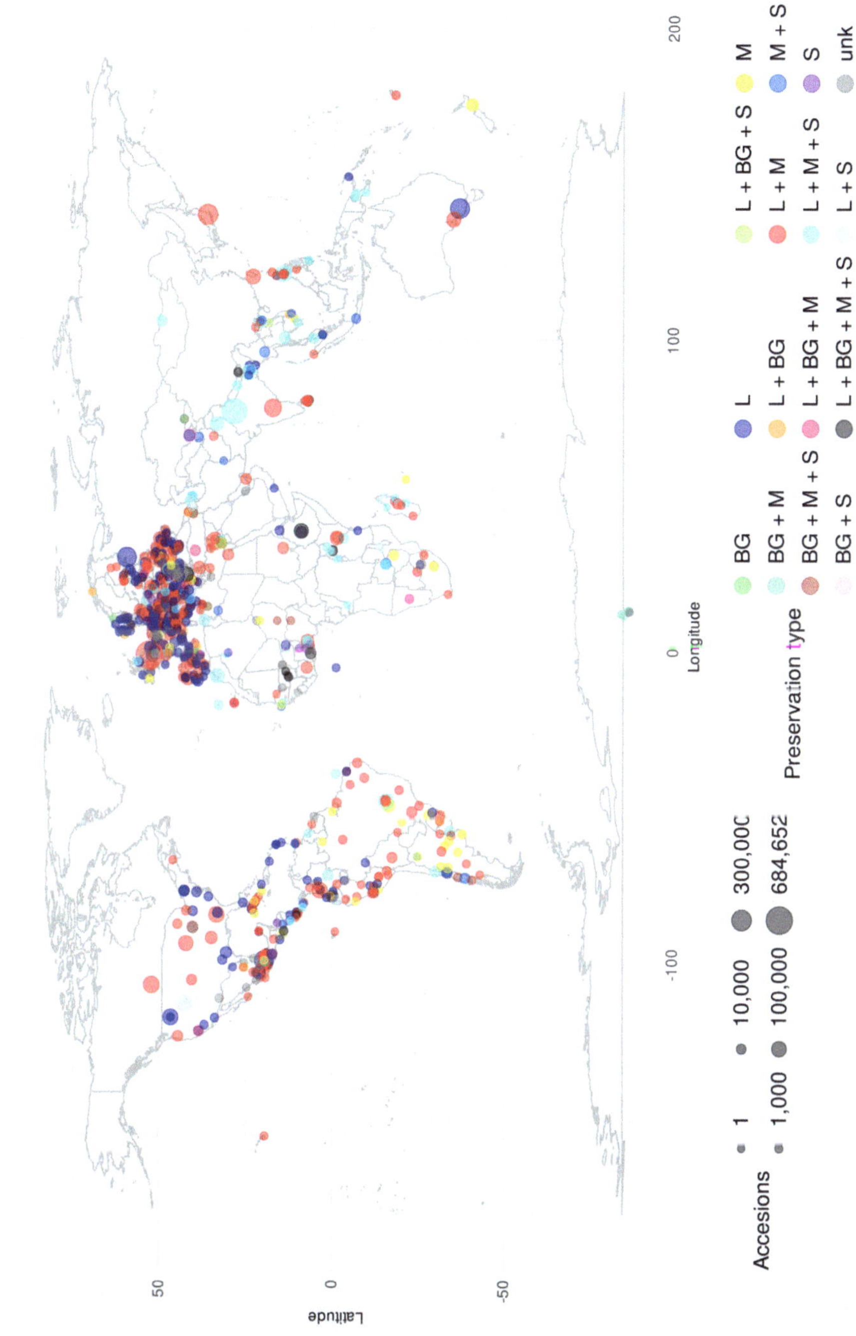

Fig. 6 Global distribution of genebanks by preservation type and regional accession (made in R core team) [13, 63]. Distribution of gene banks by preservation type: short-term (S), medium-term (M), long-term or cryopreservation (L), and botanical gardens (BG). The size of the circles represents the number of accessions

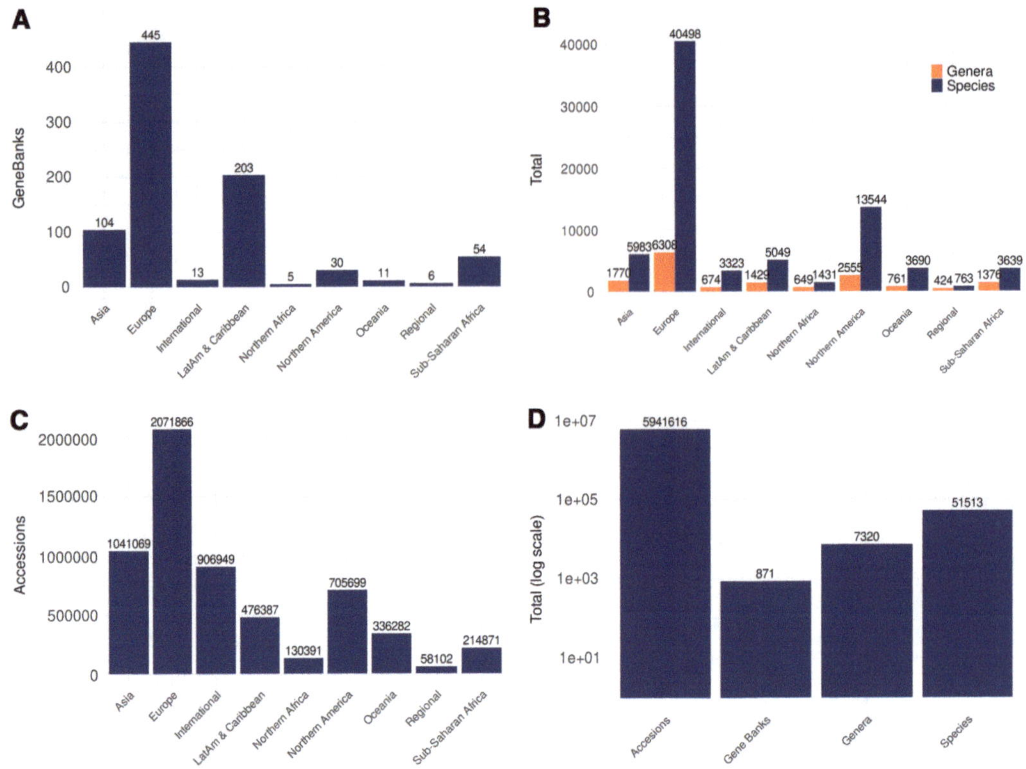

Fig. 7 Regional statistics of genebanks (**A**), preserved genera and species (**B**), number of accessions (**C**), and global totals of genebank metrics (**D**) (made in R core team) [13, 63]

the USDA-ARS at Washington State University, renowned for its extensive conservation capacity (Table 1).

In Latin America and the Caribbean, the region hosts a substantial number of germplasm banks (203) but holds a relatively low volume of accessions (476,387). Notably, Brazil and Mexico account for 73% of the conserved species and 60% of the regional accessions. In Oceania, New Zealand and Australia preserve the highest species richness; however, Australia maintains most of the accessions (89%).

Infrastructure and funding limitations constrain the contribution of germplasm banks in Africa (Fig. 7C, D). Despite its vast biodiversity, Sub-Saharan and North Africa collectively safeguard only 345,262 accessions representing 5070 species (Fig. 7B).

International and regional gene banks play a critical role in global food security through the efforts of institutions such as the International Maize and Wheat Improvement Center (CIMMYT) and the International Potato Center (CIP), which consolidate the conservation of strategic genetic material and foster international collaboration [64–67].

Ex situ conservation systems—defined as the maintenance of organisms or their germplasm outside their natural habitats—have emerged as a complementary strategy to in situ conservation mechanisms. Their primary objective is to safeguard genetic resources through the storage and propagation of germplasm collections that capture the variability for preservation [61]. Thus, Gene banks facilitate both the conservation and sustainable use of PGRs, supporting current agricultural demands and ensuring future resource availability.

3.1 Botanical Gardens

Around the World, Botanical gardens conserve plant biodiversity through living collections, fulfilling purposes related to research, conservation, exhibition, and education [68]. These institutions preserve approximately 30% of the world's plant species diversity [6] and play a critical role in the reintroduction of species that have gone extinct in the wildlife [15]. However, maintaining ex situ collections in botanical gardens requires significant resources, with labor costs representing nearly 50% of total operational expenses [69].

The genetic variation of plant populations-maintained ex situ could decline markedly over successive generations due to factors such as high rates of inbreeding, genetic drift, and the limited genetic source derived from the initial collection of specimens [15]. To mitigate these effects, several strategies have been developed, including the collection of larger numbers of individuals from wild populations, planned crossings between genetically diverse specimens, and the strategic exchange of plant material among botanical gardens [70].

Nevertheless, the exchange and conservation of living specimens result in the risk of introducing phytopathogens or viruses, highlighting the need for rigorous biosecurity measures [71]. Recommended practices include the use of disinfectant mats, quarantining new accessions, continuous phytosanitary monitoring, isolation of contaminated specimens, and the dissemination of biosecurity protocols.

Botanical gardens function as vital reservoirs of plant genetic resources, actively contributing to research, education, and ecological restoration initiatives [68]. Their roles are enhanced through shared access to accession data, national and international collaborations, and the integration of advanced biotechnological tools into conservation programs [59]. These efforts are fundamental for ensuring the preservation of plant genetic diversity for future generations.

3.2 Seed Banks

Seed banks were initially established to maintain seed viability during transport and, consequently, to secure the productivity of important crop varieties. Over time, the systematic collection and storage of seeds evolved into the principal method of ex situ plant conservation. In the United States, early seed banks—prior to 1967—were primarily associated with breeding programs managed by the Office of Foreign Seed and Plant Introduction, founded in 1898 [40].

Seed preservation through seed banks leads the long-term storage of high levels of plant genetic diversity, while optimizing both cost and spatial efficiency [72]. Today, seed banks constitute the primary strategy for conserving PGRs globally [68]. Approximately 59% of long-term and 33% of medium-term conserved accessions are stored in seed banks [13].

Orthodox seeds—those possess the strength of dehydration to moisture levels between 3% and 7% and storage at low temperatures (−18°C to −20°C) without losing viability—are ideal candidates for seed bank conservation. Examples include *Zea mays* L., *Phaseolus vulgaris* L., and *Glycine max* (L.) Merr. Conversely, species producing recalcitrant seeds can only be stored at temperatures ≥1°C and typically remain viable for less than 1 year [73].

The standard process for seed conservation involves four steps: (I) collecting a wide sample of wild seeds to capture genetic diversity; (II) drying the seeds to a moisture content of 3–7% to prevent ice crystal formation; (III) storing seeds at −18 to −20°C in airtight containers; and (IV) recovering or germinating seeds as needed. Since seed viability gradually declines over time, it is crucial to standardize optimal storage durations to maintain acceptable viability rates, typically around 85% [73, 74].

3.3 In Vitro Method for Conservation Plant Biodiversity

Among the current biotechnological methods for the conservation of PGRs are in vitro conservation, DNA banks, and cryopreservation [5]. Plant tissue culture (PTC) is based on the principle of totipotency, involving the exposure of explants—cells, tissues, organs (for cloning), or seeds—to a culture medium enriched with macronutrients, micronutrients, and plant hormones under controlled photoperiod and aseptic conditions [75]. Within the context of germplasm conservation, PTC offers multiple advantages over conventional techniques: it allows immediate access to and exchange of germplasm, protects against natural disasters, facilitates the production of pathogen-free plants, optimizes space and costs, supports mass micropropagation, and enables short- and medium-term preservation of genetic diversity [76].

The initial stages of PTC involve the collection of plant material and the establishment of in vitro cultures. Developing a species-specific protocol is essential, using methods established for taxonomically related species as references. Critical considerations include the species' biology and conservation status, with resources such as Plants of the World Online (https://powo.science.kew.org/) [77] and The IUCN Red List of Threatened Species (https://www.iucnredlist.org/) [78] serving as valuable information sources.

Species that are difficult to propagate, possess recalcitrant or non-viable seeds, exhibit low fertility, or are endangered and poorly suited to traditional ex situ conservation methods are prime candidates for in vitro conservation and cryopreservation [76]. Notable examples include orchids [79, 80], triploid cultivars of the *Musa* genus [81], *Aloe* species [82], and mahogany *(Swietenia macrophylla)* [83].

The establishment of a germplasm bank via PTC (Fig. 8) encompasses six key steps: (I) collection and quarantine of plant material; (II) selection of explants; (III) in vitro establishment; (IV) micropropagation; (V) selection of a minimal growth protocol; and (VI) eventual utilization of the germplasm. The initial step, involving collection and quarantine, ensures that healthy and contamination-free material enters the culture system. Explant selection is dictated by the conservation goal: seeds or tissue samples from diverse accessions are preferred for maximizing genetic diversity, whereas specific tissues such as leaves, buds, or stem segments are selected when conserving specific genotypes [76, 84, 85].

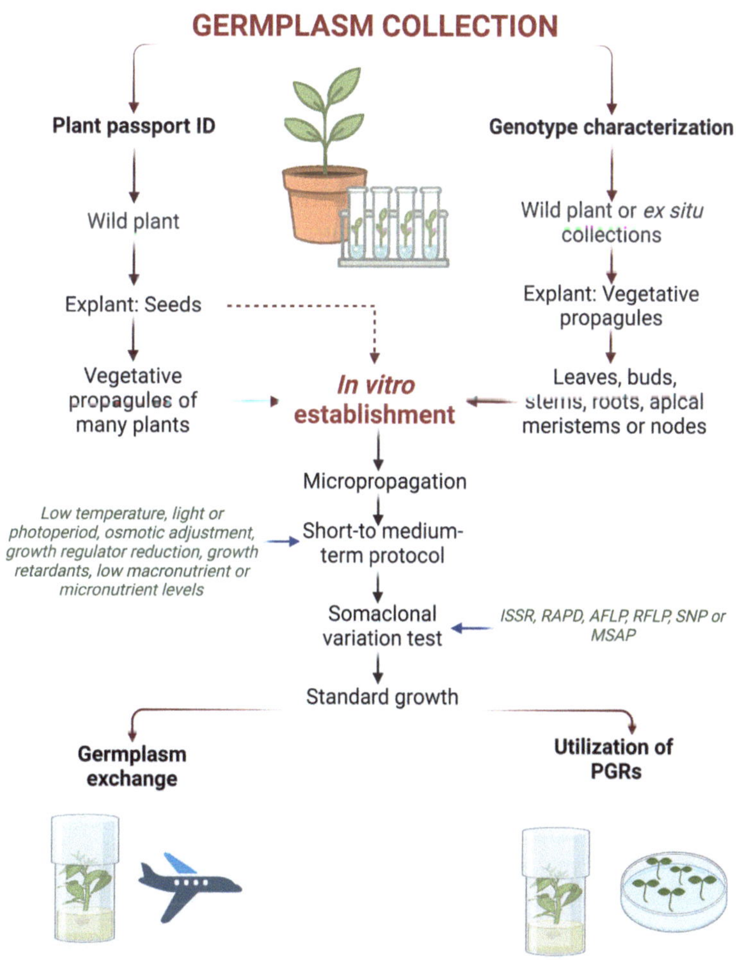

Fig. 8 Workflow for establishing an in vitro germplasm bank

During the establishment phase, explants undergo surface disinfection and sterilization protocols before being cultured aseptically on appropriate media, often supplemented with antibiotics and antimicrobials [86]. For complete plants, pretreatment with fungicides and bactericides is recommended to further minimize contamination risks [87].

Following successful in vitro establishment, micropropagation is conducted on standardized species-specific or related protocols (Fig. 9). This step leads to the rapid multiplication of individuals and the efficient use of space and resources [75]. Regular assessments using molecular markers are essential to monitor and maintain genetic stability and to determine the permissible number of subcultures [88]. Subsequently, the selection of a minimal growth protocol involves testing various growth-retardant strategies through randomized or factorial experimental designs to optimize conditions for slowed explant development [89].

Short- and medium-term conservation focuses on inhibiting explant growth and extending the intervals between subcultures, allowing maintenance for periods ranging from several months to years [10, 12]. Key parameters for a successful protocol include genetic stability, high survival rates, and effective growth inhibition [90].

The in vitro conservation and genetic improvement of germplasm require capturing the widest possible genetic base within the target populations [91]. Consequently, germplasm preserved via PTC could serve two primary applications: (I) plant breeding, by selecting

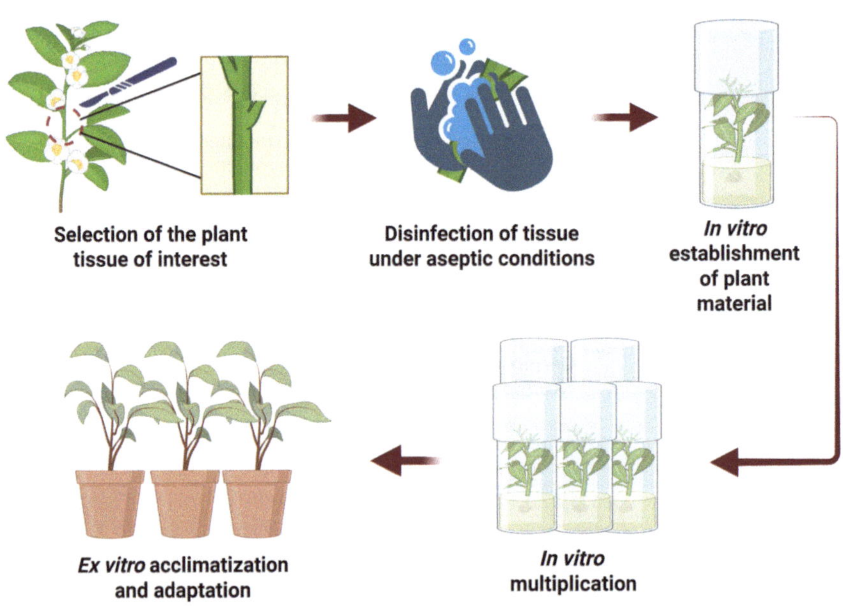

Fig. 9 Steps of in vitro plant tissue culture

and mass-propagating elite genotypes [85, 92]; and (II) ecological restoration, through the reintroduction of genetically diverse plant material into natural environments [93].

A range of specialized basal media is available to accommodate different plant groups and tissue types. The Murashige and Skoog (MS) medium remains the most widely employed, while Knudson C medium (for epiphytic orchids), Woody Plant Medium (WPM), and Driver and Kuniyuki Woody Plant Medium (DKW) are used for specific species. Growth retardants such as paclobutrazol (PBZ), ancymidol, abscisic acid (ABA), and chlormequat chloride (CCC) are commonly applied to inhibit growth [80, 94–96]. Additionally, osmotic agents like polyethylene glycol (PEG), mannitol, sucrose, and sorbitol have been used for growth reduction [97, 98]. However, the use of these agents may induce somaclonal variation, necessitating the development of complementary strategies such as reduced light intensity, lowered macro- and micronutrient concentrations, controlled temperatures, and modified container atmospheres to minimize variation.

Various combinations of techniques have been evaluated for the conservation of species of agricultural, medicinal, industrial, ornamental, and ecological importance (Table 2). Despite their potential, in vitro techniques are subjected to several limitations, including the recalcitrance of certain species to regeneration [132, 133], somaclonal variation [134], genotype-specific conservation responses [110], vulnerability to contamination, and sensitivity to infrastructural failures such as power outages. Additionally, PTC requires specialized facilities and highly skilled personnel.

Nevertheless, plant tissue culture remains an indispensable complementary strategy within comprehensive PGRs conservation programs. Other emerging biotechnological approaches include the in vitro regeneration and conservation of synthetic seeds [135] and cryopreservation techniques for long-term storage [136].

3.4 Synthetic Seeds

Synthetic seeds are artificially created structures designed to encapsulate plant tissues including somatic embryos, shoots, apices, or cells with the objective of regenerating complete plants under either in vitro or ex vitro conditions [137]. Their composition typically includes a nutrient source (such as MS medium), growth regulators, a gelling agent, and additional components such as antibiotics or fungicides [138]. Among the principal advantages of synthetic seed technology are its ease of handling, storage, and transport, alongside its potential applications in mass propagation and short- to medium-term germplasm conservation [135, 139].

Two main types of synthetic seeds exist: hydrated and desiccated. The distinction between them lies in the dehydration process applied to desiccated seeds, typically conducted which is typically carried out overnight under aseptic conditions, often within a laminar flow chamber [140].

Table 2

Short- and medium-term in vitro conservation methods for ecological (EP), food (FP), industrial (IP), medicinal (MP), ornamental (OP), and woody plants (WP)

Species	Plant category	Tissue	Methods/condition	Storage period	Agents/modifications	Reference
Alcantarea nahoumii (Leme) J.R.Grant	EP	Seedlings	MS + 22 μmoL m^{-2} s^{-1} PPFD + 12 h light	2 years	Light and photoperiod change	[99]
Fragaria ananassa cv. Festival	FP	Runner segments and runner tips	¼ MS	270 days	Basal media	[100]
Colocasia esculenta (L.) Schott	FP	Apical meristems	MS + 2 mg L^{-1} ancymidol	168 days	Growth retardants	[101]
Solanum tuberosum L.	FP	Shoot apices	MS + 2 mg L^{-1} PAC + 50 mg L^{-1} de AgNPs	180 days	Growth retardants and nanotechnology	[95]
Vitis vinifera L.	FP	Nodal explants	¾ MS + 3 % sucrose + 3.5 % glucose + 2.5 % mannitol + 5°C + darkness	360 days	Osmotic agents, low temperature and darkness	[102]
Allium sativum L.	FP and MP	Garlic clove explant	MS + 68.46 g L^{-1} sucrose + 36.43 g L^{-1} sorbitol + 5°C + darkness	1 year	Osmotic agents, low temperature and darkness	[103]
Agave potatorum Zucc.	IP	Shoots	MS + 3 mg L^{-1} PBZ	180 days	Growth retardants	[104]
Gossypium barbadense L.	IP	Cotyledonary nodes	MS (1/2 macronutrients) + 2 mg L^{-1} $AgNO_3$	120 days	Basal media	[105]
Elaeis guineensis var. Tenera	IP	Plants from zygotic embryos	½ MS + 20°C + 12 h light	1 year	Basal media and low temperature	[106]
Elaeis oleifera (Kunth) Cortés	IP	Plants from zygotic embryos	½ MS + 20°C + 12 h light	1 year	Basal media and low temperature	[106]
Saccharum spp. Hybrid	IP	Seedlings	½ MS + 10 mg L^{-1} sucrose + 18°C	2 years	Low temperature and basal media	[107]

Species	Plant category	Tissue	Methods/condition	Storage period	Agents/modifications	Reference
Vanilla planifolia × *Vanilla pompona*	IP and FP	Nodal segmets	MS + 3 mg L^{-1} ABA + blue LED light	168 days	Light change	[108]
Vanilla planifolia Andrews	IP and FP	Nodal segmets	MS + 3 mg L^{-1} ABA + 22°C	120 days	Growth retardants	[90]
Passiflora edulis Sims 'UENF Rio Dourado'	IP and FP	Nodal explants	¼ MSM + 10 g L^{-1} sucrose + 20°C	180 days	Low temperature and basal media	[109]
Zingiber officinale Roscoe	MP	Rhizome shoots	MS + 90 g L^{-1} sucrose + 0.1 mg L^{-1} NAA + 1 mg L^{-1} BA	1 year	Osmotic agents	[110]
Lavandula angustifolia cv. 'Sineva'	MP	Bud meristems	¼ DM + 0.2 mg L^{-1} CCC + 60 g L^{-1} sucrose + 6°C + 1.25 μmol m^{-2} s^{-1} PPFD	2 years	Basal media, growth retardants, low temperature and light	[111]
Arnica montana L.	MP	Stem segments	½ MS + 20 g L^{-1}sucrose + 30.0 g L^{-1} sorbitol + 20 μmol m^{-2} s^{-1} PPFD	180 days	Low light, basal media and osmotic agents	[112]
Mentha X piperita	MP	Shoot apices	MS + 10°C + 2–4 μmol m^{-2} s^{-1} PPFD + 16 h light	2 years	Low light and temperature	[113]
Valeriana officinalis L.	MP	Axillary buds	¼ MS + 0.1 mg L^{-1} BAP	84 days	Basal media	[114]
Capparis spinosa L.	MP	Seedlings	MS + 3% sorbitol	119 days	Osmotic agents	[115]
Curcuma sp.	MP	Rhizome bud	MS + 24.6 μM 2iP	379 days	Plant growth regulator	[116]
Anthurium andraeanum Linden ex André	OP	Shoots	MS + 0.5 mg L^{-1} sorbitol	240 days	Osmotic agents	[117]
Melocactus paucispinus Heimen & R. Raul	OP	Apical segment of the cladodes	MS + 75 g L^{-1} sucrose	1 year	Osmotic agents	[118]

(continued)

Table 2
(continued)

Species	Plant category	Tissue	Methods/condition	Storage period	Agents/modifications	Reference
Melocactus glaucescens Buining & Brederoo	OP	Apical segment of the cladodes	MS + 75 g L^{-1} sucrose	1 year	Osmotic agents	[118]
Ranunculus asiaticus L.	OP	Shoots	DKW + 2 mg L^{-1} kinetina + 60 g L^{-1} sucrose + 6°C + darkness	270 days	Osmotic agents, low temperature and darkness	[119]
Turbinicarpus schmiedickeanus subsp. flaviflorus (G. Frank & A.B. Lau) Glass & R.A. Foster	OP	Shoots	MS + 30 g L^{-1} mannitol	1 year	Osmotic agents	[120]
Dianthus ingoldbyi Turrill	OP	Axillary buds	MS + 0.5 mg L^{-1} NAA + 4°C + darkness	180 days	Low temperature	[121]
Lilium davidii Duch. ex Elwes	OP	Seedlings	¼ MS + 3 mg L^{-1} ABA + −2°C + darkness	15 months	Basal media, growth retardants, low temperature and darkness	[122]
Nandina domestica Thunb	OP	Shoots	DKW + 60 g L^{-1} sucrose + 4°C + darkness	180 days	Osmotic agents, low temperature and darkness	[123]
Heliconia Champneiana cv. Splash	OP	Shoots	MS + 0.1 mg L^{-1} IBA + blue light	12 weeks	Light change and growth retardants	[124]
Laelia anceps Lindl.	OP and EP	Seedlings	MS + 2 mg L^{-1} PBZ	168 days	Growth retardants	[79]

Species	Plant category	Tissue	Methods/condition	Storage period	Agents/modifications	Reference
Cedrela fissilis Vell.	WP and EP	Shoot tips, cotyledonary and epicotyl nodal segments	MS + 2.5 mM de BAP	180 days	Encapsulation	[125]
Tetraclinis articulata (Vahl) Mast.	WP and EP	Shoots	SH + 4°C + darkness	196 days	Low temperature and darkness	[126]
Sequoia sempervirens (D. Don) Endl.	WP and EP	Apical shoots	QL + 4°C + darkness	1 year	Low temperature and darkness	[127]
Hancornia speciosa Gomes	WP and EP	Seedlings	MS + 15 g L^{-1} sucrose + 5 g L^{-1} sorbitol	120 days	Osmotic agents	[128]
Olea europaea L.	WP and FP	Shoots	MS + 4°C + 8 h light	240 days	Low temperature and light	[129]
Mangifera indica L.	WP and FP	Immature embryos	½ Ms + 100 mg L^{-1} Cysteine + 0.5 mg L^{-1} GA + 22 µmol m^{-2} s^{-1} PPFD + 12 h light	360 days	Low temperature and light, basal media and growth regulators	[130]
Citrus × *Jambhiri* Lush.	WP and FP	Microcuttings	WPM + 25 g L^{-1} sucrose + 22°C + 20 µmol m^{-2} s^{-1} + 12 h light	360 days	Low temperature and light	[131]

Gelation of the encapsulating matrix is achieved through the interaction of sodium alginate (usually at concentrations of 3–4%) and calcium chloride (at 75–100 mM) [141, 142]. Furthermore, agents incorporated into the synthetic seeds to enhance viability, germination, and protection include plant growth regulators such as auxins and cytokinins, which promote root and shoot development, respectively. Sucrose is often added as the principal carbon source [143]. Additional supplements may comprise fungicides, bactericides, fertilizers, or beneficial microorganisms such as mycorrhizal fungi, all of which contribute to contamination prevention and improved ex vitro germination rates [144].

The optimal storage duration and temperature for synthetic seeds are highly species-dependent and influenced by the specific preparation methods employed. Hung and Trueman (2012) demonstrated that synthetic seeds derived from eucalyptus shoots (*Corymbia torelliana* × *C. citriodora*) could be stored at 25°C for up to 8 weeks while maintaining regeneration rates between 50% and 84% [145]. In contrast, synthetic seeds of blackberry (cultivar "Thornfree") stored at 4°C retained high regeneration rates for up to 120 days when maintained in MS medium; however, deviations from this temperature or modifications to the regeneration substrate significantly reduced viability [146]. Interestingly, synthetic seeds of *Phalaenopsis* showed greater regeneration when stored at 25°C for up to 180 days, whereas storage at 4°C severely compromised their viability [142].

These findings underscore the necessity of developing species-specific protocols to define optimal storage conditions that maximize the viability, germination, and subsequent growth of synthetic seeds.

3.5 Cryopreservation

Techniques for the rapid multiplication and production of abundant plant material via tissue culture methods are critical prerequisites for the successful establishment of in vitro conservation and cryopreservation programs [147].

Cryopreservation is widely regarded as the ideal strategy for the long-term ex situ conservation of biodiversity. Defined as the process by which biological material is frozen and stored in liquid nitrogen (LN) at extremely low temperatures, typically around −196°C [148, 149], cryopreservation induces a complete cessation of metabolic activity within the preserved material [62], thereby enabling its storage for theoretically indefinite periods [150].

This technology, when optimized, offers safe, long-term preservation of biological materials without compromising their biological function or inducing genetic alterations [151]. Consequently, cryopreservation strengthens the stability and propagation potential of germplasm collections. Its principal advantage lies in its broad applicability to a diverse range of genotypes and tissues, including pollen, seeds, meristems, dormant buds, cell suspensions, embryogenic cultures, somatic and zygotic embryos, and callus tissue [152]. Additionally, cryotherapy, a derivative application of cryopreservation, can also be employed to eradicate plant pathogens such as phytoplasmas, viruses, and bacteria [153, 154].

Meristematic regions, characterized by active cell division, high cytoplasmic density, and minimal vacuolization, are typically selected for cryopreservation due to their increased resistance to freezing-induced damage [155]. The ultralow temperatures effectively halt all metabolic and biochemical activities [156].

Cryopreservation protocols aim to eliminate freezable water from cells through either physical or osmotic dehydration, followed by ultrarapid cooling. These protocols can be broadly classified into two types: classical cryopreservation, which involves cooling in the presence of ice formation, and vitrification-based approaches, where ice formation is avoided [10].

Vitrification refers to the direct transition of cellular water from the liquid phase to an amorphous, glassy state, without forming ice crystals [157]. This mechanism is critical for preserving hydrated cells, tissues, and organs at liquid nitrogen temperatures [158], as it minimizes tissue collapse, solute concentration effects, and pH alterations during dehydration [159]. However, achieving vitrification is technically challenging because water naturally tends to crystallize even at temperatures slightly below freezing. To counteract this tendency, highly concentrated cryoprotectant solutions are employed [160].

Common cryoprotectants include dimethyl sulfoxide (DMSO), polyethylene glycol (PEG), sucrose, sorbitol, and mannitol. These agents, acting either osmotically or through membrane penetration (as in the case of DMSO), protect cellular integrity during cryopreservation. They are typically incorporated into a solution or medium in which cells are suspended and then processed through programmable freezing devices. As external ice formation begins, solute concentrations rise, facilitating water efflux from cells and rapid replacement with cryoprotectants [150].

Sakai et al. introduced an effective cryoprotective formulation known as Plant Vitrification Solution 2 (PVS2), consisting of 30% (w/v) glycerol, 15% (w/v) ethylene glycol, and 15% (w/v) DMSO in a culture medium supplemented with 0.4 M sucrose at pH 5.8. PVS2 is an approximately 8 M solution that can supercool below −70°C and vitrify into a stable glass around −115°C after rapid cooling [161]. Upon warming, PVS2 exhibits a glass transition at −115°C, devitrification near −75°C, and melting around −36°C [162].

Successful cryopreservation requires the prevention of lethal intracellular ice formation during cooling. Thus, appropriate tissue dehydration prior to immersion in liquid nitrogen is imperative [163]. Additional critical factors influencing cryopreservation results include preconditioning, hardening, osmoprotection before cryopreservation, unloading after thawing, as well as parameters such as exposure time to PVS solutions, cryoprotectant concentration, temperature, explant size, age, and physiological condition [164].

Prior to the 1990s, conventional slow-freezing protocols were predominantly used [160]. Akira Sakai, a pioneer in the field, reported the survival of mulberry twigs subjected to liquid nitrogen exposure [165]. However, it took nearly two decades for cryopreservation to be firmly established as a research field. Advances have since transitioned from slow freezing methods to vitrification-based techniques [156].

Classical cryopreservation approaches are primarily suited for undifferentiated tissues such as cell suspensions and calluses. In contrast, differentiated tissues including seeds, embryonic axes, meristems, and zygotic or somatic embryos, achieve better survival rates with specialized methods such as encapsulation-dehydration (ED), vitrification, encapsulation-vitrification, desiccation, preculture, preculture-desiccation, and droplet-vitrification [150]. For detailed descriptions of these techniques, the reader is referred to Chap. 30 of this volume.

3.5.1 Slow Freezing

Classical cryopreservation methods, encompassing slow freezing and simple freezing techniques, were developed during the 1970s and 1980s. In the slow freezing approach, biological samples, including meristems, embryos, and cells are placed in cryotubes and treated with cryoprotectants such as dimethyl sulfoxide (DMSO), glycerol, ethylene glycol, and sucrose. Subsequently, the samples are precooled on ice at −7°C subjected to controlled freezing at a rate of 0.3–0.5°C per minute down to −40°C using a programmable freezer, facilitating progressive dehydration. Upon reaching the target temperature, the samples are immersed in liquid nitrogen (LN) for storage. Recovery of the plant material involves rapid thawing followed by re-establishment in appropriate culture media [163].

Despite their proven efficacy, these classical methods present operational complexities, primarily due to the need for sophisticated and costly programmable freezing equipment [10].

The principle underlying classical cryopreservation is based on inducing freeze-dehydration through slow cooling. As the temperature declines, extracellular ice formation precedes intracellular freezing, creating a concentration gradient that drives water efflux from the cytoplasm and vacuoles to the extracellular space, where it crystallizes. Consequently, the extent of cellular dehydration is determined by the cooling rate and the prefreezing temperature employed prior to immersion in liquid nitrogen [159].

Classical cryopreservation techniques have been successfully applied to undifferentiated plant culture systems, such as cell suspensions and calluses, across a wide range of species [166, 167]. Additionally, they have demonstrated efficacy for the cryopreservation of shoot tips from cold-tolerant plant species [168].

3.5.2 Techniques Based on Vitrification

Since the late 1980s, several highly efficient vitrification-based procedures have been developed for the cryopreservation of plant materials [162, 169–172]. These advances have been particularly significant for the preservation of complex structures such as meristems and somatic embryos, which could not be effectively cryopreserved using conventional slow-freezing protocols [173–175]. Moreover, when vitrification protocols are applied under optimized conditions, the integrity of the entire meristematic structure is largely preserved, enabling direct regrowth without the formation of intermediate calluses [176, 177]. This is critical for clonal germplasm conservation, as it minimizes the risk of somaclonal variation associated with callus-mediated regeneration [178, 179].

Further purifications in cryopreservation techniques have led to the development of approaches such as droplet–vitrification (D–V), vitrification using aluminum cryoplates (V-cryoplate), and air dehydration with cryoplates (D-cryoplate). These methods involve the direct immersion of biological samples into liquid nitrogen (LN) for ultra-rapid cooling, followed by ultra-rapid warming in a rewarming solution. The use of aluminum foil strips or cryoplates facilitates extremely high cooling (approximately 4000–5000°C min^{-1}) and warming rates (approximately 3000–4500°C min^{-1}), which contribute to high post-thaw regrowth rates [180].

As outlined in the 2020 Plant Conservation Report, cryopreservation repositories (cryobanks) have been established globally for a range of species, including *Manihot esculenta*, *Solanum tuberosum*, *Malus domestica*, *Pyrus communis*, *Coffea arabica*, *Morus alba* L., and *Allium sativum*, employing various cryopreservation protocols [181]. These efforts serve as a foundation for future PGRs conservation initiatives.

The development and dissemination of simplified and standardized protocols, supported by appropriate infrastructure and trained personnel, are critical steps toward overcoming current limitations in PGRs cryopreservation. Such advancements will facilitate the widespread implementation of cryogenic techniques in research institutions, laboratories, and cryobanks, contributing significantly to the preservation of plant biodiversity and food security [156].

Assessment of genetic stability following cryopreservation is essential to ensure the fidelity of conserved germplasm. Various molecular markers, including Inter-Simple Sequence Repeats (ISSR), Simple Sequence Repeats (SSR or microsatellites), Single Nucleotide Polymorphisms (SNP), Amplified Fragment Length Polymorphisms (AFLP), Random Amplified Polymorphic DNA (RAPD), Restriction Fragment Length Polymorphisms (RFLP), and Methylation Sensitive Amplified Polymorphisms (MSAP), have been employed for this purpose [88, 134, 182].

A bibliometric analysis conducted on the Scopus platform using the keywords "cryopreservation" and "plants" identified

3092 publications between 2000 and 2025, with a publication peak observed in 2021 (226 articles) (Fig. 10). However, a decline in publication numbers has been noted since 2022.

Keyword co-occurrence analysis reveals that the most highly cited topics include "cryopreservation," "cryoprotectant agent," "vitrification," "physiology," and "controlled" (Fig. 11]. This pattern indicates a strong research focus on cryoprotectant agents, as well as related biochemical evaluations involving antioxidant activity (189 occurrences), molecular biology analyses (291 occurrences), and studies addressing temperature-induced physiological changes (133 occurrences).

In the broader context of cryobiology, elucidating the physiological and molecular mechanisms that enable plant survival at extremely low temperatures remains a central research goal. Understanding plant responses to dehydration, freezing, and exposure to cryoprotectant agents provides critical insights for refining cryopreservation strategies. This knowledge is vital for ensuring successful recovery and long-term viability of conserved materials [183].

The co-occurrence map generated from the Scopus bibliometric data using the search terms "cryopreservation," "plants," and "biochemistry" or "molecular biology" highlights the multidisciplinary nature of research in this field (Fig. 12). Topics range from genetic stability and gene expression studies to species-specific cryobiological investigations, underlining the pivotal role of transcriptomics and other omics approaches in understanding cryopreservation impacts at the molecular level.

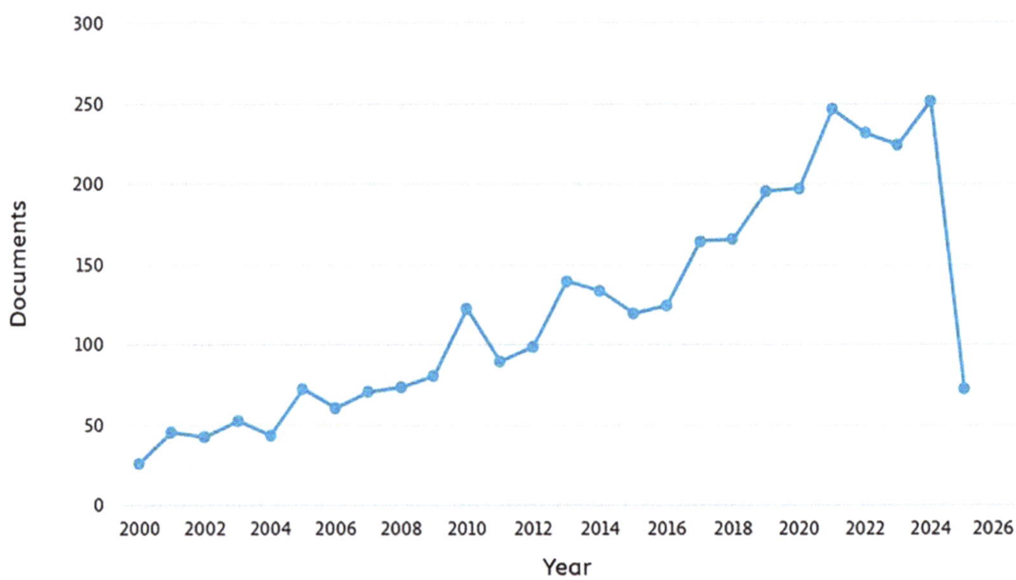

Fig. 10 Total number of publications indexed in Scopus (2000–2025) using the keywords "Cryopreservation" AND "Plants"

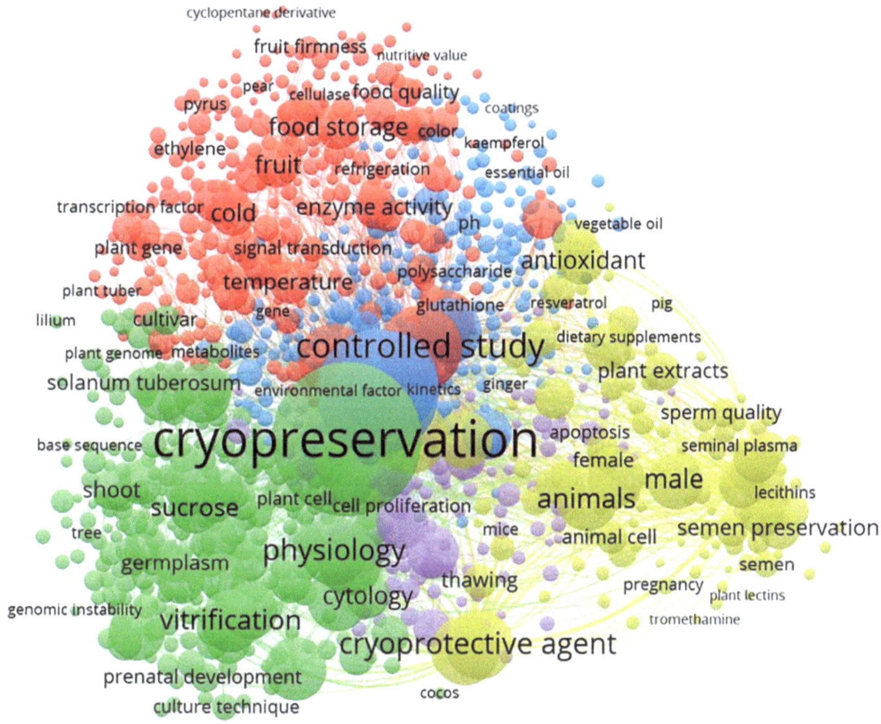

Fig. 11 Term co-occurrence network of Scopus publications (2000–2025) on plant cryopreservation using the keywords "Cryopreservation" AND "Plants"

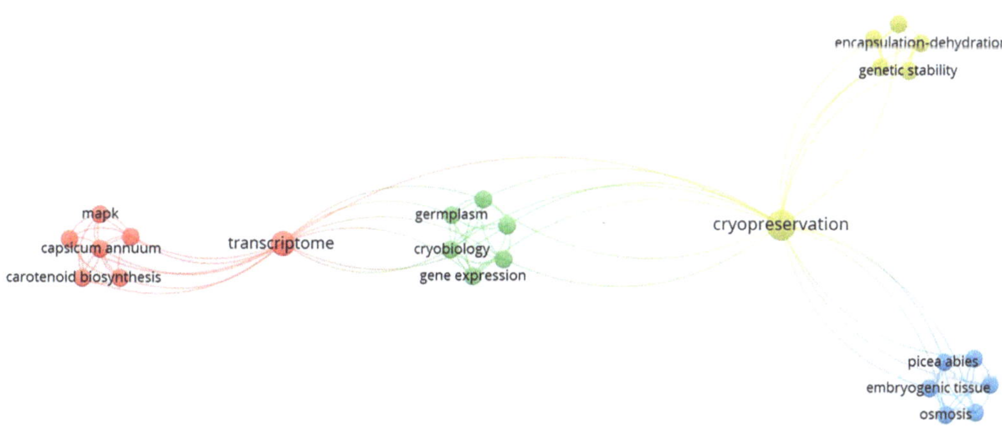

Fig. 12 Thematic network of cryopreservation-related plant research using the keywords "Criopreservation" AND "Plants" AND ("Biochemistry" OR "Molecular biology")

Overall, cryopreservation represents a cornerstone technique for the long-term conservation of plant species. Although significant advances have been made, the bibliometric analysis underscores the persistent knowledge gaps regarding the detailed molecular responses to cryogenic stress. Future research integrating molecular biology and biochemical approaches will be crucial to further optimize cryopreservation protocols and enhance the conservation of plant genetic resources.

4 Conclusion

The conservation of plant resources involves both in situ and ex situ strategies, representing a complex undertaking that demands multidisciplinary integration. These approaches are complementary in preserving biological diversity, facilitating the sustainable use of plant genetic resources of cultural, economic, and ecological significance.

Successful implementation of these strategies depends on several critical factors, including species biology, genetic stability, conservation methodologies, operational costs, and public policy frameworks. The relevance of plant resource conservation extends beyond local contexts and has global implications, as these resources underpin agriculture, food security, human health, scientific research, and cultural identity. Furthermore, they play an essential role in climate change mitigation and biodiversity preservation.

In situ conservation supports the continuation of evolutionary processes within natural ecosystems, while ex situ conservation offers a safeguard against catastrophic events and abrupt environmental changes. When integrated, these approaches provide a comprehensive safety net that enhances the survival, accessibility, and long-term sustainability of plant resources for future generations.

Looking ahead, conservation efforts must adopt more intelligent, adaptive, and participatory strategies. This requires interdisciplinary frameworks that integrate scientific rigor with traditional knowledge and community engagement. To ensure the effectiveness and resilience of these approaches, sustained political will, long-term funding, and an inclusive vision that recognizes the intrinsic value of plant life beyond its utilitarian functions are needed. In this context, emerging technologies such as genomics, artificial intelligence, and remote sensing platforms offer promising tools for advancing the precision and efficiency of plant resource management.

References

1. Halewood M, Chiurugwi T, Sackville Hamilton R, Kurtz B, Marden E, Welch E, Michiels F, Mozafari J, Sabran M, Patron N, Kersey P, Bastow R, Dorius S, Dias S, McCouch S, Powell W (2018) Plant genetic resources for food and agriculture: opportunities and challenges emerging from the science and information technology revolution. N Phytol 217(4):1407–1419. https://doi.org/10.1111/nph.14993

2. Hoban S, Bruford MW, Funk WC, Galbusera P, Griffith MP, Grueber CE, Heuertz M, Hunter ME, Hvilsom C, Kalamujic Stroil B, Kershaw F, Khoury CK, Laikre L, Lopes-Fernandes M, MacDonald AJ, Mergeay J, Meek M, Mittan C, Mukassabi TA, O'Brien D, Ogden R, Palma-Silva C, Ramakrishnan U, Segelbacher G, Shaw RE, Sjögren-Gulve P, Veličković N, Vernesi C (2021) Global commitments to conserving and monitoring

genetic diversity are now necessary and feasible. Bioscience 71(9):964–976. https://doi.org/10.1093/biosci/biab054
3. Engels JMM, Ebert AW (2021) A critical review of the current global ex situ conservation system for plant agrobiodiversity. I. History of the development of the global system in the context of the political/legal framework and its major conservation components. Plants 8:1557. https://doi.org/10.3390/plants10081557
4. Díaz M, Concepción ED, Morales MB, Alonso JC, Azcárate FM, Bartomeus I, Bota G, Brotons L, García D, Giralt D, Gutiérrez JE, López-Bao JV, Mañosa S, Milla R, Miñarro M, Navarro A, Olea PP, Palacín C, Peco B, Rey PJ, Seoane J, Suárez-Seoane S, Schöb C, Tarjuelo R, Traba J, Valera F, Velado-Alonso E (2021) Environmental objectives of Spanish agriculture: scientific guidelines for their effective implementation under the Common Agricultural Policy 2023-2030. Ardeola 68(2):445–460. https://doi.org/10.13157/arla.68.2.2021.fo1
5. Priyanka V, Kumar R, Dhaliwal I, Kaushik P (2021) Germplasm conservation: instrumental in agricultural biodiversity—a review. Sustainability 13(12):6743. https://doi.org/10.3390/su13126743
6. Mounce R, Smith P, Brockington S (2017) Ex situ conservation of plant diversity in the world's botanic gardens. Nat Plants 3(10):795–802. https://doi.org/10.1038/s41477-017-0019-3
7. Białoskórska M, Rucińska A, Boczkowska M (2024) Molecular mechanisms underlying freezing tolerance in plants: implications for cryopreservation. Int J Mol Sci 25(18):10110. https://doi.org/10.3390/ijms251810110
8. González-Arnao MT, Engelmann F (2013) Introducción a la conservación ex situ de los recursos genéticos vegetales. In: González-Arnao MT, Engelmann F (eds). Crioconservación de plantas en América Latina y el Caribe, IICA, San José
9. Mestanza-Ramón C, Henkanaththegedara SM, Vásconez Duchicela P, Vargas Tierras Y, Sánchez Capa M, Constante Mejía D, Jimenez Gutierrez M, Charco Guamán M, Mestanza-Ramón P (2020) In-situ and ex-situ biodiversity conservation in ecuador: a review of policies, actions and challenges. Diversity 12(8):315. https://doi.org/10.3390/d12080315
10. Cruz-Cruz CA, González-Arnao MT, Engelmann F (2013) Biotechnology and conservation of plant biodiversity. Resources 2:73–95. https://doi.org/10.3390/resources2020073
11. Mahanayak B (2024) Ex-situ and in-situ conservation of wild life. World J Biol Pharm Health Sci 18(03):277–282. https://doi.org/10.30574/wjbphs.2024.18.3.0371
12. Sánchez-Chiang N, Jiménez VM (2010) Técnicas de conservación in vitro para el establecimiento de bancos de germoplasma en cultivos tropicales. Agron Mesoam 21(1):193–205
13. WIEWS (2022) WIEWS-World Information and Early Warning System on Plant Genetic Resources for Food and Agriculture. https://www.fao.org/wiews/. Accessed 10 Apr 2025
14. Griffiths KE, Balding ST, Dickie JB, Lewis GP, Pearce TR, Grenyer R (2015) Maximizing the phylogenetic diversity of seed banks. Conserv Biol 29(2):370–381. https://doi.org/10.1111/cobi.12390
15. Abeli T, Dalrymple S, Godefroid S, Mondoni A, Müller JV, Rossi G, Orsenigo S (2020) Ex situ collections and their potential for the restoration of extinct plants. Conserv Biol 34(2):303–313. https://doi.org/10.1111/cobi.13391
16. Kallow S, Mertens A, Janssens SB, Vandelook F, Dickie J, Swennen R, Panis B (2022) Banana seed genetic resources for food security: status, constraints, and future priorities. Food Energy Secur 11(1):e345. https://doi.org/10.1002/fes3.345
17. Mirioba JN, Emitaro W, Obwanga B, Gaya H, Leley N, Otuoma J, Maina JM, Kawaka F (2024) Orchid species diversity across a forest disturbance gradient in west Mau forest, Kenya. PLoS One 19(8):e0307887. https://doi.org/10.1371/journal.pone.0307887
18. Saddoud Debbabi O, Miazzi MM, Elloumi O, Fendri M, Ben Amar F, Savoia M, Sion S, Souabni H, Mnasri SR, Abdelaali SB, Jendoubi F, Mangini G, Famiani F, Taranto F, Montemurro C, Msallem M (2020) Recovery, assessment, and molecular characterization of minor olive genotypes in Tunisia. Plants (Basel) 9(3):382. https://doi.org/10.3390/plants9030382
19. Ouaja M, Bahri BA, Aouini L, Ferjaoui S, Medini M, Marcel TC, Hamza S (2021) Morphological characterization and genetic diversity analysis of Tunisian durum wheat (*Triticum turgidum* var. durum) accessions. BMC Genom Data 22:1–17. https://doi.org/10.1186/s12863-021-00958-3
20. Lupien SL, Dugan FM, Ward KM, O'Donnell K (2017) Wilt, crown, and root rot of common rose mallow (*Hibiscus moscheutos*) caused by a novel Fusarium sp. Plant Dis 101(2):354–358. https://doi.org/10.1094/pdis-05-16-0717-re

21. Clement SL, Hellier BC, Elberson LR, Staska RT, Evans MA (2007) Flies (Diptera: Muscidae: Calliphoridae) are efficient pollinators of *Allium ampeloprasum* L.(Alliaceae) in field cages. J Econ Entomol 100(1):131–135. https://doi.org/10.1603/0022-0493(2007)100[131:fdmcae]2.0.co;2
22. Cunha Alves AA, Azevedo VCR (2018) Embrapa network for Brazilian plant genetic resources conservation. Biopreserv Biobank 16(5):350–360. https://doi.org/10.1089/bio.2018.0044
23. Dinato NB, Imaculada Santos IR, Zanotto Vigna BB, de Paula AF, Fávero AP (2020) Pollen cryopreservation for plant breeding and genetic resources conservation. Cryo Lett 41(3):115–117
24. Kumar S, Archak S, Tyagi RK, Kumar J, VK V, Jacob SR, Srinivasan K, Radhamani J, Parimalan R, Sivaswamy M, Tyagi S, Yadav M, Kumari J, Deepali, Sharma S, Bhagat I, Meeta M, Bains NS, Chowdhury AK, Saha BC, Bhattacharya PM, Kumari J, Singh MC, Gangwar OP, Prasad P, Bharadwaj SC, Gogoi R, Sharma JB, Kumar GMS, Saharan MS, Bag M, Roy A, Prasad TV, Sharma RK, Dutta M, Sharma I, Bansal KC (2016) Evaluation of 19,460 wheat accessions conserved in the Indian national genebank to identify new sources of resistance to rust and spot blotch diseases. PLoS One 11(12):e0167702. https://doi.org/10.1371/journal.pone.0167702
25. Kumari J, Lakhwani D, Jakhar P, Sharma S, Tiwari S, Mittal S, Avashthi H, Shekhawat N, Singh K, Mishra KK, Singh R, Yadav MC, Singh GP, Singh AK (2023) Association mapping reveals novel genes and genomic regions controlling grain size architecture in mini core accessions of Indian National Genebank wheat germplasm collection. Front Plant Sci 14:1148658. https://doi.org/10.3389/fpls.2023.1148658
26. Kumar S, Jacob SR, Mir RR, Vikas VK, Kulwal P, Chandra T, Kaur S, Kumar U, Kumar S, Sharma S, Singh R, Prasad S, Singh AM, Kumari J, Saharan MS, Bhardwaj SC, Prasad M, Kalia S, Singh K (2022) Indian wheat genomics initiative for harnessing the potential of wheat germplasm resources for breeding disease-resistant, nutrient-dense, and climate-resilient cultivars. Front Genet 13:834366. https://doi.org/10.3389/fgene.2022.834366
27. Malaviya DR, Raman H, Dear BS, Raman R, Roy AK, Kaushal P, Chandra A, Hughes SJ (2019) Genetic diversity and lineage based on SSR markers of two genomic resources among trifolium collections held within the australian pastures Genebank. Open J Genet 9(01). https://doi.org/10.4236/ojgen.2019.91001
28. Smith RW, Harris CA, Cox K, McClements D, Clark SG, Hossain Z, Humphries AW (2021) A history of Australian pasture genetic resource collections. Crop Pasture Sci https://doi.org/10.1071/CP20336
29. Westengen OT, Lusty C, Yazbek M, Amri A, Asdal Å (2020) Safeguarding a global seed heritage from Syria to Svalbard. Nat Plants 6(11):1311–1317. https://doi.org/10.1038/s41477-020-00802-z
30. Hansson M, Youssef HM, Zakhrabekova S, Stuart D, Svensson JT, Dockter C, Stein N, Waugh R, Lundqvist U, Franckowiak J (2024) A guide to barley mutants. Hereditas 161(11). https://doi.org/10.1186/s41065-023-00304-w
31. Otim MH, Alibu S, Asea G, Abalo G, Sserumaga JP, Adumo S, Alupo J, Ochen S, Tefera T, Bruce AY, Beyene Y, Meisel B, Tende R, Nang'ayo F, Baguma Y, Mugo S, Oikeh SO (2022) Performance of Bt maize event MON810 in controlling maize stem borers Chilo partellus and Busseola fusca in Uganda. Crop Prot 156:105945. https://doi.org/10.1016/j.cropro.2022.105945
32. IUCN (2021) Nature 2030: one nature, one future. International Union for Conservation of Nature. Available via IUCN Library Portal. https://portals.iucn.org/library/sites/library/files/documents/WCC-7th-001-En.pdf. Accessed 10 Apr 2025
33. IUCN-CEM (2022) The IUCN Red List of Ecosystems. Version 2022-1. https://iucn.org/. Accessed 10 Apr 2025
34. Goettsch B, Hilton-Taylor C, Cruz-Piñón G, Duffy JP, Frances A, Hernández HM, Inger R, Pollock C, Schipper J, Superina M, Taylor NP, Tognelli M, Abba AM, Arias S, Arreola-Nava HJ, Baker MA, Bárcenas RT, Barrios D, Braun P, Butterworth CA, Búrquez A, Caceres F, Chazaro-Basañez M, Corral-Díaz R, Perea MV, Demaio PH, Barros WAD, Durán R, Faúndez Yancas L, Felger RS, Fitz-Maurice B, Fitz-Maurice WA, Gann G, Gómez-Hinostrosa C, Gonzales-Torres LR, Griffith MP, Guerrero PC, Hammel B, Heil KD, Hernández-Oria JG, Hoffmann M, Ishihara MI, Kiesling R, Larocca J, León-de la Luz JL, Loaiza SCR, Lowry M, Machado MC, Majure LC, Martínez Ávalos JG, Martorell C, Maschinski J, Méndez E, Mittermeier RA, Nassar JM, Negrón-Ortiz V, Oakley LJ, Ortega-Baes P, Ferreira ABP, Pinkava DJ, Porter JM, Puente-Martinez R, Gamarra JR, Pérez PS, Martínez ES, Smith M, Sotomayor M, del CJM, Stuart SN, Muñoz JLT, Terrazas T, Terry M, Trevisson M, Valverde T, Van

Devender TR, Véliz-Pérez ME, Walter HE, Wyatt SA, Zappi D, Zavala-Hurtado JA, Gaston KJ (2015) High proportion of cactus species threatened with extinction. Nat Plants 1(10):1–7. https://doi.org/10.1038/nplants.2015.142
35. Havens K, Kramer AT, Guerrant EO (2014) Getting plant conservation right (or not): the case of the United States. Int J Plant Sci 175(1):3–10. https://doi.org/10.1086/674103
36. Corlett RT (2016) Plant diversity in a changing world: status, trends, and conservation needs. Plant Divers 38(1):10–16. https://doi.org/10.1016/j.pld.2016.01.001
37. Zegeye H (2017) In situ and ex situ conservation: complementary approaches for maintaining biodiversity. IJRES 4(1):1–12
38. Jin X, Ren H (2020) In situ and ex situ conservation of Chinese plants. In: Ren H (ed) Conservation andreintroduction of rare and endangered plants in China. Springer, Singapore
39. Heywood VH (2014) An overview of in situ conservation of plant species in the Mediterranean. Flora Mediterr 24:5–24. https://doi.org/10.7320/FlMedit24.001
40. Braverman I (2014) Conservation without nature: the trouble with in situ versus ex situ conservation. Geoforum 51:47–57. https://doi.org/10.1016/j.geoforum.2013.09.018
41. Rotach P (2005) In situ conservation methods. In: Geburek T, Turok J (eds) Conservation and Management of Forest Genetic Resources in Europe. Arbora, Zvolen
42. Maxted N, Kell S, Ford-Lloyd B, Dulloo E, Toledo Á (2012) Toward the systematic conservation of global crop wild relative diversity. Crop Sci 52(2):774–785. https://doi.org/10.1111/j.1748-1090.2012.00192.x
43. Heywood VH (2014) An overview of in situ conservation of plant species in the Mediterranean. Flora Mediterr 24:5–24. https://doi.org/10.7320/FlMedit24.005
44. Jones KR, Venter O, Fuller RA, Allan JR, Maxwell SL, Negret PJ, Watson JE (2018) One-third of global protected land is under intense human pressure. Science 360(6390):788–791. https://doi.org/10.1126/science.aap9565
45. Heywood VH (2019) Conserving plants within and beyond protected areas–still problematic and future uncertain. Plant Divers 41(2):36–49. https://doi.org/10.1016/j.pld.2018.10.001
46. Velazco SJE, Bedrij NA, Rojas JL, Keller HA, Ribeiro BR, De Marco P (2022) Quantifying the role of protected areas for safeguarding the uses of biodiversity. Biol Conserv 268:109525. https://doi.org/10.1016/j.biocon.2022.109525
47. Maron M, Simmonds JS, Watson JE (2018) Bold nature retention targets are essential for the global environment agenda. Nat Ecol Evol 2(8):1194–1195. https://doi.org/10.1038/s41559-018-0595-2
48. Cao M, Peng L, Liu S (2015) Analysis of the network of protected areas in China based on a geographic perspective: current status, issues and integration. Sustainability 7(11):15617–15631. https://doi.org/10.3390/su71115617
49. Bejatović G, Ristić J (2016) Modern concept of tourism development and normative protection of National Parks. Tour Int Sci Conf Vrnjačka Banja 1(2):483–500
50. Bell J, Stockdale A (2015) Evolving national park models: the emergence of an economic imperative and its effect on the contested nature of the "national" park concept in Northern Ireland. Land Use Policy 49:213–226. https://doi.org/10.1016/j.landusepol.2015.08
51. Rhama B, Timang JH, Palangka JR, Raya KP (2020) The meta-analysis of Ecotourism in National Parks. Afr J Hosp 9(1):1–17
52. Rastandeh A, Brown DK, Zari MP (2018) Site selection of urban wildlife sanctuaries for safeguarding indigenous biodiversity against increased predator pressures. UFUG 32:21–31. https://doi.org/10.1016/j.ufug.2018.03.019
53. Innes J, Burns B, Sanders A, Hayward MW (2015) The impact of private sanctuary networks on reintroduction programs. In: Armstrong DP, Hayward MW, Moro D, Seddon PJ (eds). Advances in reintroduction biology of Australian and New Zealand fauna. CSIRO Publishing, Canberra
54. UNESCO (2024) What are biosphere reserves? https://www.unesco.org/es/mab/wnbr/about Accessed 28 Oct 2024
55. Ruiz-Mallén I, Corbera E, Calvo-Boyero D, Reyes-García V, Brown K (2015) How do biosphere reserves influence local vulnerability and adaptation? Evidence from Latin America. Glob Environ Change 33:97–108
56. Van Cuong C, Dart P, Hockings M (2017) Biosphere reserves: attributes for success. J Env Manag 188:9–17. https://doi.org/10.1016/j.jenvman.2016.11.069
57. Birader K (2023) Genetic diversity and the adaptation of species to changing environments. J Biodivers Endanger Species 11(03):474
58. Pence VC, Meyer A, Linsky J, Gratzfeld J, Pritchard HW, Westwood M, Bruns EB (2022) Defining exceptional species—a conceptual framework to expand and advance *ex*

situ conservation of plant diversity beyond conventional seed banking. Biol Conserv 266:109440. https://doi.org/10.1016/j.biocon.2021.109440
59. Ulian T, Sacandé M, Hudson A, Mattana E (2017) Conservation of indigenous plants to support community livelihoods: the MGU–Useful Plants Project. JEPM 60(4):668–683. https://doi.org/10.1080/09640568.2016.1166101
60. Egan LM, Hofmann RW, Ghamkhar K, Hoyos-Villegas V (2021) Prospects for Trifolium improvement through germplasm characterisation and pre-breeding in New Zealand and beyond. Front Plant Sci 12:653191. https://doi.org/10.3389/fpls.2021.653191
61. Gutiérrez CB, Koch ZL (2015) Conservación de germoplasma ex situ: Protocolos y estrategias para la mantención de un banco in vitro. Cienc Investig For 21(1):69–82. https://doi.org/10.52904/0718-4646.2015.433
62. Morales MMB (2015) Conservación in vitro: una perspectiva para el manejo de los Recursos Fitogenéticos. Rev Investig Agraria Ambient 6(1):67–82. https://doi.org/10.22490/21456453.1264
63. R Core Team (2024) R: A Language and Environment for Statistical Computing_. R Foundation for Statistical Computing, Vienna, Austria. https://www.R-project.org/
64. Ortiz R, Taba S, Tovar VHC, Mezzalama M, Xu Y, Yan J, Crouch JH (2010) Conserving and enhancing maize genetic resources as global public goods–a perspective from CIMMYT. Crop Sci 50(1):13–28. https://doi.org/10.2135/cropsci2009.06.0297
65. Lüttringhaus S, Pradel W, Suarez V, Manrique-Carpintero NC, Anglin NL, Ellis D, Hareau G, Jamora N, Smale M, Gómez R (2021) Dynamic guardianship of potato landraces by Andean communities and the genebank of the International Potato Center. CABI Agric Biosci 2(45):1–16. https://doi.org/10.1186/s43170-021-00065-4
66. Lindqvist-Kreuze H, Bonierbale M, Grüneberg WJ, Mendes T, De Boeck B, Campos H (2024) Potato and sweetpotato breeding at the International Potato Center: approaches, outcomes and the way forward. Theor Appl Genet 137(1):12. https://doi.org/10.1007/s00122-023-04515-7
67. King J, Dreisigacker S, Reynolds M, Bandyopadhyay A, Braun HJ, Crespo-Herrera L, Crossa J, Govindan V, Huerta J, Ibba MI, Robles-Zazueta CA, Saint Pierre C, Singh PK, Singh RP, Achary VMM, Bhavani S, Blasch G, Cheng S, Dempewolf H, Flavell RB, Gerard G, Grewal S, Griffiths S, Hawkesford M, He X, Hearne S, Hodson D, Howell P, Kamali MRJ, Karwat H, Kilian B, King IP, Kishii M, Kommerell VM, Lagudah E, Lan C, Montesinos-Lopez OA, Nicholson P, Pérez-Rodríguez P, Pinto F, Pixley K, Rebetzke G, Rivera-Amado C, Sansaloni C, Schulthess U, Sharma S, Shewry P, Subbarao G, Tiwari TP, Trethowan R, Uauy C (2024) Wheat genetic resources have avoided disease pandemics, improved food security, and reduced environmental footprints: a review of historical impacts and future opportunities. Glob Chang Biol 30(8):e17440. https://doi.org/10.1111/gcb.17440
68. Breman E, Ballesteros D, Castillo-Lorenzo E, Cockel C, Dickie J, Faruk A, O'Donnell K, Offord CA, Pironon S, Sharrock S, Ulian T (2021) Plant diversity conservation challenges and prospects—the perspective of botanic gardens and the Millennium Seed Bank. Plants 10(11):2371
69. Philpott M, Pence VC, Bassüner B, Clayton AS, Coffey EED, Downing JL, Edwards CE, Folgado R, Ligon JJ, Powell C, Ree JF, Seglias AE, Sugii N, Zale PJ, Zeldin J (2022) Harnessing the power of botanical gardens: evaluating the costs and resources needed for exceptional plant conservation. APPS 10(5):e11495. https://doi.org/10.1002/aps3.11495
70. Cibrian-Jaramillo A, Hird A, Oleas N, Ma H, Meerow AW, Francisco-Ortega J, Griffith MP (2013) What is the conservation value of a plant in a botanic garden? Using indicators to improve management of ex situ collections. Bot Rev 79:559–577. https://doi.org/10.1007/s12229-013-9120-0
71. Hayden K (2020) Botanic gardens and plant pathogens: a risk-based approach at the Royal Botanic Garden Edinburgh. Sibbaldia 18:127–139. https://doi.org/10.24823/Sibbaldia.2020.293
72. Hoyle GL, Sommerville KD, Liyanage GS, Worboys S, Guja LK, Stevens AV, Crayn DM (2023). Seed banking is more applicable to the preservation of tropical montane flora than previously assumed: a review and cloud forest case study. Glob Ecol Conserv e02627. https://doi.org/10.1016/j.gecco.2023.e02627
73. De Vitis M, Hay FR, Dickie JB, Trivedi C, Choi J, Fiegener R (2020) Seed storage: maintaining seed viability and vigor for restoration use. Restor Ecol 28(S3):S249–S255. https://doi.org/10.1111/rec.13174
74. Tyagi RK, Agrawal A (2015) Revised genebank standards for management of plant genetic resources. Indian J Agric Sci 85(2):157–165. https://doi.org/10.56093/ijas.v85i2.46437
75. Phillips GC, Garda M (2019) Plant tissue culture media and practices: an overview. In Vitro

Cell Dev Biol Plant 55:242–257. https://doi.org/10.1007/s11627-019-09983-5
76. Imarhiagbe O, Osazee JO, Aiwansoba RO, Shittu OH (2016) In vitro germplasm collection and storage: a review. In Vitro Int J Biol Res 1(1):09–14
77. POWO (2025) Plants of the World Online. Facilitated by the Royal Botanic Gardens, Kew. https://powo.science.kew.org/ Accessed 10 April 2025.
78. IUCN (2025) The IUCN Red List of Threatened Species. https://www.iucnredlist.org/ Accessed 10 Apr 2025
79. Ramírez-Mosqueda MA, Cruz-Cruz CA, Atlahua-Temoxtle J, Bello-Bello JJ (2019) In vitro conservation and regeneration of Laelia anceps Lindl. S Afr J Bot 121:219–223. https://doi.org/10.1016/j.sajb.2018.11.010
80. Cruz-Cruz CA, González-Arnao MT, Bautista-Aguilar JR, Ramírez-Mosqueda MA (2022) In vitro short-term storage of *Stanhopea tigrina* Bateman ex Lind. S Afr J Bot 151:334–338. https://doi.org/10.1016/j.sajb.2022.10.014
81. Rodrigues PHV, Oliveira EL, Demetrio CA, Ambrosano GB, Piedade SMS (2022) Effects of different light spectra on the slow-grown in vitro storage and quality of banana plantlets cv. Prata Catarina (AAB). Plant Cell Tiss Organ Cult 150(2):479–485
82. Berhe BD, Sbhatu DB, Munawar TM, Gebreyohannes G (2023) Aloe monticola Reynolds: a refugee of the mountains—contributing towards its conservation through in vitro propagation. Heliyon 9(12). https://doi.org/10.1016/j.heliyon.2023.e22955
83. Pereira CD, Bernini CS, Jantsch MR, Medeiros RA, de MOURA LC (2021) Germination and in vitro propagation of Brazilian mahogany (*Swietenia macrophylla* King). Nativa 9(5):595–599. https://doi.org/10.31413/nativa.v9i5.12129
84. Nasiruddin M, Islam AR (2018) In vitro slow-growth conservation for two genotypes of *Solanum tuberosum* L. Bangladesh J Bot 47(3):369–380. https://doi.org/10.3329/bjb.v47i3.38654
85. Pathirana C, Attanayake U, Dissanayake U, Gamlath L, Ketipearachchi K, Madhujith T, Eeswara J (2020) Establishment of a micropropagation protocol for elite accessions of Bael (*Aegle marmelos* (L.) Corr.), a tropical hardwood species. Adv Agric 2020(1):8840386. https://doi.org/10.1155/2020/8840386
86. Nadha HK, Salwan R, Kasana RC, Anand M, Sood A (2012) Identification and elimination of bacterial contamination during in vitro propagation of *Guadua angustifolia* Kunth. Pharmacogn Mag 8(30):93. https://doi.org/10.4103/0973-1296.96547
87. Lara-Ascencio M, Andrade-Rodríguez M, Guillén-Sánchez D, Sotelo-Nava H, Villegas-Torres OG (2021) Establishment of in vitro aseptic culture of *Philodendron xanadu* Croat. Rev Ciênc Agron 52(2):e20197034. https://doi.org/10.5935/1806-6690.20210024
88. Hussain H, Nisar M (2020) Assessment of plant genetic variations using molecular markers: a review. J Appl Biol Biotechnol 8(5):99–109. https://doi.org/10.7324/JABB.2020.80514
89. El-Hawaz RF, Adelberg J, Naylor-Adelberg J, Eisenreich R, Van der Meij J (2019) The effect of slow-growth strategy on a production of Petunia× hybrida Vilm. microcuttings. In Vitro Cell Dev Biol Plant 55:433–441. https://doi.org/10.1007/s11627-019-09996-0
90. Bautista-Aguilar JR, Iglesias-Andreu LG, Martínez-Castillo J, Ramírez-Mosqueda MA, Ortiz-García MM (2021) In vitro conservation and genetic stability in *Vanilla planifolia* Jacks. HortScience 56(12):1494–1498. https://doi.org/10.21273/HORTSCI16118-21
91. Singh C, RAI P, Rajkumar BK (2017) Recent advances in conservation of plant genetic resources. ARTOAJ 7:1–2. https://doi.org/10.19080/ARTOAJ.2017.07.555718
92. Rai MK, Kalia RK, Singh R, Gangola MP, Dhawan AK (2011) Developing stress tolerant plants through in vitro selection—an overview of the recent progress. EEB 71(1):89–98. https://doi.org/10.1016/j.envexpbot.2010.10.021
93. Sarasan V (2010) Importance of in vitro technology to future conservation programmes worldwide. Kew Bull 65:549–554. https://doi.org/10.1007/s12225-011-9250-7
94. Buldakov SA (2021) Use of growth inhibitor chlormequat chloride in potato culture in vitro. E3S Web Conf 285:03003. https://doi.org/10.1051/e3sconf/202128503003
95. Baltazar-Bernal O, Spinoso-Castillo JL, Mancilla-Álvarez E, Muñoz-Márquez RA, Bello-Bello JJ (2023) In vitro Conservation and Regeneration of Potato (*Solanum tuberosum* L.): role of Paclobutrazol and Silver Nanoparticles. Horticulturae 9(6):676. https://doi.org/10.3390/horticulturae9060676
96. Noorrohmah S, Sari L, Ermayanti TM (2023) The effects of ancymidol and sucrose concentrations on growth of breadfruit Artocarpus altilis (Park.) Fosberg: shoot culture for in vitro preservation. AIP Conf Proc 2606(1). https://doi.org/10.1063/5.0118388
97. Martínez-Santos E, Cruz-Cruz CA, Spinoso-Castillo JL, Bello-Bello JJ (2021) *In vitro* response of vanilla (*Vanilla planifolia* Jacks. ex Andrews) to PEG-induced osmotic stress. Sci Rep 11(1):22611. https://doi.org/10.1038/s41598-021-02207-0

98. Cioloca M, Tican A, Bădărău C, Bărăscu N, Popa M (2021) In vitro medium term conservation of sweet potato genotypes using mannitol and sorbitol. Rom Agric Res 38:123–132. https://doi.org/10.59665/rar3813
99. dos Santos Silva SS, de Souza EH, Nepomuceno CF, Souza FVD, de Carvalho Costa MAP (2020) Micropropagation and in vitro conservation of *Alcantarea nahoumii* (Bromeliaceae), an endemic and endangered species of the Brazilian Atlantic Forest. Acta Scientiarum. Biol Sci 42:1–10. https://doi.org/10.4025/actascibiolsci.v42i1.52940
100. Kadhim ZK, Abdulhussein MA (2021) Minimal media strength for in vitro conservation of strawberry (*Fragaria ananassa*) cultures. Basrah J Agric Sci 34(2):1–9. https://doi.org/10.37077/25200860.2021.34.2.01
101. Mancilla-Álvarez E, Ramírez-Mosqueda MA, Arano-Avalos S, Núnez-Pastrana R, Bello-Bello JJ (2019) In vitro techniques to the conservation and plant regeneration of malanga (*Colocasia esculenta* L. Schott). HortScience 54(3):514–518. https://doi.org/10.21273/HORTSCI13835-18
102. AlMousa R, Hassan N (2022) Effect of osmotic regulators on in vitro conservation of grape. J Genet Environ Resour Conserv 10(1):60–66
103. Tirado B, Gómez-Rodríguez VM, Cruz-Cárdenas CI, Zelaya-Molina LX, Ramírez-Vega H, Sandoval-Cancino G (2023) In vitro conservation of Mexican garlic varieties by minimal growth. Plants (Basel) 12(23):3929. https://doi.org/10.3390/plants12233929
104. Spinoso-Castillo JL, Pérez-Sato JA, Schettino-Salomón SS, Bello-Bello JJ (2022) An alternative method for medium-term in vitro conservation of different plant species through gibberellin inhibitors. In Vitro Cell Dev Biol Plant 58(4):606–614. https://doi.org/10.1007/s11627-022-10263-y
105. Delgado-Paredes G, Rojas-Idrogo C, Vásquez-Díaz C, Esquerre-Ibañez B, Bazán-Sernaqué P, Custodio Ayasta P, Silva ZD (2024) Collection,fiber color characterization and germplasm conservation of native cotton *Gossypium Barbadense* L. Chil J Agric Anim Sci Ex Agro-Cienc 40(1):1–12. https://doi.org/10.29393/CHJAAS40-1CFGF70001
106. Camillo J, Scherwinski-Pereira JE (2015) In vitro maintenance, under slow-growth conditions, of oil palm germplasm obtained by embryo rescue. Pesqui Agropecu Bras 50(05):426–429. https://doi.org/10.1590/S0100-204X2015000500010
107. Banasiak M, Snyman SJ (2017) Exploring in vitro germplasm conservation options for sugarcane (*Saccharum* spp. hybrids) in South Africa. In Vitro Cell Dev Biol Plant 53:402–409
108. Bautista-Aguilar JR, Iglesias-Andreu LG, Flores-Estévez N, Sánchez-Coello NG (2024) Effect of LED lighting on minimal growth for in vitro conservation of vanilla genotypes. Propag Ornam Plants 24(1):13–19
109. Generoso AL, Carvalho VS, Sales RA, Brito NL, Viana AP, Pereira TNS (2023) New evidence for in vitro conservation of nodal segments of the passion fruit 'UENF Rio Dourado' (*Passiflora edulis* Sims). Acta Scientiarum Agron 45:e59498. https://doi.org/10.4025/actasciagron.v45i1.59498
110. Tyagi RK, Agrawal A, Yusuf A (2006) Conservation of Zingiber germplasm through in vitro rhizome formation. Sci Hortic 108(2):210–219. https://doi.org/10.1016/j.scienta.2006.01.018
111. Brailko V, Ivanova N, Zhdanova I, Mitrofanova O (2020) Morphological and anatomical features of narrow-leaved lavender plants with prolonged conservation under in vitro genebank. BIO Web Conf 24:5. https://doi.org/10.1051/bioconf/20202400015
112. Petrova M, Zayova E, Geneva M, Dimitrova L, Vitkova A, Stanilova M (2021) Multiplication and conservation of threatened medicinal plant Arnica montana L. by in vitro techniques. Agric Conspec Sci 86(1):57–65
113. Martín C, Senula A, González I, Acosta A, Keller EJ, González-Benito ME (2013) Genetic identity of three mint accessions stored by different conservation procedures: field collection, in vitro and cryopreservation. Genet Resour Crop Evol 60:243–249. https://doi.org/10.1007/s10722-012-9830-x
114. Syahid SF, Palindung LS (2023) In vitro conservation of *Valeriana officinalis* L. through minimal growth. E3S Web Conf 373:03025. https://doi.org/10.1051/e3sconf/202337303025
115. Al-Mahmood HJ, Shatnawi MA, Shibli RA, Makhadmeh IM, Abubaker SM, Shadiadeh AN (2012) Clonal propagation and medium-term conservation of *Capparis spinosa*: a medicinal plant. J Med Plants Res 6(22):3826–3836
116. Tyagi RK, Yusuf A, Dua P, Agrawal A (2004) In vitro plant regeneration and genotype conservation of eight wild species of Curcuma. Biol Plant 48:129–132. https://doi.org/10.1023/B:BIOP.0000024289.68669.ef
117. Hanan MA, Sherif SS (2013) Conservation plantlets of *Anthurium andraeanum* (Lindeu ex Andre) by using in vitro. Middle East J 02(2):36–43
118. Torres-Silva G, Schnadelbach AS, Bezerra HB, Lima-Brito A, Resende SV (2021) In vitro conservation and genetic diversity of threatened

species of Melocactus (Cactaceae). Biodivers Conserv 30(4):1067–1080. https://doi.org/10.1007/s10531-021-02132-8
119. Benelli C, Ozudogru EA, Lambardi M, Dradi G (2011) In vitro conservation of ornamental plants by slow growth storage. In: VII International Symposium on In Vitro Culture and Horticultural Breeding, vol 961, pp 89–93. https://doi.org/10.17660/ActaHortic.2012.961.8
120. Pérez-Molphe-Balch E, Pérez-Reyes ME, De La Rosa MDL (2012) In vitro conservation of Turbinicarpus (Cactaceae) under slow growth conditions. Haseltonia 2012(17):51–57. https://doi.org/10.2985/1070-0048-17.1.6
121. Arda H, Dayan S, Kartal Ç, Güler N (2016) In vitro conservation of critically endangered Dianthus ingoldbyi Turrill under slow growth conditions. Trak Univ J Nat Sci 17(1):47–54
122. Yun-peng D, Wen-yuan L, Ming-fang Z, Heng-bin H, Gui-xia J (2012) The establishment of a slow-growth conservation system in vitro for two wild lily species. Afr J Biotechnol 11(8):1981–1990. https://doi.org/10.5897/AJB11.2868
123. Ozudogru A, da SILVA DPC, Ergun KAYA, Dradi G, Paiva R, Lambardi M (2013) In vitro conservation and cryopreservation of *Nandina domestica*, an outdoor ornamental shrub. Not Bot Horti Agrobo 41(2):638–645. https://doi.org/10.15835/nbha4129335
124. Rodrigues PHV, Arruda F, Forti VA (2018) Slow-grown in vitro conservation of *Heliconia champneiana* cv. Splash under different light spectra. Sci Agric 75(2):163–166. https://doi.org/10.1590/1678-992x-2016-0394
125. Da Costa Nunes E, Benson EE, Oltramari AC, Sibila Araujo P, Righetto Moser J, Viana AM (2003) In vitro conservation of *Cedrela fissilis* Vellozo (Meliaceae), a native tree of the Brazilian Atlantic Forest. Biodivers Conserv 12:837–848. https://doi.org/10.1023/A:1022492226341
126. Juan-Vicedo J, Serrano-Martínez F, Cano-Castillo M, Casas JL (2022) In vitro propagation, genetic assessment, and medium-term conservation of the coastal endangered species *Tetraclinis articulata* (Vahl) Masters (Cupressaceae) from adult trees. Plants 11(2):187. https://doi.org/10.3390/plants11020187
127. Ozudogru EA, Kirdok E, Kaya E, Capuana M, De Carlo A, Engelmann F (2011) Medium-term conservation of redwood (*Sequoia sempervirens* (D. Don.) Endl.) in vitro shoot cultures and encapsulated buds. Sci Hortic 127(3):431–435. https://doi.org/10.1016/j.scienta.2010.10.013
128. de Oliveira KS, Ibrahim Aloufa MA (2022) Slow growth in vitro culture for conservation of *Hancornia speciosa* gomes. Floresta 52(1). https://doi.org/10.5380/rf.v52i1.69209
129. Lambardi M, Benelli C, De Carlo A, Fabbri A, Grassi S, Lynch PT (2002) Medium- and long- term in vitro conservation of olive germplasm (*Olea Europaea* L.). Acta Hortic 586:109–112. https://doi.org/10.17660/ActaHortic.2002.586.14
130. Souza FV, Reboucas CC, Souza EH, Peixoto CP (2020) In vitro conservation of mango (*Mangifera indica* L.) Ubá and Carlota cvs. through culturing immature embryos. An Acad Bras Ciênc 92(suppl 1):e20190400. https://doi.org/10.1590/0001-3765202020190400
131. Carvalho MDJDSD, Souza ADS, Santos EB, Soares Filho WDS, Ledo CADS, Costa EMR, Souza FVD (2022) In vitro conservation of 'Florida Rough'lemon plants. Ciênc Rural 52(12):e20210276. https://doi.org/10.1590/0103-8478cr20210276
132. Hossein Pour A, Aydin M, Haliloglu K (2020) Plant regeneration system in recalcitrant rye (*Secale cereale* L.). Biologia 75:1017–1028. https://doi.org/10.2478/s11756-019-00395-9
133. Boboc P, Cantor M, Cordea MI, Cătană C (2022) Plant regeneration protocol for recalcitrant passionflower (*Passiflora quadrangularis* L.). Horticulturae 8(4):337. https://doi.org/10.3390/horticulturae8040337
134. González-Benito ME, Ibáñez MÁ, Pirredda M, Mira S, Martín C (2020) Application of the MSAP technique to evaluate epigenetic changes in plant conservation. Int J Mol Sci 21(20):7459. https://doi.org/10.3390/ijms21207459
135. Kader A, Sinha SN, Ghosh P (2022) A strategy for development of genetically stable synseeds of *Azadirachta indica* A. Juss. (Neem) suitable for in vitro storage. Plant Cell Tiss Organ Cult 151(1):47–58. https://doi.org/10.1007/s11240-022-02329-x
136. Nagel M, Pence V, Ballesteros D, Lambardi M, Popova E, Panis B (2024) Plant cryopreservation: principles, applications, and challenges of banking plant diversity at ultralow temperatures. Annu Rev Plant Biol 75. https://doi.org/10.1146/annurev-arplant-070623-103551
137. Reddy MC, Murthy KSR, Pullaiah T (2012) Synthetic seeds: a review in agriculture and forestry. AJB 11(78):14254–14275. https://doi.org/10.5897/AJB12.770
138. Garg R, Maheshwari S (2023) Synthetic seed technology, application and future trends. EPH Int J Agric Environ Res 9(1):1–10. https://doi.org/10.53555/eijaer.v9i1.67
139. Kamińska M, Gołębiewski M, Tretyn A, Trejgell A (2018) Efficient long-term

conservation of *Taraxacum pieninicum* synthetic seeds in slow growth conditions. Plant Cell Tiss Organ Cult 132:469–478
140. Abbas MK, Mahood HE, Alhasan AS (2022) Production of synthetic seeds in vegetable crops: a review. IOP Conf Ser 1060(1):012099. https://doi.org/10.1088/1755-1315/1060/1/012099
141. Gantait S, Kundu S, Yeasmin L, Ali MN (2017) Impact of differential levels of sodium alginate, calcium chloride and basal media on germination frequency of genetically true artificial seeds of *Rauvolfia serpentina* (L.) Benth. ex Kurz. J Appl Res Med Aromat Plants 4:75–81. https://doi.org/10.1016/j.jarmap.2017.01.005
142. Mahdavi Z, Daylami SD, Fadavi A, Vahdati K (2023) Artificial seed production of Phalaenopsis orchid: effect of encapsulation materials, temperature, light spectra, and storage period. Plant Cell Tiss Organ Cult 155(3):797–808. https://doi.org/10.1007/s11240-023-02600-9
143. Pasternak TP, Steinmacher D (2024) Plant growth regulation in cell and tissue culture in vitro. Plants (Basel) 13(2):327. https://doi.org/10.3390/plants13020327
144. Yang H, Li NQ, Gao JY (2023) A novel method to produce massive seedlings via symbiotic seed germination in orchids. Front Plant Sci 14:1114105. https://doi.org/10.3389/fpls.2023.1114105
145. Hung CD, Trueman SJ (2012) Alginate encapsulation of shoot tips and nodal segments for short-term storage and distribution of the eucalypt *Corymbia torelliana* × *C. citriodora*. Acta Physiol Plant 34:117–128. https://doi.org/10.1007/s11738-011-0810-0
146. Regni L, Micheli M, Del Pino AM, Facchin SL, Rabica E, Camilloni L, Cesarini A, Proietti P (2024) Blackberry synthetic seeds storage: effects of temperature, time, and sowing substrate. Plant Cell Tiss Organ Cult 158(1):17. https://doi.org/10.1007/s11240-024-02816-3
147. Agrawal A, Singh S, Malhotra EV, Meena DPS, Tyagi RK (2019) In vitro conservation and cryopreservation of clonally propagated horticultural species. In: Rajasekharan PE, Rao VR (eds). Conservation and utilization of horticultural genetic resources. Springer, Singapore
148. Wang MR, Bi W, Shukla MR, Ren L, Hamborg Z, Blystad DR, Saxena PK, Wang QC (2021) Epigenetic and genetic integrity, metabolic stability, and field performance of cryopreserved plants. Plants (Basel) 10(9):1889. https://doi.org/10.3390/plants10091889
149. Białoskórska M, Rucińska A, Boczkowska M (2024) Molecular mechanisms underlying freezing tolerance in plants: implications for cryopreservation. Int J Mol Sci 25(18):10110. https://doi.org/10.3390/ijms251810110
150. Kalaiselvi R, Rajasekar M, Gomathi S (2017) Cryopreservation of plant materials—a review. IJCS 5(5):560–564
151. Harding K (2004) Genetic integrity of cryopreserved plant cells: a review. Cryo Lett 25(1):3–22
152. Engelmann F (2004) Plant cryopreservation: progress and prospects. In Vitro Cell Dev Biol Plant 40:427–433. https://doi.org/10.1079/IVP2004541
153. González-Arnao MT, Méndez-Chávez M, Vásquez-Hernández S, Núñez-Pastrana R, Cruz-Cruz CA (2020) Osmo-and cryotherapy of sugarcane (*Saccharum* spp. L.) shoot-tips infected with sugarcane mosaic virus (SCMV). VirusDisease 31:497–502. https://doi.org/10.1007/s13337-020-00636-5
154. Wang QC, Panis B, Engelmann F, Lambardi M, Valkonen JT (2009) Cryotherapy of shoot tips: a technique for pathogen eradication to produce healthy planting materials and prepare healthy plant genetic resources for cryopreservation. Ann Appl Biol 154:351–363. https://doi.org/10.1111/j.1744-7348.2008.00308.x
155. García-Águila L, de Feria M, Acosta K (2007) Aspectos básicos de la conservación in vitro de germoplasma vegetal. Biotecnol Vegetal 7(2):67–79
156. Benelli C (2021) Plant cryopreservation: a look at the present and the future. Plants (Basel) 10(12):2744. https://doi.org/10.3390/plants10122744
157. Fahy GM, MacFarlane DR, Angell CA, Meryman HT (1984) Vitrification as an approach to cryopreservation. Cryobiology 21(4):407–426. https://doi.org/10.1016/0011-2240(84)90079-8
158. Sakai A, Thinh NT, Takagi H (2000) Cryopreservation of in vitro grown apical meristems of some vegetatively propagated tropical monocots by vitrification. In: Engelmann F, Takagi H (eds). Cryopreservation of tropical plant germplasm. Current research progress and application. IPGRI, Italy
159. González-Arnao MT, Juárez J, Ortega C, Navarro L, Durán-Vila N (2003) Cryopreservation of ovules and somatic embryos of citrus using the encapsulation-dehydration technique. Cryo Letters 24(2):85–94
160. Tanaka D, Akimoto-Kato A, Kusaba M, Taniguchi K, Niino T, Maki S, Matsumoto T (2016) Cryopreservation of chrysanthemum shoot tips by D cryo-plate method. J Jpn Soc Agric Technol Mgt 23:1–7
161. Sakai A, Kobayashi S, Oiyama I (1991) Cryopreservation of nucellar cells of navel orange (*Citrus sinensis* Osb.) by a simple freezing

method. Plant Sci 74(2):243–248. https://doi.org/10.1016/0168-9452(91)90052-A
162. Sakai A, Kobayashi S, Oiyama I (1990) Cryopreservation of nucellar cells of navel orange (*Citrus sinensis* Osb. var. brasiliensis Tanaka) by vitrification. Plant Cell Rep 9:30–33. https://doi.org/10.1007/BF00232130
163. Matsumoto T, Dhungana SA, Kunitake H, Niino T, Yamamoto SI, Fukui K, Tanaka D (2017) Cryopreservation of blueberry shoot tips derived from in vitro and current shoots using D cryo-plate technique. Plant Biotechnol (Tokyo) 34(1):1–5. https://doi.org/10.5511/plantbiotechnology.16.1231b
164. Zamecnik J, Faltus M, Bilavcik A (2021) Vitrification solutions for plant cryopreservation: modification and properties. Plants (Basel) 10(12):2623. https://doi.org/10.3390/plants10122623
165. Sakai A (1956) Survival of plant tissue at super low temperatures. Low Temp Sci Ser B 14:17–23
166. Kartha KK, Engelmann F (1994) Cryopreservation and germplasm storage. In: Vasil IK, Thorpe TA (eds). Plant cell and tissue culture. Springer, Dordrecht
167. Withers LA, Engelmann F (1998) In vitro conservation of plant genetic resources. In: Hammerschlag FA, Litz RE (eds). Biotechnology of perennial fruit crops. CAB International
168. Reed BM, Uchendu E (2008) Controlled rate cooling. In: Reed BM (ed). Plant cryopreservation: a practical guide. Springer, New York
169. Langis R, Schnabel B, Earle ED, Steponkus PL (1989) Cryopreservation of *Brassica campestris* L. cell suspensions by vitrification. CryoLetters 10(6):421–428
170. Towill LE (1990) Cryopreservation of isolated mint shoot tips by vitrification. Plant Cell Rep 9:178–180. https://doi.org/10.1007/BF00232174
171. Uragami A, Sakai A, Kato T (1989) Cryopreservation of asparagus (*Asparagus officinalis* L.) cell cultures by vitrification. Plant Cell Rep 8(4):225–228. https://doi.org/10.1007/BF00232529
172. Fabre J, Dereuddre J (1990) Encapsulation-dehydration: a new approach to cryopreservation of Solanum shoot-tips. CryoLetters 11(6):413–426
173. Panis B, Lambardi M (2006) Status of cryopreservation technologies in plants (crops and forest trees). In: Ruane J, Sonnino A (eds) The role of biotechnology in exploring and protecting agricultural genetic resources, pp 43-54. FAO, Rome
174. González-Arnao MT, Moreira T, Urra C (1996) Importance of pregrowth with sucrose and vitrification for the cryopreservation of sugarcane apices using encapsulation-dehydration. CryoLetters 17(3):141–148
175. González-Arnao MT, Urra C, Engelmann F (1998) Cryopreservation of encapsulated sugarcane apices: effect of storage temperature and storage duration. CryoLetters 20(5):347–352
176. Matsumoto T, Sakai A, Yamada K (1994) Cryopreservation of in vitro-grown apical meristems of wasabi (*Wasabia japonica*) by vitrification and subsequent high plant regeneration. Plant Cell Rep 13(7):442–446. https://doi.org/10.1007/bf00231963
177. Hirai D, Sakai A (1999) Cryopreservation of in vitro-grown axillary shoot tip meristems of mint (*Mentha spicata* L.) by encapsulation-vitrification. Plant Cell Rep 19(2):150–155. https://doi.org/10.1007/s002990050725
178. Takagi H (2000) Recent developments in cryopreservation of shoot apices of tropical species. In: Engelmann F, Takagi H (eds). Cryopreservation of tropical plant germplasm. Current research progress and application. IPGRI, Italy
179. Sakai A (2004) Plant cryopreservation. In: Fuller BJ, Lane N, Benson EE (eds) Life in the frozen state. CRC Press, Boca Raton
180. Wang MR, Lambardi M, Engelmann F, Pathirana R, Panis B, Volk GM, Wang QC (2021) Advances in cryopreservation of in vitro-derived propagules: technologies and explant sources. Plant Cell Tiss Organ Cult 144:7–20. https://doi.org/10.1007/s11240-020-01770-0
181. Sharrock S (2020) Plant Conservation Report 2020: a review of progress in implementation of the Global Strategy for Plant Conservation 2011–2020. Secretariat of the Convention on Biological Diversity, Montréal, Canada and Botanic Gardens Conservation International, Richmond, UK
182. Ferreira MS, Rocha AJ, Nascimento FS, Oliveira WDS, Soares JMS, Rebouças TA, Lino LSM, Haddad F, Ferreira CF, Santos-Serejo JA, Fernández JS, Amorim EP (2023) The role of somaclonal variation in plant genetic improvement: a systematic review. Agronomy 13(3):730. https://doi.org/10.3390/agronomy13030730
183. Martinez-Montero ME, Harding K (2015) Cryobionomics: evaluating the concept in plant cryopreservation. In: Barh D, Khan M, Davies E (eds). PlantOmics: the omics of plant science. Springer, New Delhi

Chapter 3

In Situ Conservation of Agricultural Biodiversity and Generation of Projects in Ejidos and Communities in Mexico

Jorge Cadena-Iñiguez, Jorge David Cadena-Zamudio, Rosalinda González-Santos, María Isabel Iñiguez-Luna, Daniel Alejandro Cadena-Zamudio and Kazuo N. Watanabe

Abstract

Agricultural biodiversity genetic resources are tied to social groups through ancestral uses, form part of traditional knowledge, and create a sense of belonging and identity. In Mexico, the "ejido" exists as a form of social property, defined as: "communal land on the outskirts of a settlement designated for services such as grain storage or livestock grazing." In this context, many genetic resources are found on these properties, which often show low levels of development. The signing of the Nagoya Protocol requires the characterization, distinction, and conservation of agricultural species and their wild relatives. For this reason, a methodology was designed for the in situ conservation of agricultural biodiversity, focusing on local agroecosystems, with special emphasis on the geographical areas of local and indigenous communities in Mexico. The methodological foundation includes engagement, awareness raising, partnerships, resource inventories, and the creation of community-based social and bioprospecting projects to build value chains, with the involvement of higher education and research institutions. This strategy integrates sociocultural, economic, and biological variables that guide its operation and ensure the fair and equitable distribution of benefits to the cultural heirs reflected in each plant resource. It also aligns with Mexico's international conservation goals.

Keywords Wild relatives, Bioprospecting, Value chains, In situ conservation, Agricultural biodiversity, Plant genetic resources, Nagoya Protocol, Local agroecosystem

1 Introduction

The 2011-2020 Strategic Plan for Biodiversity under the Convention on Biological Diversity (CBD) consists of a global ten-year action framework through which countries commit to protecting biodiversity and enhancing the benefits it provides for human well-being. It consists of a vision, a mission, and 20 global goals known as the Aichi Targets [1].

Mexico's National Biodiversity Strategy (ENBioMEX) has a mission, a vision for 2030, and 14 guiding principles. The Action

Marco A. Ramírez-Mosqueda et al. (eds.), *Conservation of Plant Genetic Resources*, Methods in Molecular Biology, vol. 3011, https://doi.org/10.1007/978-1-0716-5130-8_3,

Plan consists of six strategic pillars, 24 lines of action, and 160 specific actions. Among its main pillars are knowledge and education, communication and environmental culture, conservation and restoration, sustainable use and management, and addressing pressure factors that aim to increase actions that positively impact biodiversity and reduce the direct causes of its loss. The integration and governance pillar strengthens the implementation of these actions through coordination among stakeholders and sectors, harmonization of the legal framework, and the promotion of integration and cooperation [2]. The second strategic of ENBioMEX, related to conservation and restoration (2.0), includes in situ conservation as one of its lines of action. This is defined as the process of protecting an endangered animal or plant species in its natural habitat, whether by intervening in the habitat or protecting the species from predators. Agricultural biodiversity, a branch of general biodiversity, encompasses all crop varieties. This term includes all components of biodiversity (genetic, species, and ecosystems) that are essential for food and agriculture, supporting the agroecosystems where agriculture takes place. The FAO (2019) highlights that agricultural biodiversity is fundamental for humanity as it ensures food production in diverse environments, maintains food availability stability under adverse conditions such as droughts and pests, and provides a wide variety of nutritious foods [3]. Moreover, wild edible plants are a vital food source, especially in poor regions of the world, and represent a crucial gene source for breeding programs.

The above has implications that go beyond taxonomy, particularly when dealing with local genetic resources of bioprospective importance. Having a detailed characterization (morphological, chemical, genetic, and biological) of a biological variant allows for its precise distinction and facilitates the execution of Material Transfer Agreements (MTA) within the framework of the Nagoya Protocol (NP). This also aids in the design of fair and equitable benefit-sharing (ABS) models for the use or exploitation of genetic resources [4].

In Mexico, agricultural genetic resources are closely linked to social groups through both ancestral and current uses. These resources are part of traditional knowledge or local know-how, which has created a sense of ancestral belonging, granting them identity [5]. The "ejido" system, which emerged after the Mexican Revolution, represents a form of rural collective land ownership, where ejidos have legal personality and their own assets. Common-use ejidal lands (CUEL) form the economic backbone of these communities and are managed collectively [6]. In ejidos and indigenous communities, genetic resources are considered common property, and many of them have been improved in situ through traditional practices. These resources are associated not only with food use but also with a wide range of local knowledge and uses [7]. This communal perception of agricultural biodiversity, along with ethnic identity and governance systems, should be a key variable in designing in situ conservation strategies in Mexico. Based on the above, the aim of this protocol is to implement an in situ conservation strategy for genetic resources identified as components of Mexico's agricultural biodiversity through local projects in ejidos and communities. This conservation model is based on the local agroecosystem approach, with a special focus on the geographical areas of local and indigenous communities.

2 Materials

2.1 Inventory Survey and Initial Diagnosis

1. Cartography and georeferencing: handheld GPS or smartphone with GNSS; topographic maps; compass.
2. Field sampling: measuring tapes (≥50 m); transect ropes; stakes; flagging tape/adhesive; quadrats (1 × 1 m and 10 × 10 m frames); clinometer.
3. Data recording: waterproof notebooks; clipboard; printed forms and/or tablet with digital forms; permanent markers; Tyvek labels.
4. Documentation: camera (or smartphone) and batteries; drone (optional).
5. Botanical collection: disinfectable pruning shears and knives; gloves; paper bags and Whirl-Pak bags; 2 mL tubes; indicator silica gel; 70% alcohol; cooler with cold packs.
6. Site characterization (optional): soil auger; bags; portable pH meter; conductivity meter; soil moisture meter.

2.2 Ethnobotanical Analysis and Participatory Workshops

1. Workshops/interviews: voice recorders; informed-consent forms; semi-structured guides; flip chart/markers; projector (if electricity is available); tables/chairs; bilingual materials (local language—Spanish); refreshments.
2. Field support: GPS to locate referenced sites; participatory-mapping materials (kraft paper; pushpins/stickers).
3. Definition and operation of CRA and creation of local infrastructure (4 and 8).
4. Demarcation and signage: posts; chain-link fencing or boundary markers; paint/plaques; informational signs with rules of use.
5. Community nursery: seedbeds/trays; substrates (soil-sand-organic matter mix); perlite/vermiculite; polyethylene bags; watering cans or drip irrigation; shade netting (30%–50%); thermo-hygrometers.
6. Ethnobotanical garden: local seedlings; UV-resistant tags; plaques with scientific/common name and use.

7. Community herbarium: plant presses; blotting paper; herbarium cardboard; PVA glue; gummed tape; labels; −20°C freezer for insect control; metal cabinets.
8. Community seed bank: sieves/screens; cleaning trays; dehumidifier; hygrometers/thermometers; indicator silica gel; airtight glass jars or trilaminate envelopes; heat sealer; analytical balance; dry chamber (e.g., 15°C and 15%–25% RH, or desiccator); 4°C refrigerator and −20°C freezer; T/RH dataloggers.
9. Documentation systems: laptop; UPS; external drive; scanner; database software (PostgreSQL/Access) and templates for passport data and records.

2.3 Monitoring and Evaluation

1. Marking and tracking: marking tape; numbering stamps; UV-resistant labels; microchips or RFID tags (optional); standardized notebooks/templates.
2. Measurement: DBH tapes (for trees); dendrometric rulers; permanent quadrats; GPS.
3. Automated recording: weather stations (or T/RH/precipitation dataloggers); soil moisture meters.
4. Software: QGIS; R/Excel for indicators and time series; data repository with version control.

2.4 Training and Education of Local Managers

1. Educational material: printed/bilingual manuals; practice kits for seed collection, drying, and storage; reference samples; posters; projector/speakers.
2. Workshop logistics: community halls; flip charts; consumables (stationery); travel expenses.

2.5 Governance, Regulations, and ABS

1. Documentation and forms: assembly minutes; printed community regulations; CRA agreements; Prior Informed Consent (PIC) forms; Mutually Agreed Terms (MAT); Material Transfer Agreement (MTA); Access and Benefit-Sharing (ABS) records.
2. Document management: filing cabinets/safe; stamps; bilingual logbooks; digital backup system (folders with copies).

2.6 Community Databases

1. Hardware/software: community laptop; printer-scanner; external backup; database and spreadsheet platforms; passport-data templates for accessions.

2.7 Safety and Logistics

1. PPE: nitrile gloves; boots; hat; goggles; sunscreen; rain gear; first-aid kit.
2. Communications and support: radios or local telephone; headlamps; reflective vests.
3. Transport and storage: backpacks/stackable plastic boxes; coolers; fuel; vehicle maintenance.

3 Methods

General note. Maintain traceability with unique codes (CRA-Site-Species-Batch/Individual-Date) and record each step in the Database (DB).

3.1 Permits, Design, and Safety

1. Permits and ethics. Obtain assembly agreements and Prior Informed Consent (PIC); prepare Mutually Agreed Terms (MAT) and a Material Transfer Agreement (MTA) if material will be transferred (*see* **Note 1**).
2. Prioritization. Identify priority species/varieties using ecological, cultural, and risk criteria, with documented justification.
3. Local team. Appoint Local Managers and sub-process leads (inventory, seed bank, monitoring, Access and Benefit-Sharing — ABS) and log their roles in the DB (*see* **Note 2**).
4. Safety. Assess risks (weather, access, fauna) and prepare Personal Protective Equipment (PPE) and first-aid kit.
5. Phase closure. Draft an assembly minute naming responsible persons; without this document, do not proceed (*see* **Note 3**).

3.2 Delimiting Community Restriction Areas (CRA)

1. Community meeting. Present objectives, scope, and preliminary rules of the community-conservation model. Introduce the Local Agroecosystem (LoAgro) as the operational unit for preventing genetic erosion and organizing conservation activities. Within this context, CRA are identified and delimited as spaces for managing, conserving, and improving agrobiodiversity (Fig. 1).
2. Preliminary polygons. Use a Geographic Information System (GIS; e.g., QGIS) and GPS to outline conservation and management zones.
3. Field verification. Walk boundaries and adjust for ecological connectivity, accessibility, and tenure.
4. Signage. Install stakes, flagging tape, and signs displaying rules of use.
5. Registration. Record coordinates, areas, and responsible persons in the DB (*see* **Note 4**).

3.3 Floristic-Ethnobotanical Inventory (Design and Survey)

1. Sampling design. Define transects and/or quadrats for each environment (crop, edge, fallow, forest). Plant Resources (PR) result from the interaction among Ecological Environment (EE), local Culture (C), perceived Useful Vegetation (V), and historical Tradition (T) (Fig. 2).
2. Botanical survey. For each site, record common and scientific names (if known), phenology, relative abundance, coordinates, and a photograph.

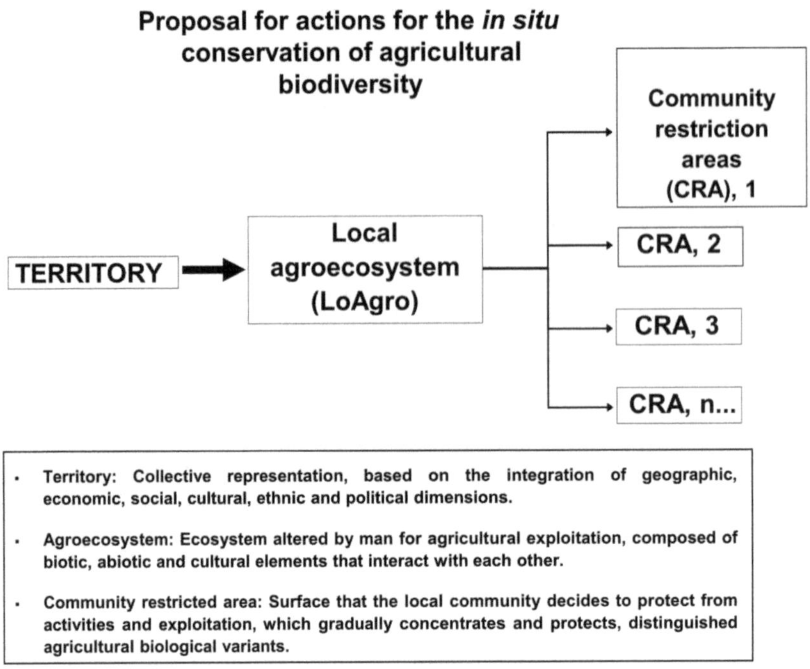

Fig. 1 Proposal for actions for in situ conservation of agricultural biodiversity in areas of social property in Mexico, under the Local Agroecosystem system and community restriction areas (modified from ref. [2])

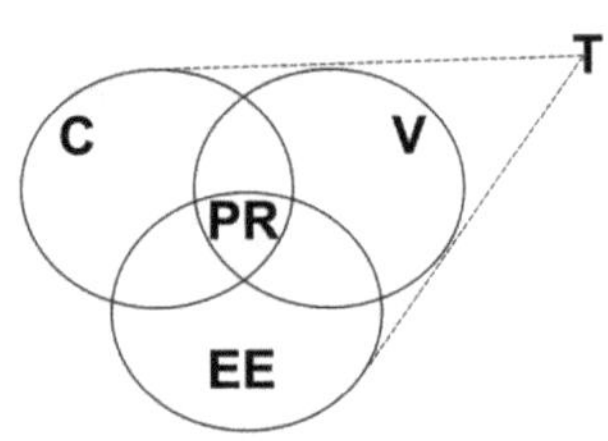

Fig. 2 Spatial relationship of the concept of plant resource, according to the variables of ecological environment, culture, vegetation and time

3. Interviews/workshops. Use semi-structured guides on uses, management, and seasonality with consent (*see* **Note 1**). This phase identifies the subset of plant resources actually used by the community, a selection modulated by ecological and cultural factors that can expand (e) or restrict (r) the useful-species set. Sustained restriction may lead to genetic erosion (Fig. 3).
4. Vouchers. Collect reference specimens when appropriate; press and label them for the community or partner herbarium (*see* **Note 5**).
5. Taxonomic cleaning. Validate names (genus, species, authority) and standardize synonyms in the DB; use double entry and reconciliation (*see* **Note 6**).

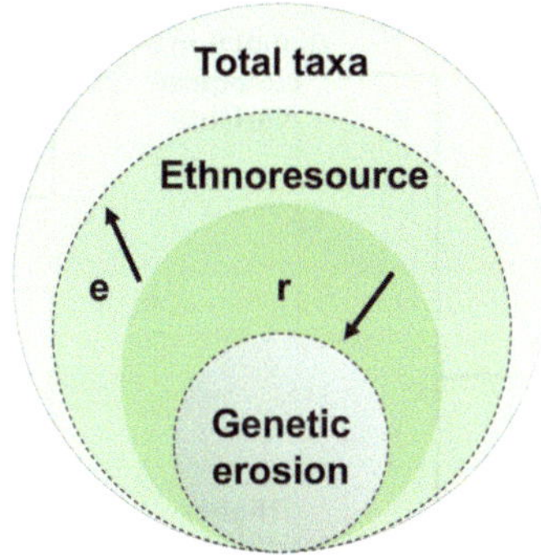

Fig. 3 Variables for the expansion and restriction of taxa based on the area of daily action of a human group (Taken from Cuevas, 1994)

3.4 Collecting Material for Genetic Analysis (Optional/by Agreement)

1. Selection. Sample 20–30 individuals per population, spaced to minimize relatedness.
2. Collection. Make clean cuts with disinfected shears; place tissue in tubes with indicator silica gel (*see* **Note 7**).
3. Labelling. Assign a unique code per sample and note collector, time, and micro-habitat; link to the DB (*see* **Note 2**).
4. Preservation. Transport in a cooler and store at −20°C (or −80°C, as required).
5. Shipment. Prepare a manifest and MTA; coordinate with the receiving laboratory.

3.5 Establishing the Nursery and Ethnobotanical Garden

1. Site. Choose an area with nearby water and good drainage; install 30%–50% shade netting.
2. Substrates. Prepare a soil-sand-organic mix (or commercial substrate); sterilize if needed.
3. Propagation. Sow or take cuttings per species; record batch, origin, and method; aim for >70% establishment (*see* **Note 8**).
4. Management. Adjust irrigation per species, perform mechanical weed control, and enforce phytosanitary quarantine (*see* **Note 9**).
5. Reintroduction pathway. Plan transplants to CRA with post-establishment monitoring and UV-resistant labels (*see* **Note 9**).

3.6 Community Seed Bank (Cleaning, Drying, Packaging, Storage)

1. Reception. Verify batches (origin, date, species) and register in DB; pre-clean to remove debris.
2. Fine cleaning. Use sieves/screens and manual winnowing until ≥95% purity (or agreed standard).
3. Controlled drying. Place in a dry chamber (e.g., 15°C; relative humidity 15%–25%) or a desiccator with silica gel until constant mass (*see* **Note 10**).

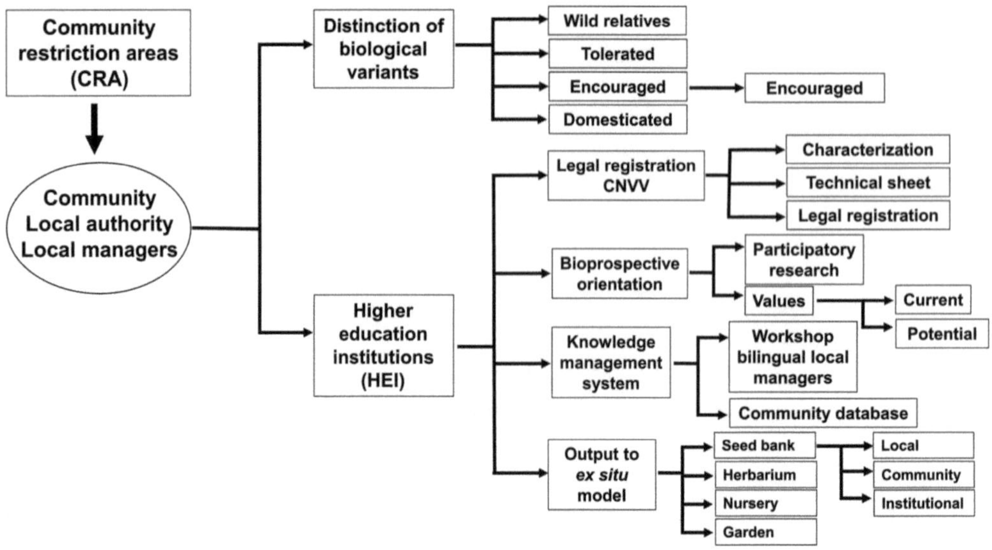

Fig. 4 Actions to be developed in the areas of community restriction within the local agroecosystem (modified from ref. [2])

4. Conditioning. Pack in trilaminate envelopes or airtight jars and heat-seal; include a moisture indicator (*see* **Note 11**).
5. Labelling and passport. Write species (in italics) + authority, batch code, origin, date, seed count, estimated RH, and responsible person; link to DB (*see* **Note 2**).
6. Storage. Short-term: 4°C; long-term (orthodox seeds): −20°C; monitor T/RH with dataloggers (*see* **Note 12**).
7. Replenishment. Set viability thresholds for recollection or regeneration (*see* **Note 13**). Promote orderly germplasm exchange under formal agreements (MTA) and full traceability (Fig. 4).

3.7 Viability and Germination Tests

1. Sampling. Take a representative subsample per batch (e.g., 50–100 seeds or agreed %).
2. Test conditions. Germinate on paper/filter or inert substrate at species-specific temperature and photoperiod; pre-moisten uniformly (*see* **Note 14**).
3. Criteria. Define normal/abnormal seedlings and maximum test duration per species (*see* **Note 14**).
4. Decision. Calculate germination % and mean time; if below the threshold, activate recollection/regeneration (*see* **Note 13**).

3.8 Community DB and Document Management

1. Structure. Create tables: Species (taxonomy/authority), Sites (CRA/coords), Batches (passport), Uses (ethnobotany), Monitoring (indicators), ABS (agreements).

2. Validation. Use controlled vocabularies and unique keys; double entry for inventories and passports (*see* **Note 15**).
3. Backup. Make weekly copies (external drive and cloud/partner institution).
4. Reports. Produce availability lists, maps, and data sheets for assemblies (*see* **Note 16**).

3.9 Indicator Monitoring (Biological, Sociocultural, Economic, Governance)

1. Selection. Choose measurable, comparable indicators (e.g., species richness, survival, participation, local income, rule compliance).
2. Permanent plots. Establish, georeference, and mark with UV-resistant labels; log coordinates in DB (*see* **Note 17**).
3. Frequencies. Richness/cover: annual; participation/training: semi-annual; cultural identity: biennial.
4. Recording. Measure, photograph, and capture data; note events (weather, disturbances).
5. Analysis. Update the indicator dashboard and discuss management adjustments in assembly (*see* **Note 17**).

3.10 Governance and Operating Rules

1. Management council. Form community representation (including women and youth) and define documented roles. Training Technicians and Local Managers is recommended; they facilitate in-situ conservation through participatory processes, technical support, and territorial management. The Social Intervention Model (SIM) links community, institutional, and economic actors, fostering local value networks and territorial sustainability (Fig. 5).
2. Rules. Set allowable extraction, no-take zones, sanctions, and conflict-resolution mechanisms.
3. Transparency. Post seed-bank transactions and agreements on a community board; conduct an annual social audit.
4. Community fund. Operate an account with clear rules for benefit use; prevent conflicts of interest (*see* **Note 18**).

3.11 ABS Access to Genetic Resources and Benefit-Sharing

1. Determination. Identify ownership and whether material is in situ or from the seed bank. For resources on social property land, follow the Nagoya Protocol, applying PIC, MAT, and MTA. The operational flow is shown in (Fig. 6).
2. Instruments. Prepare and sign PIC, MAT, and MTA before any transfer (*see* **Note 1**).
3. Benefits. Agree on monetary/non-monetary benefits and sharing criteria, prioritizing the provider community (*see* **Note 19**).
4. Registration. Digitize agreements and link them to batches/species in the DB.

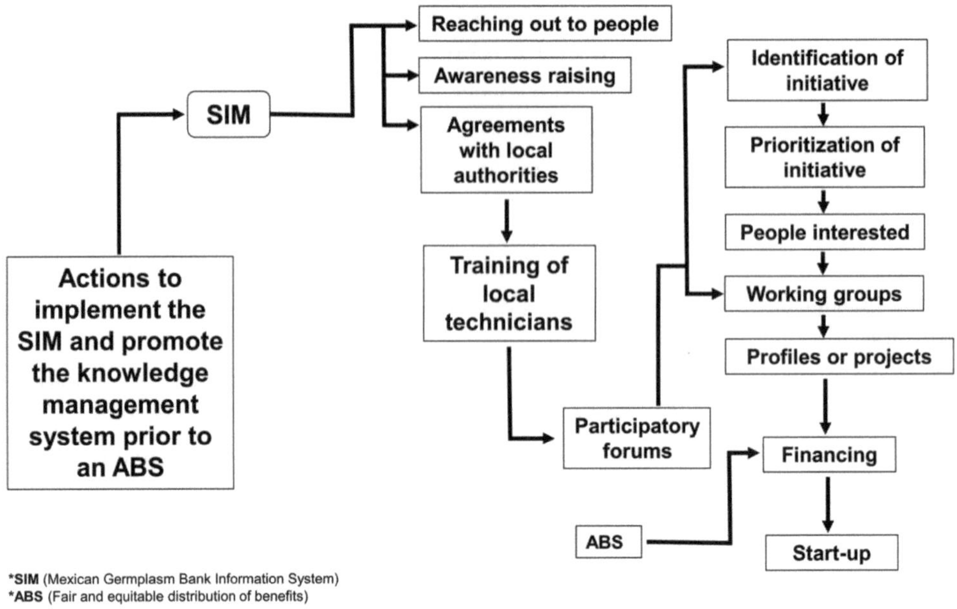

Fig. 5 Community operating protocol for the fair and equitable distribution of benefits (ABS) applying the Social Intervention Model (SIM), which promotes local development, avoids conflicts over the distribution of monetary benefits, and promotes local projects (modified from ref. [2])

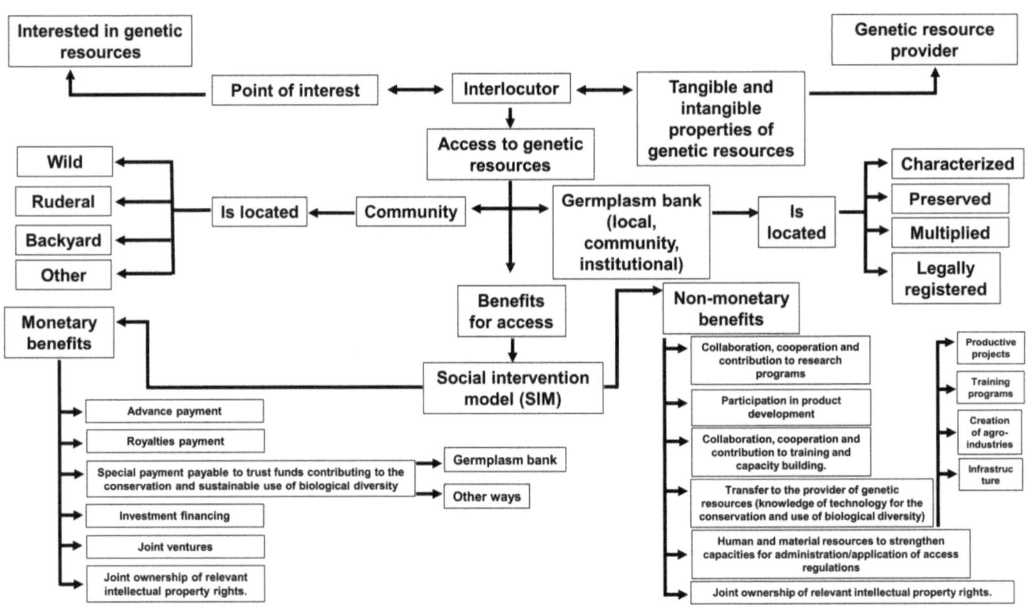

Fig. 6 Overview of the route to operate access to a genetic resource, applying the Social Intervention Model (SIM) for the fair and equitable distribution of benefits (ABS) for local and indigenous communities within the framework of the Nagoya Protocol (modified from ref. [2])

3.12 Institutional Articulation (Biobanks, National Network, HEI [Higher Education Institution], NGO [Non-Governmental Organization])

1. Agreements. Draft letters and accords for characterization, safeguarding, and training.
2. Exchange. Formalize entries/exits with MTA and complete passports; attach the minimum-data set (*see* **Note 20**).
3. Training. Schedule annual workshops (collection, drying, storage, ABS, DB).

3.13 Safety, Bioethics, and Waste Management

1. PPE. Use gloves, goggles, boots, hat, and raincoat as required; carry antiseptic, gauze, and bandages (*see* **Note 21**).
2. Hygiene. Disinfect tools between samples/sites; manage lab waste per regulations.
3. Emergencies. Define evacuation routes and meeting points; maintain radio or satellite-phone communication when needed.

3.14 Data Management and Analysis

1. Curation. Check taxonomy, coordinates, passport data, and document changes (*see* **Note 22**).
2. Indicators. Calculate metrics (richness, survival, germination %, participation) and trends.
3. Reporting. Issue semi-annual and annual reports with management recommendations and visualizations (*see* **Note 17**).

4 Notes

1. PIC/MAT/MTA made easy. Use pre-filled templates with mandatory fields and a checklist (signatures, dates, annexes). Save a PDF for each agreement; never collect or transfer material without these documents.
2. Traceability and DB. Encode hierarchically and use printed labels (optional: QR). Every physical label must exist in the DB with the same code; forbid code reuse.
3. Assembly traffic-light. Do not advance without the minute and assigned responsibilities. Before closing, verify objectives, responsible persons, dates, and minimum resources.
4. Error-free polygons. Capture coordinates in the field and validate in GIS (QGIS). Attach a photo of each sign and a sketch with control points.
5. Reliable vouchers. Press specimens in the field when possible; include diagnostic parts. Freeze 48 h at –20°C for insect control before mounting.
6. Fast double entry. Two people enter the inventory independently; reconcile with a different report or script.

7. Zero contamination. Change gloves per sample and disinfect shears (70% alcohol or brief flame). Carry wipes and indicator silica gel.
8. >70% establishment. Weekly: live plants/planted plants × 100. If <70%, adjust irrigation/substrate or use more lignified cuttings.
9. Labels that last. Use UV-resistant labels and double-tie in the nursery. Never mix batches; create physical pens to separate materials.
10. Simple drying. If no dry chamber, use airtight containers with silica gel and indicator; weigh trays every 24–48 h until mass is constant.
11. Perfect seal. Test the sealer; add a silica sachet and an RH indicator inside each package.
12. Orthodox vs. recalcitrant. Only orthodox seeds tolerate desiccation and −20°C. For recalcitrant/intermediate seeds, document exceptions and consider living collections.
13. Replenishment thresholds. Set per species (e.g., 85% initial; re-test yearly; recollect if <70%). Record the decision and date in the DB.
14. Meaningful germination. Moisten substrate evenly; define a "normal seedling" with reference photos and species-specific duration.
15. DB validation. Use controlled vocabularies and unique keys; reject records that break rules.
16. Robust plots. Mark plot corners with numbered rods/plates; record azimuth and a panoramic photo.
17. Clear dashboard. Build a dashboard (R/Excel) with traffic-light indicators and trend graphs; discuss at every assembly.
18. Conflict-free governance. List functions and signatures per role; for the community fund, set spending rules and keep a public log.
19. Benefit sharing. Document criteria (variant contribution, costs, vulnerability). Prioritize the community contributing most diversity.
20. Minimum exchange package. Attach full passport, moisture %, recent viability %, lot photos, and a signed MTA copy.
21. Practical PPE. Include a lightweight raincoat, headlamp, radio, permanent marker, and antiseptic spray for surfaces.
22. Change control. Use versioning (v 1.0, v 1.1...). For each change record date, responsible person, and justification.

References

1. Parks L, Tsioumani E (2023) Transforming biodiversity governance? Indigenous peoples' contributions to the Convention on biological diversity. Biol Conserv 280:109933
2. Cadena Iñiguez J, González Santos R, Cuevas Sánchez J, Riviello Flores MDLL, Ruiz Posadas LDM (2020) La Conservación in situ de la Biodiversidad Agrícola, y la generación de proyectos en ejidos y comunidades de México.
3. Food, Nations AOotU (2019). The state of the world's biodiversity for food and agriculture. FAO, Rome
4. Tvedt MW, Fauchald OK (2011) Implementing the Nagoya Protocol on ABS: a hypothetical case study on enforcing benefit sharing in Norway. J World Intellect Prop 14(5):383–402
5. Orozco-Hernández ME, García-Fajardo B, Álvarez-Arteaga G, Mireles-Lezama P (2017) Tendencias del sector agrícola, Estado de México. Quivera Rev estud territ 19(1): 99–121
6. Cadena-Iñiguez J, Morales-Flores F, Martínez-Becerra A, Trejo-Téllez B, Hernández-Rosas F, López-Romero G, Figueroa-Rodríguez K (2016) MODELO DE INTERVENCION SOCIAL (MIS) Y SU APLICACION AL DESARROLLO DE EMPRESAS RURALES. Agro Productividad 9(S1):72–76
7. Cadena-Iñiguez J (2010) Formación de gestores locales como estrategia para favorecer el relevo intergeneracional en ejidos. Agro Productividad 3(3):14–22

Chapter 4

Germination Protocol for Cedar Walnut Seeds (*Juglans pyriformis* Liebm.) Using Thermal Shock

Juan M. Pichardo-González, Esmeralda J. Cruz-Gutiérrez, Martín Quintana-Camargo, Edgar Torres-García, María G. Pérez-Reynozo and Luis Alberto Gómez-Reyes

Abstract

Juglans pyriformis Liebm. (Juglandaceae) is a species endemic to Mexico, which inhabits the mesophyll mountain forest and is highly valued for the quality of its wood. Therefore, it has been overexploited and its population has declined; it is currently listed as endangered. Therefore, it is important to implement conservation measures. One conservation strategy could be to collect seeds in tree for the establishment of forest plantations. Another strategy is to conserve them in vitro germplasm banks through the propagation and establishment of vegetative explants. Under natural and nursery conditions, *Juglans pyriformis* seeds require constant humidity to germinate, which is very late and slow due to the physical barrier posed by the hard pericarp. Likewise, when seeds of this species are germinated in laboratory, pre-germination treatments must be applied to promote germination and obtain plants for the propagation and conservation this species using in vitro techniques. The application of a thermal shock by subjecting the seed to 30 s in liquid nitrogen and immediately after another 30 s in boiling water is an effective treatment to promote germination in seeds of this species.

Keywords Cedar walnut, Germination, Stratification, Dormancy

1 Introduction

Juglans pyriformis Liebm. (Juglandaceae) is an endemic species of Mexico, characteristic of the mesophyll mountain forest, valued for the quality of its wood, which is used in the elaboration of fine furniture and musical instruments. Therefore, it has been overexploited and is currently listed as threatened [1]. However, the decrease and fragmentation of the mesophyll mountain forest, as well as the overexploitation of this species, are strongly impacting its populations, which have decreased. This is affecting the genetic variability of the species in a negative way [2], so it is important to implement actions for its conservation.

Marco A. Ramírez-Mosqueda et al. (eds.), *Conservation of Plant Genetic Resources*, Methods in Molecular Biology, vol. 3011,
https://doi.org/10.1007/978-1-0716-5130-8_4,

Endemic species such as *Juglans pyriformis* have been little studied, and today their endangered status requires generating such knowledge, so that from it strategies for their conservation can be proposed, as a key element of the mesophyll mountain forest in Mexico [3]. An in situ conservation strategy for threatened species such as *Juglans pyriformis* may be the selection of seed-producing trees, which guarantees natural regeneration, collecting seed of known origin for the establishment of conservation plantations, and maintaining the forest structure [4]. Another strategy is through their conservation in vitro germplasm banks, which are sites for the conservation of genetic resources under controlled laboratory conditions and that involve various in vitro cultivation and storage techniques. In these germplasm banks, the aim is to maximize the diversity of specimens collected from populations in the field or in their center of origin. The collection unit is maintained under controlled conditions, and these can be vegetative explants from plants of the species [5, 6].

The fruit of *Juglans pyriformis* is a globose pseudodrupe of 3.5–4.5 cm in diameter, consisting of a rough nut surrounded by a fleshy indehiscent shell, which rots easily to release the seed. The endocarp of this species is very hard and ends up separating into two valves during germination [7]. Under natural and nursery conditions, the seed of *Juglans pyriformis* requires constant humidity to germinate, being this process very late and slow due to the physical barrier represented by the hard pericarp [8]. Similarly, when the seed of this species is germinated at the laboratory level, it is necessary to apply pre-germination treatments to promote germination and obtain plants to propagate and conserve this species through in vitro techniques. An effective treatment to promote germination in *Juglans pyriformis* seeds is the application of a thermal shock by subjecting the seed to 30 s in liquid nitrogen and immediately after another 30 s in boiling water [9].

2 Materials

2.1 Plant Material

1. Cedar walnut seeds (*Junglas pyriformis* Liebm.). The quantity of seeds will depend on the quantity available.

2.2 Tap Water for High-Temperature Treatment of Seeds

1. Tap water (H_2O) (3 L or more depending on the quantity of seeds to be treated). The water is poured into 1000 mL beakers.

2.3 Liquid Nitrogen for Low-Temperature Treatment of Seeds

1. Liquid nitrogen (2 L or more depending on the quantity of seeds to be treated). The nitrogen is poured into liquid nitrogen (LN) containers.

2.4 Type of Substrates

1. Peat moss, agrolite, and vermiculite.

2.5 Materials and Equipment

1. Liquid nitrogen containers.
2. Heating plate.
3. Glassware: 500–1000 mL beakers, stirring rod, Erlenmeyer flasks, 3–5 L buckets or containers (for planting seeds).
4. Black polyethylene bags or plastic trays or containers at least 20 cm high for sowing seeds.
5. Seed germination chamber or seed germination room with controlled conditions.

3 Methods

3.1 Generation of High-Temperature Conditions

1. Empty tap water into a 1000 mL beaker to approximately 800 mL, which is placed on the hot plate.
2. Turn on the hotplate and set it to 100°C. Wait for the water to boil.

3.2 Generation of Low-Temperature Conditions

1. Empty liquid nitrogen into a container to 3/4 of its capacity (liquid nitrogen has a temperature of −196°C).

3.3 Thermal Shock Treatment of Juglans Pyriformis Nuts

1. Using tweezers, grasp a *Juglans pyriformis* nut and place it in the container with liquid nitrogen for 30 s.
2. Using the same tongs, remove the nut from the liquid nitrogen and place it in boiling water at 100°C for 30 s to create a thermal shock (Fig. 1) (*see* **Note 1**).

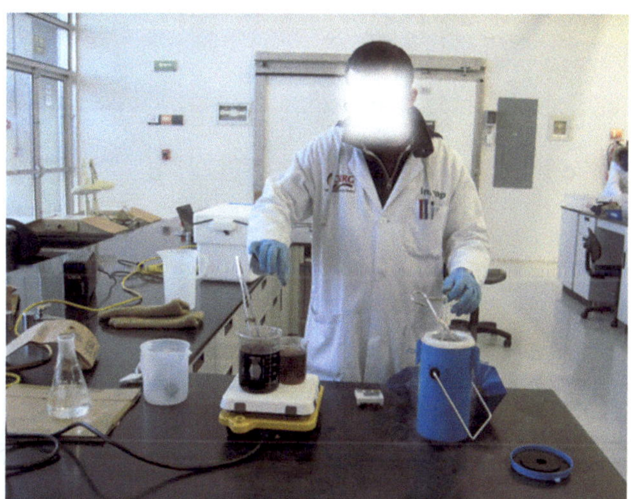

Fig. 1 Seeds of *Junglas pyiriformis* Liebm. immersed in boiling water and immediately afterward immersed in liquid nitrogen

3. Remove the nut from the liquid nitrogen and place it in a 3–5 L bucket or container.
4. Continue to process all the nuts until the treatment is complete.

3.4 Preparation of the Substrate for Sowing the Nuts for Germination and Obtention of Juglans Pyriformis Seedlings

1. In a container, pour three parts of peat moss, one part agrolite, and one part vermiculite (peat moss, agrolite, and vermiculite, in a 3:1:1 ratio).
2. Mix the substrates.
3. Add tap water and continue mixing the moistened substrates. Continue adding water until the substrate mix reaches "field capacity," meaning the mix is moist, but without excess moisture. At this point, the substrate is ready for sowing (*see* **Note 2**).

3.5 Sowing the Walnuts in a Suitable Germination Substrate

1. In a container, preferably a large one, empty the moistened substrate mix into the container, leaving about 10 cm of soil free at the top (*see* **Note 3**).
2. Sow the heat-shocked *Juglans pyriformis* walnuts, distributing them evenly across the tray (Fig. 2a, b).
3. Cover the sown walnuts with a 2 to 2.5 cm layer of moistened substrate.

Fig. 2 Sowing *Junglas pyriformis* seeds in substrate. (**a**) Establishing of sowing of the seeds in trays with substrate; (**b**) distribution of the sown seeds in the tray; (**c**) emergence of a *Junglas pyriformis* seedling; (**d**) *Juglans pyriformis* plants ready for in vitro propagation

3.6 Germination of Juglans Pyriformis Seeds in a Germination Chamber with Controlled Conditions

1. Place the trays or containers with the sown nuts in a germination chamber programmed at a temperature between 25°C and 30°C, with a relative humidity of 70% and a photoperiod of 16 h of light and 8 h of darkness.
2. Periodically irrigate the trays with the sown nuts when the substrate is lacking moisture.
3. Count the seedlings that have emerged in the substrate (Fig. 2c), water them, and monitor their growth in the room until they have sufficient tissue for propagation and in vitro conservation.

4 Notes

1. To ensure uniformity in thermal shock, process the seeds in small batches (three to four seeds) and time exactly 30 s in each phase, avoiding any delay between liquid N_2 and boiling water.
2. Before sowing, perform a surface disinfection with 0.5%–1% sodium hypochlorite for 2–3 min, rinsing three times with distilled water to reduce fungal and bacterial contamination.
3. The substrate must be moistened to actual field capacity (when squeezed by hand, one to two drops of water come out, but without dripping) to avoid waterlogging or water deficit during germination.

References

1. Acosta-Hernández CC, Luna-Rodríguez M, Octavio-Aguilar P, Morales-Romero Z, Galindo-González J, Noa-Carrazana JC, Vázquez-Torres SM, Iglesias-Andreu LG (2011) Efecto del aprovechamiento forestal sobre la variación morfológica de *Juglans pyriformis* Liebm. Rev Chapingo Ser Cienc For Ambiente 17(3):379–388
2. Acosta-Hernández CC, Luna-Rodríguez M, Noa-Carrazana JC, Galindo-González J, Vázquez-Torres MS, Morales-Romero Z, Iglesias-Andreu L (2011) Caracterización morfológica y dasométrica de la especie amenazada Juglans pyriformis Liebm. Rev Chapingo Ser Cienc For Ambiente 17(1):59–67
3. Acosta-Hernández CC, Ortiz-Muñoz E, Cornú-García JD, Perusquía Chávez JS, Linares-Márquez P (2019) Biología y silvicultura del cedro-nogal (*Juglans pyriformis*) para su conservación en México. *Monografías de la Asociación Chelonia, Volumen XV*. Asociación Chelonia, Madrid, p 76
4. Ortiz ME, Acosta-Hernández CC, Linares MP, Morales RZ, Rebolledo CV (2016) Selección de árboles semilleros de *Juglans pyriformis* Liebm. en poblaciones naturales de Coatepec y Coacoatzintla, Veracruz. Rev Mex Cienc For 7(38):43–58
5. Rao NK (2004) Plant genetic resources: advancing conservation and use through biotechnology. Afr J Biotechnol 3:136–145
6. Wang YL, Fan MJ, Liaw SL (2005) Cryopreservation of in vitro–grown shoot tips of papaya (*Carica papaya* L.) by vitrification. Bot Bull Acad Sin 46:29–34
7. Niembro RA (2002) Imágenes de árboles y arbustos del Estado de Veracruz. INECOL, A.C. Formato en CD inédito
8. Pedraza RA (2003) Germinación en condiciones de vivero y campo del nogal (*Juglans pyriformis* Liebm). Amaranto 16(3):2–11
9. Pichardo-González JM, Cruz-Gutiérrez EJ, Quintana-Camargo M, Torres-García E, Gómez-Reyes LA (2015) Rompimiento de la latencia de semillas de Junglas pyriformis, mediante choque térmico, Memorias del Congreso Internacional SOMEREFO. Ixtapan de la Sal, Estado de México, p. 218

Check for updates

Chapter 5

Evaluation and Short-Term Conservation of *Thrinax radiata* (Arecaceae) seeds

Dafne V. Bacab-Caamal, Ramon Souza-Perera and Ricardo Alvarez-Espino

Abstract

Physiological characterization and quality evaluation of seeds stored in germplasm banks is a challenge in seeds that present complex mechanisms of seed dormancy, underdeveloped embryos and/or abundant endosperm. Neutral red staining is proposed as a complementary method to standard viability and germination tests for cellular viability evaluation of *Thrinax radiata* (Arecaceae) seeds.

Keywords Arecaceae, Histology, Neutral red staining, Seed longevity, Seed viability, Storage behavior

1 Introduction

Thrinax radiata Lodd. ex Schult. & Schult. f. (Aracaceae), commonly called Chit palm, is a monocot that can grow up to 8 meters tall, grows near coastal areas [1], and has been reported from Florida to Nicaragua. In Mexico, its presence is exclusive to the Yucatan Peninsula. Its spherical seeds are known to occupy more than two-thirds of the total volume of the drupe (fruit), and it has been assumed that seed storage has negative effects on viability and germination [2]. Although *T. radiata* is not endemic to Mexico, the Mexican Official Standard classifies this wild species as an endangered species [3], so its collection and storage in germplasm banks has been promoted for years.

The standards suggested by the International Seed Testing Association for seed assessment rarely consider native wild species [4] in which seed quantity is regularly limited and seed morphology and physiology are highly variable and complex [5–7]. Protocols for the initial evaluation of seeds of wild species and their monitoring during storage are scarce and often must be modified according to the species studied [8]. For this purpose, viability tests are often complementary to seed germination tests.

Marco A. Ramírez-Mosqueda et al. (eds.), *Conservation of Plant Genetic Resources*, Methods in Molecular Biology, vol. 3011,
https://doi.org/10.1007/978-1-0716-5130-8_5,

2,3,5-Triphenyl tetrazolium chloride (TTC) staining is a common method to verify seed viability. However, this technique has limitations since results are sometimes difficult to interpret, mainly in species with underdeveloped embryos, as well as in species where the endosperm tends to be stained. In addition, for some types of seeds these methods are insufficient to testing quality, for example, in those seeds with deep morphological and physiological dormancy types, and/or in those seeds that have been stored for long periods in germplasm banks. During long-term seed storage, some physiological and biochemical mechanisms deteriorate cell metabolism, negatively affecting viability and germination. Among the main effects at the cellular level are damage to the physical integrity of membranes, resulting from the generation of free radicals and lipid peroxidation, which leads to high permeability, and therefore, cells decrease the retention of soluble compounds [9]. Likewise, DNA fragmentation [10] and decreased respiration rate are common features in aged, albeit orthodox, seeds [11]. Therefore, for these seeds, additional protocols are needed to confirm viability and maintenance of tissues and membranes.

Neutral red is a lipophilic dye of eurohydine that is protonated at acid pH, producing a reddish color; however, it can change to yellow if the pH ranges from 6.8 to 8. In plant cells, it is used to identify lysosomes, nucleus and vacuoles [12], even though, because the pH of the apoplast is about 5, this extracellular space also tends to stain. Metabolic and biochemical changes that occur at the cellular level and interfere with the membrane, wall or tonoplast decrease the absorption of neutral red, allowing this dye to serve as an indicator of cellular viability. For years, this staining has been used as a standard test of viability and cytotoxicity in living cells, mainly in cell cultures, because viable cells tend to absorb and actively retain the dye [13]. In plants, Gonzalez et al. [14] tested its efficiency to determine the viability in onion parenchyma. Here, we establish a protocol for the initial evaluation of *T. radiata* seeds; to establish storage behavior and for monitoring in genebanks and also, we proposed a complementary test using neutral red staining, over histological cuts of fixed tissue, based on the integrity of the membrane as indicative of cell deterioration in the endosperm and embryo cells to determine the viability of seed accessions even after several years of storage.

2 Materials

1. All solutions are prepared with sterile distilled water.

2.1 Seed Desiccation and Storage Behavior

1. Silica gel with orange moisture indicator.
2. Airtight glass jars.

3. 2,3,5-Triphenyl tetrazolium chloride (TTC): 1% solution in water (*see* **Note 1**).
4. Petri dishes (90 mm) with Whatman No. 1 filter paper, soaked with distilled water and sealed with a layer of Parafilm.

2.2 Histochemical Analysis

1. 50-mL falcon tubes.
2. Glass desiccator.
3. Watchmaker's forceps.
4. High-profile microtome blades.
5. 1.5-mL Eppendorf tubes.
6. Embedding molds and block holders.
7. Brushes and pencils.
8. Micropipettes and tips 1000-μL and 50-μL.
9. Glass jar with silica dry gel.
10. Transparent nail varnish.
11. Slides and coverslips.
12. Scalpel and aluminum foil.
13. Pipettes and disposable handkerchiefs.

2.3 Equipment

1. Growth Chamber (model Binder 720 KBW E5.1).
2. Water bath.
3. Vacuum pump.
4. Analytical balance.
5. Extraction bell.
6. Heating plate.
7. Rotary microtome.
8. Light field Optical Microscope.
9. Freezer.

2.4 Solutions

FAA fixative. Take 10 mL of formaldehyde, 50 mL of ethanol, 5 mL of glacial acetic acid and add water up to a volume of 100 mL. Mix by immersion. Store at 4°C.

Poly-L-lysine 10%. Take 10 mL of poly-L-lysine and add 90 mL of water. Mix by immersion. Store at 4°C.

Ethanol gradients. Take 30 mL, 50 mL, 70 mL, 85 mL, and 96 mL of ethanol. For each case, adjust with water to the final volume of 100 mL. Mixing by immersion

Inlay medium: methacrylate resin (*see* **Note 2**).

Neutral red. Dissolve 0.5 g of neutral red in 50 mL of water. Dissolve 0.5 g of NaCl in 10 mL of water. Mix both solutions and filters.

80% glycerol. Take 8 mL of glycerol and 2 mL of water, mix by immersion.

3 Methods

1. All procedures are performed at room temperature unless otherwise indicated.
2. The fixation and the embedding process of the plant material should be performed under an extraction bell.
3. The proposed methodology illustrates the steps taken for the analysis of each storage time in fresh and accession samples of *Thrinax radiata* seeds.
4. The seed accessions come from the GermoLab-CICY seed bank and are stored at 4°C.

3.1 Seed Desiccation and Storage Behavior

1. Collect mature *Thrinax radiata* (white color) fruits from the plants, separate the seeds from the fleshy fruit under running water, and let it air dry for 3–5 days (*see* **Note 3**).
2. Use a sample of ten seeds to estimate the average moisture content of fresh seeds. Individually weigh each seed inside an aluminum cone. Estimate the moisture content (MC) using a modified method of "constant low temperature," placing the seeds in an electric oven for 24 h and estimating the water loss through the difference between fresh weight and dry weight.
3. Dry the remaining seeds to 4.4%–5% in an airtight glass jar with silica gel (seed to silica gel mass ratio of approximately 1:1), regularly weighing labeled seeds ($n = 10$). Calculate the decrease in moisture content based on the weight of the fresh seeds. Store a seed lot under laboratory conditions (22°C, 50–60 relative humidity), and store another seed lot at −20°C in a freezer.
4. Take a sample of sixty stored seeds from both conditions after 1, 3, 12, and 24 months of storage and perform viability and germination tests.
5. Soak the seed sample in distilled water for 24 h. Mechanically scarify thirty seeds by removing the operculum with a scalpel with the help of a stereoscopic microscope.
6. Immerse the scarified seeds in tetrazolium chloride at 35°C for 16 h in dark conditions. After this time, remove the tetrazolium solution and wash the seeds three times.

7. Perform a longitudinal cut to the seeds in the operculum area and assess the viability of the seeds by observing the pink to reddish staining of the embryos. Estimate the viability percentage of each seed lot, maintaining moisture with distilled water while the seeds are checked.
8. Sow thirty no-scarified seeds in Petri dishes ($n = 6$) with filter paper and distilled water. Seal Petri dishes with parafilm to decrease water evaporation. Incubate in a growth chamber at 25–35°C alternating temperature with a photoperiod of 12 h/ 12 h light/darkness. Check the Petri dishes every 3 days to count the number of germinated seeds, considering a germinated seed when observing the emergence of the whitish cotyledonary petiole. Estimate the germination time of 50% of the seeds sown and the final germination percentage.
9. Determine the optimal seed conservation conditions based on the results of viability and germination tests. In the present study, *T. radiata* seeds were tolerant to desiccation, but sensitive to freezing at −20°C. On the other hand, the seeds maintained high percentages of viability and germination capacity stored at least for 24 months at laboratory temperature (*see* **Note 4**).
10. After establishing the evaluation and conservation protocol for *T. radiata* using fresh seeds, perform viability and germination tests to evaluate accessions of seeds that have been stored for long periods of time. For this protocol, seeds stored for 2 years, 11 years and 24 years were evaluated.

3.2 Seed Fixing with FAA

1. In a 50 mL falcon tube, six seeds are embedded in 30 mL of FAA solution (*see* **Note 5**).
2. Falcon tubes are placed without cap in a glass desiccator connected to a pump to make vacuum.
3. Carefully cover the desiccator and turn on the pump until it reaches 19.685 in Hg. Simultaneously, the pump is turned off and the valve is closed to keep the vacuum inside the desiccator for 15 min.
4. After this time, the valve slowly opens to release the vacuum.
5. The fixator is changed with a new FAA solution and a second vacuum round is reapplied under the same conditions.
6. The falcon tube is closed and stored at 4°C for 48 h.

3.3 Embedding Seeds in Methacrylate Resin

1. To process the seeds, we use a modified method to the one suggested by the manufacturing specifications of the "JB-4 Embedding kit."
2. FAA fixator is a decanted, falcon type tube with seeds inside, filled with 30 mL of 30% ethanol.
3. The uncovered falcon tube is again brought to the glass desiccator and subjected to a vacuum round, applying a pressure of 20 in Hg for 5 min.
4. At the end of this period, the 30% ethanol is replaced by a new solution and waits 2.5 h (*see* **Note 6**).
5. Again the 30% ethanol solution is decanted and replaced by 50% ethanol. Steps 2 through 4 are repeated with 50%, 70%, 85%, 96%, and 100% ethanol.
6. 12 mL of pre-infiltration solution is prepared in a light-protected tube (*see* **Note 7**) where 6 mL of JB-4 solution A (monomer) is poured and 6 mL of absolute ethanol added to it. It is mixed by immersion and left at room temperature until use.
7. On the other hand, the absolute alcohol is decanted to leave only the seeds in the falcon tube. The seeds are carefully transferred with the help of Watchmaker's forceps (without applying much pressure) to the falcon tube containing the pre-infiltration solution. The seeds are stored for 72 h in dark conditions.
8. The infiltration solution is prepared by weighing 0.125 g of benzoyl peroxide and transferring to a light-protected falcon tube, slowly adding 10 mL of JB4-Solution A and mixing by vigorous immersion for 5 to 7 min, until you observe that benzoil has completely dissolved.
9. The pre-infiltration solution is decanted from the falcon tube and replaced by the infiltration solution. Again, the glass desiccator is used to apply 7.87 in Hg of pressure for 3 min. Over this time, the falcon tube is closed and the seeds must remain in darkness for a week.
10. 10 mL of infiltration solution is prepared again (see step 7).
11. With the help of watchmaker's forceps, each seed is taken and with a scalpel a small cut is made parallel to the seed coat (to make it easier to position) and up to two seeds are placed for each embedding mold.
12. 1 mL of infiltration solution is transferred to a light-protected 1.5 mL Eppendorf tube, 40 μL of JB4-solution B (Accelerator) is added. This will be our embedding solution and is mixed by immersion vigorously for 30 s (*see* **Note 8**).

13. With the help of the micropipette, 500 μL of embedding solution is taken and slowly decanted over the seeds oriented into the inlay mold. The block holder is then placed, and a slight pressure is applied with the hand for about 1 min (*see* **Note 9**).
14. The 500 μL of the remaining embedding solution shall be taken and poured into the block holder hole. Immediately the block with resin is covered to avoid exposure to light and is stored to 4°C where it remains for 24 h (*see* **Note 10**).
15. After this period, the embedded tissue is dismounted and with the help of a scalpel remove the excess of methacrylate around the seeds (*see* **Note 11**).
16. To store at room temperature in a silica gel bottle for at least 1 day (*see* **Note 12**).

3.4 Microtome Cutting and Tissue Collection

1. Prior to cutting, the slides should be loaded to prevent tissue washing during staining. To do this, the slides must be fully immersed in the 10% poly-L-lysine solution for at least 4 h and a maximum of 24 h (*see* **Note 13**), then allowed to dry and stored in a dust-free place until use.
2. The supports are placed on the microtome, always making sure that they are as fixed as possible.
3. The high-profile blade should be placed in the place indicated (*see* **Note 14**).
4. The sample will begin to be roughed with a thickness of approximately 20 μm, when reaching the area of interest (full picture: embryo and endosperm), the thickness of the cut is reduced to 5 μm (*see* **Note 15**).
5. After obtaining the cut, it is collected from the surface of the microtome with the help of tweezers (taking it from a corner) and immediately placed in the water of the water bath at 45°C, where it is left to stretch for 2 to 3 min.
6. With the help of a brush, the histological cut is oriented and collected on the slide (*see* **Note 16**). It is briefly observed under the microscope to see the integrity (*see* **Note 17**) and immediately dried on a heating plate at 45°C for approximately 5 min.
7. The cuts are placed in a storage box, always correctly labelling the sample (*see* **Note 18**).

3.5 Neutral Red Staining

1. Select the histological section that will be stained, cover it completely with distilled water, leave it for two minutes and then decant it.
2. The slide is carried to a light-protected Petri box, placed in a horizontal position and taken approximately 200 μL for each histological cut of the neutral red solution. All the tissue should be covered by the solution. It is allowed to dye for 10 min.

3. At the end of the period, the neutral red dye is decanted and washed with distilled water (*see* **Note 19**).
4. Immediately (without allowing the tissue to dehydrate) two or three drops of 80% glycerol are placed on the tissue and the coverslip is carefully placed (*see* **Note 20**).
5. A layer of clear nail varnish is placed around the coverslip to prevent glycerol from spilling onto the microscope stage plate.
6. Let dry for one or two minutes and immediately proceed to be observed under the light-field optical microscope (*see* **Notes 21 and 22**).
7. Viable cells of both endosperm and embryo retain a red and intense coloration in the middle sheets, cell walls and vacuoles, while non-viable cells fail to retain the dye (see Fig. 1).
8. At the time of observation, micrographs can be taken for a later more thorough analysis of the cell.

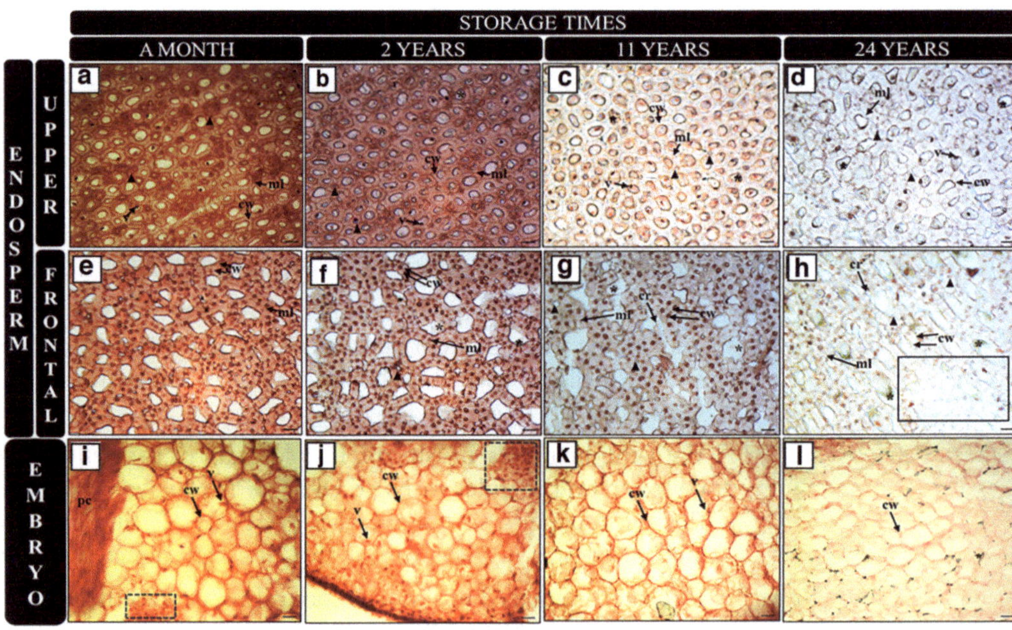

Fig. 1 Determination of cell viability in *Thrinax radiata* seeds fixed using neutral red staining (red-orange). A comparison is shown using (**a–d**) transverse and (**e–h**) longitudinal sections of the endosperm, as well as (**i–l**) a segment of the frontal profile of the embryo; as a function of four different seed storage times (a month; 2, 11, and 24 years). In this technique, the loss of seed viability is reflected in the retention of the dye in the endosperm and embryo cells; it was found that in seeds of 11 years of storage, viability is already considerably affected, while at 20 years, cell viability has been completely lost. All micrographs were stained with neutral red. In all cases, scale bars indicate 50 μm. Simple pits (arrowhead), cell wall (cw), middle lamella (ml), cell rupture (cr), vacuole (v), compound accumulation (*), meristematic areas (dashed rectangle inside), unstained tissue (black rectangle inside)

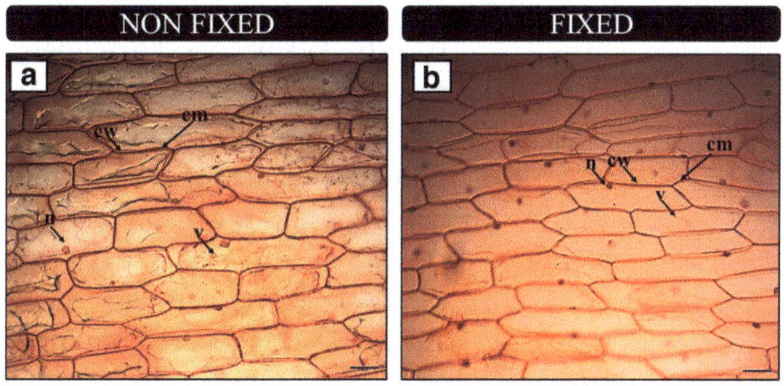

Fig. 2 Comparison of parenchyma cells staining with neutral red (red-orange). (**a**) Showing a section of onion staining in vivo and (**b**) showing a tissue section 24 h after fixed with FAA under vacuum (similar conditions using *Thrinax radiata* seeds). Although in (**b**) a slight decrease in staining intensity is observed; in both (**a**) and (**b**) the cell walls and membranes, nucleus and vacuole are stained, characteristics that indicate cell viability. Scale bars indicate 100 μm. Cell wall (cw), cell membrane (cm), vacuole (v), nucleus (n)

3.6 Checking the Effectiveness of Neutral Red Staining

1. To verify that the fixative does not alter the response of the cells to the neutral red dye. The following could be carried out:
2. Fix a section of the onion's epidermis with FAA fixative (as indicated in Sect. 3.1).
3. Dye with neutral red (as shown in Sect. 3.5).
4. Cut an onion's epidermis sections and dye with neutral red (as shown in Sect. 3.5).
5. Compare both tissues under the 10× and 20× light field optical microscope.
6. In this final analysis, it was observed that it is effective in determining viability in fixed cells, provided that the observation is immediate, no longer than 10 min after washing and assembly with glycerol at 80% (see Fig. 2).

4 Notes

1. After multiple trials, we have determined that for different wild species, TTC at 1% concentration is useful for seed viability testing. In the case of Arecaceae species, it is essential to release the operculum with a scalpel without damaging the embryo so that the solution enters correctly. Incubation in darkness between 35°C and 40°C speeds up the success of this viability test.

2. It is recommended to use the Kit JB-4 4 (Polysciences, 00226, Badener Str, Hirschberg an der Bergstr, Germany).
3. The correct selection of mature seeds and the immediate cleaning facilitates moisture management and increases the potential for maintaining the viability of the seeds.
4. The germination velocity was variable considering that *T. radiata* seeds have a primary dormancy and the storage time influenced different phases of dormancy in the seeds. Removal of the operculum increases the rate of germination, but is not recommended because it promotes fungal infection.
5. The FAA fixative is ideal for preserving cell morphology because it rapidly penetrates tissue, while the alcohol contraction effect is counteracted by the swelling effect of acetic acid. Although this fixative can be sheltered for months at 4°C, in seeds we recommended using freshly prepared fixative solution.
6. The vegetable sample can be kept at 70% ethanol concentrations overnight, provided that the material is kept in refrigeration at °C. The next day, you can continue with the subsequent process of dehydration.
7. To protect the solutions from light, it is recommended to cover with aluminum foil.
8. The immersion of the embedding solution should be vigorous to ensure that the accelerator is very well mixed with the chemical catalyst solution, because if not mixed correctly, the solidification of the resin won't be uniform and will bring disassembly and cutting problems.
9. The pressure on the embedding mold will be applied to avoid bubble formation and the presence of oxygen, which could interfere with the polymerization process.
10. The coating of the newly embedded tissue with resin can be made with aluminum foil, dark fabric, etc. This step will prevent the resin from oxidizing and acquiring a yellowish coloration that could interfere with histological stains after cutting. In addition, maintaining the polymerization process (exothermic reaction) under cold conditions (4°C) will prevent the process of overheating of the sample and will better preserve the protein activity of the tissue; however, it will slow down the solidification process of the tissue.
11. Excess methacrylate will be removed very carefully, leaving at least 2 or 3 mm around the seed. The cut is made at an angle of 45°, leaving the base area as intact as possible, to prevent problems of vibration or detachment during cutting.
12. The use of silica gel will help to extract residual moisture from the embedded sample and increase the drying speed

and hardening of the resin, which will avoid possible complications when performing the microtome cutting. Embedded tissues should be kept in silica gel until the time of use.

13. When the solution of poly-L-lysine is newly prepared, the slides are left only a few hours (until 2 or 3 h); whereas, when the solution has been used three to four times, the slides should remain for up to 24 h. Also, after two uses, precipitating waste must be decanted. The solution should be as clean as possible to avoid stains on the slides, which may cause noise at the time of observation under the microscope.
14. During the handling of the blade we must be careful, because it is easy to damage the edge and for it to produce some scratching on the sample. Likewise, we must take all safety measures during its use to avoid an accident.
15. Roughing the sample until reaching the indicated section using significantly large cutting thicknesses will reduce time and less wear to the knife. The manipulation of the microtome rotor to obtain optimal cuts of the seed must be slow, since for hard materials if done abruptly, there is a risk of a fracture in the knife, in the sample or a winding of the histological section.
16. The brush used for collection should be as thin as possible, preferably the excess fibers should be trimmed. This will prevent the histological cut from sticking to the brush at the time of its collection on the surface of the water.
17. Before placing the slide under the microscope, wipe the back side of the slide with a soft disposable handkerchief to avoid scratches on the slide and damage to the microscope.
18. It is recommended to use a graphite pencil to write on the frosted edge of the slide, always write the type of sample, the thickness of the cut and the number of slides. This will facilitate your search when selecting the histological cut to stain and its subsequent observation.
19. Washing should be done with the slide in an inclined position (45° angle), pouring the water with a continuous but smooth flow (with the help of a pipette). It is recommended to do this wash for a period of about one minute, until the water begins to fall clear.
20. If the cover is placed at an angle of 90° and then with the help of a watchmaker the angle is decreased to 45° and then slowly lowered to the slide horizontally, it is ensured that bubbles will not form that may cause noise during the analysis of the specimen.
21. To ensure correct visualization, Kohler alignment must be performed to the microscope. Whereas, to locate the

specimen first use the objective of lower magnification. Once located, carefully move the revolver to the next objective and focus using the micrometric screw.

22. Observation of the specimen once stained with neutral red should be immediately. After 15 to 20 min, the viable cells that are stained tend to stop retaining the dye, so around the sample a reddish coloration will begin to be observed, a spot that expands in the glycerol.

References

1. Duno SR, Carnevali FC, Ramírez MI, Tapia MJ, Can IL, Hernández-Aguilar S, Embray T (2024) Flora de la Península de Yucatán. http://www.cicy.mx/sitios/flora%20 digital/ Accessed 26 Dec 2024
2. Pérez E, Ceballos G, Calvo L (2005) Germinación y supervivencia de semillas de *Thrinax radiata* (Arecaceae), una especie amenazada en la Península de Yucatán. Bot Sci 77:9–20
3. SEMARNAT (2010) Norma Oficial Mexicana NOM-059-SEMARNAT- 2010 Protección ambiental-Especies nativas de México de flora y fauna silvestres-Categorías de riesgo y especificaciones para su inclusión, exclusión o cambio-Lista de especies en riesgo. Diario Oficial de la Federación, México, p.
4. International Seed Testing Association (2020) ISTA 2020 Rules. Available at https://www.seedtest.org/en/ista-rules-2019-_content---1--3410.html
5. Baskin CC, Baskin JM (eds) (1998) Seeds: ecology, biogeography, and evolution of dormancy and germination. Academic Press, San Diego, p.
6. Frischie S, Miller AL, Pedrini S, Kildisheva OA (2020) Ensuring seed quality in ecological restoration: native seed cleaning and testing. Restor Ecol 28:S239–S248. https://doi.org/10.1111/rec.13217
7. Jaganathan GK (2020) Defining correct dormancy class matters: morphological and morphophysiological dormancy in Arecaceae. Ann For Sci 77:1–6
8. Pedrini S, Dixon KW (2020) International principles and standards for native seeds in ecological restoration. Restor Ecol 28:S286–S303. https://doi.org/10.1111/rec.13155
9. Ratajczak E, Małecka A, Bagniewska-Zadworna A, Kalemba E (2015) The production, localization and spreading of reactive oxygen species contributes to the low vitality of long-term stored common beech (*Fagus sylvatica L.*) seeds. J Plant Physiol 174:147–156. https://doi.org/10.1016/j.jplph.2014.08.021
10. Pirredda M, González-Benito M, Martín C, Mira S (2020) Genetic and epigenetic stability in rye seeds under different storage conditions: ageing and oxygen effect. Plants 9(3):393. https://doi.org/10.3390/plants9030393
11. Zhang K, Zhang Y, Sun J, Meng J, Tao J (2021) Deterioration of orthodox seeds during ageing: influencing factors, physiological alterations and the role of reactive oxygen species. Plant Physiol Biochem 158:475–485. https://doi.org/10.1016/j.plaphy.2020.11.031
12. Admon A, Jacoby B, Goldschmidt E (1980) Assessment of cytoplasmic contaminations in isolated vacuole preparations. Plant Physiol 65:85–87. https://doi.org/10.1104/pp.65.1.85
13. Repetto G, del Peso A, Zurita J (2008) Neutral red uptake assay for the estimation of cell viability/cytotoxicity. Nat Protoc 3:1125–1131. https://doi.org/10.1038/nprot.2008.75
14. González M, Jernstedt J, Slaughter D, Barrett D (2010) Microscopic quantification of cell integrity in raw and processed onion parenchyma cells. J Food Sci 75(7):E402–E408. https://doi.org/10.1111/j.1750-3841.2010.01764.x

Chapter 6

Short-Term Conservation and Exploitation of Sweet Acacia (*Vachellia farnesiana* L. Wight & Arn.) Through an Efficient Scarification-Based Germination Protocol

José Luis Aguirre-Noyola, Verónica García-Iglesias and José Daniel Chávez-González

Abstract

Vachellia farnesiana (L. Wight & Arn.), known as "huizache" or sweet acacia, is a nitrogen-fixing legume with a wide distribution in arid and semi-arid zones of Mexico and America. Its potential in ecological restoration, phytoremediation and improvement of degraded soils has generated a growing interest in establishing reliable protocols for its conservation and short-term exploitation. This is an efficient germination protocol based on mechanical scarification and subsequent in vitro culture. Seeds were obtained from mature fruits, manually scarified, disinfected, germinated under sterile conditions and sown on Fahraeus medium, which does not contain nitrogen and is commonly used to evaluate symbiosis with rhizobia. Optimal incubation conditions favoring root emergence were determined over a 96-h period. Subsequently, seedlings were grown *in vitro* for initial development and morphological and physiological analysis. This standardized protocol allows rapid, controlled and reproducible germination, which facilitates the production of plants for conservation programs, sustainable management and agricultural use. The proposed methodology also constitutes a useful tool for symbiosis and genetic improvement studies in native legumes.

Keywords *Vachellia farnesiana*, Germination, Scarification

1 Introduction

Vachellia is a diverse genus within the Fabaceae family, previously classified under the subgenus *Acacia*, and currently encompasses 163 recognized species. One hundred thirty-three are distributed across Africa and the Americas, 36 in Asia, and 11 in Australia [1, 2]. In Mexico, Sweet acacia is a shrub or small tree reaching 2 to 5 m in height, and their ranges extend from cool to tropical regions [3, 4]. *Vachellia* species are often dominant as trees or shrubs in tropical and subtropical woodlands and savannas and can also be found in temperate areas worldwide [5]. The species is native to

Marco A. Ramírez-Mosqueda et al. (eds.), *Conservation of Plant Genetic Resources*, Methods in Molecular Biology, vol. 3011, https://doi.org/10.1007/978-1-0716-5130-8_6,

tropical America, ranging from the southern United States to Brazil, Colombia, and Peru. Over time, it has spread and been cultivated in many tropical and warm regions worldwide, though it remains unclear whether this expansion is primarily due to natural processes or human activity [6–8].

V. farnesiana is a shrub that spreads outwards, with a typical height ranging from 1.5 to 4 m. The bark is characterized by a smooth to finely fissured texture and a gray-brown pigmentation. The branchlets exhibit a zigzag shape and are often hairy at the apex, but as they age, they become glabrous. Additionally, prominent lenticels are a distinguishing feature of the branchlets. The leaves possess a petiole measuring 0.2–2 cm in length, which is hairy on the upper surface and contains a circular to elongated gland. The rachis ranges from 0.3 to 5.5 cm long, with hairs primarily on the upper surface and sometimes a sugary gland at the apex. The arrangement of pinnules is in 1–7 pairs, with each pair measuring 1–4 cm and bearing hair, particularly on the upper surface. The pinnules are narrow-oblong in shape, occurring in 5–23 pairs, and measuring 3–10 mm long and 0.5–2 mm wide. Minute hairs are present along the margin near the base (and occasionally near the apex), though some pinnules may be glabrous. The midvein and lateral veins are more pronounced and slightly elevated on the abaxial surface. The spinescent stipules typically measure 5–25 mm in length. The inflorescences consist of spherical heads containing 33–95 bright yellow to orange-yellow flowers, with 1–3 or more heads per leaf axil. The peduncles are 3–30 mm long and hairy. The pods are cigar-shaped, either straight or strongly curved, approximately cylindrical, and turgid. Their dimensions range from 1.5 to 8.5 cm in length and 8–17 mm in width. The color of the pods varies from dark brown to blackish, and they are glabrous. The seeds within the pods are positioned transversely or obliquely, separated by a pithy tissue (Fig. 1) [9, 10].

Fig. 1 Morphological traits of sweet acacia. (**a**) *V. farnesiana* tree with multiple spines and bipinnate leaves. (**b**) Subcylindrical fruits (pods) ending in a sharp point. Mature fruits are reddish brown and immature fruits are green. (**c**) Flowers in yellow glomerules, arising from the axils of the spiny stipules

The *V. farnesiana* shrub can fix atmospheric nitrogen, enhancing soil fertility and promoting plant growth [11]. The spread of this species can have both positive and negative economic impacts, influencing various human activities such as agriculture, ornamental, livestock production, forestry, antioxidant, anti-inflammatory, bioenergetic potential, and tourism (6, 12–14).

Recently, *V. farnesiana* has emerged as an important plant that develops in arid and semi-arid areas of Mexico and this plant is crucial in restoring degraded soils (15–19) and soils contaminated with heavy metals (20–22). Therefore, successful germination is a critical step for the application of *V. farnesiana* in ecosystem services and phytoremediation techniques (11, 17, 23–25).

2 Materials

2.1 Plant Material

1. Sweet acacia seeds. Seeds for this protocol were obtained from 8-year-old trees of *V. farnesiana* growing at Iguala de la Independencia, Guerrero (Southwestern Mexico).

2.2 Culture Medium

1. Fahraeus medium: this medium has been used to germinate and culture the seeds of several legumes [26]. The absence of nitrogen source is useful to evaluate the symbiosis between legumes and rhizobia [27].
2. Agar as a gelling agent.
3. Solutions for pH adjustment: 1 N Hydrochloric acid (HCl) and 1 N sodium hydroxide (NaOH).

2.3 Seed Disinfection Solution

1. Ethanol solution 70% v/v (in sterile distilled water).
2. Sodium hypochlorite solution 10% v/v (in sterile distilled water) (*see* **Note 1**).
3. Sodium thiosulfate solution 2% w/v (in sterile distilled water).
4. Sterile distilled water.

2.4 Equipment

1. Plant culture room: Horizontal laminar flow cabinets, glass bead sterilizer and plant growth chamber.
2. Media preparation room: analytical balance, spatula, autoclave, pH meter, hot stirrer plate, micropipette of 100–1000 µL and plastic Petri dishes (100 mm × 15 mm).
3. Sterile stainless-steel tools: Mini-diagonal cutter (12 cm), dissecting forceps without teeth (14 cm) and Lucae Bayonet forceps (21 cm).

4. Glassware: Beakers (500 mL and 1 L), screw cap bottles (1 L), measuring cylinders (50 mL, 100 mL, and 1 L) and flasks (250 mL and 500 mL).
5. Others: Foam stopper for flask, plastic film rolls, magnetic stir bar and indelible marker.

3 Methods

3.1 Seed Scarification

1. Seeds should be obtained from pods that show no signs of physical or mechanical damage or insect infestation (Fig. 2).
2. Using a sterile mini-cutter, remove the coat from the hilum of the seed until a 2–3 mm aperture is obtained. Endosperm should be visualized as shown in Fig. 3 (*see* **Note 2**).
3. Place scarified seeds in a petri dish and keep at 4°C until disinfection.

3.2 Disinfection of Scarified Seeds

1. Working in a flow cabinet, place 50 scarified seeds in a flask (250 mL), add 100 mL of sterile distilled water and shake for 1 min. Avoid losing seeds, remove the solution.
2. Add 100 mL of 70% ethanol, shake for 1 min and remove the solution.
3. Add 100 mL of sterile distilled water, shake for 1 min and remove the solution.
4. Add 100 mL of sodium hypochlorite solution 10%, shake for 5 min and remove the solution.
5. Add 100 mL of sterile distilled water, shake for 1 min and remove the solution.

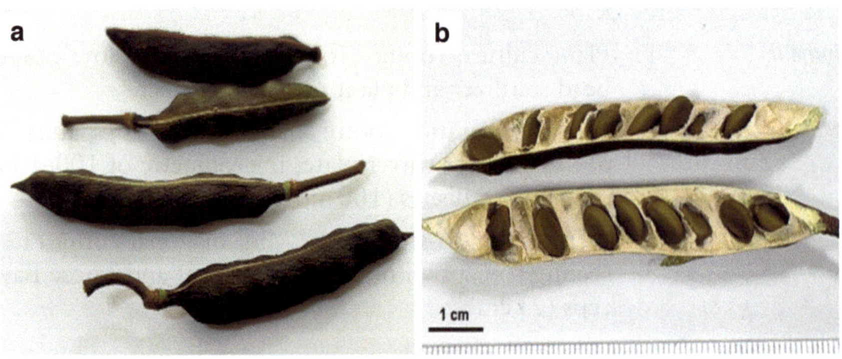

Fig. 2 Healthy fruits of sweet acacia. (**a**) Mature pods without cracks or signs of insect or fungal damage. (**b**) Seeds are disposed in two oblique rows inside a mature pod

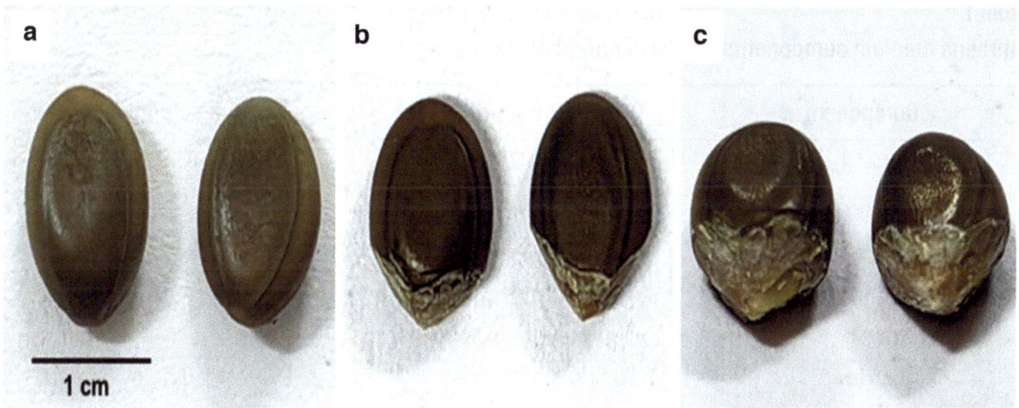

Fig. 3 Sweet acacia seeds. (**a**) Kidney-shaped, smooth, brown seeds. (**b**) Seeds with mechanical scarification at the hilum end. (**c**) Color difference between the seed coat and endosperm

6. Add 100 mL of sodium thiosulfate solution 2%, shake for 5 min and remove the solution (*see* **Note 3**).
7. Add 100 mL of sterile distilled water, shake for 1 min and remove the solution.
8. Repeat the previous step two more times.
9. With the sterile Lucae Bayonet forceps, take the disinfected seeds from the flask and place them in a petri dish.

3.3 Culture Media

1. Perform the preparation of sterile stock solutions of the components of the Fahraeus medium (semi-solid): according to the formulation in Table 1.
2. Add 940 mL of distilled water and 10 mL of the stock solutions (100×) of MgSO4, KH_2PO_4, Na_2HPO_4, and the micronutrient solution to a 1 L beaker (*see* **Note 4**).
3. Add 7.5 g of agar to the mixture.
4. Stir with a magnetic stir bar on a hot stirrer plate (90°C) until the agar is dissolved.
5. Adjust to pH 7 with NaOH or HCl solutions as appropriate.
6. Transfer the mixture to a bottle with a screw cap.
7. Sterilize by autoclaving for 15 min at 121°C with a pressure of 15 lb.
8. Allow the mixture to warm (approximately 50°C) and add 10 mL of the $CaCl_2$ and Fe-citrate stock solutions (*see* **Note 5**).

Table 1
Fahraeus medium components for germination protocols

Component	Working solution (1×)	Stock solution (100×)
$CaCl_2$	0.132 g/L	13.2 g/L
$MgSO_4 \cdot 7H_2O$	0.12 g/L	12 g/L
KH_2PO4	0.10 g/L	1 g/L
$Na_2HPO_4 \cdot 2H_2O$	0.075 g/L	7.5 g/L
	Micronutrient solution	
Fe-citrate	0.005 g/L	0.5 g/L
$MnCl_2 \cdot 4H_2O$	0.07 mg/L	7 mg/L
$CuSO_4 \cdot 5H2O$,	0.07 mg/L	7 mg/L
$ZnCl_2$	0.07 mg/L	7 mg/L
H_3BO_3	0.07 mg/L	7 mg/L
$Na_2MoO_4 \cdot 2H_2O$	0.07 mg/L	7 mg/L

9. Once all the components of the Fahraeus medium have been mixed, disperse 25 mL in Petri dishes and allow to solidify in a flow cabinet.
10. Store the Fahraeus medium plates at 4°C until use.

3.4 Seed Germination

1. Working in a flow cabinet, pick up with a sterile dissecting forceps 10 disinfected scarified seeds.
2. Spread the seeds on Fahraeus medium plates as in Fig. 4.
3. Wait 5 min for the seeds to adhere to the medium.
4. Incubate the plates at 28°C in complete darkness in a plant growth chamber. This step will allow the seed hydration.
5. Inspect the seeds at 96 h and stop incubation if radicle emergence (1–2 cm length) is observed (Fig. 5).
6. If necessary, quantify the percentage germination from data from 10 plates of Fahraeus medium.

3.5 In Vitro Culture of Sweet Acacia Plants

1. Working in a flow cabinet, add 230 mL of freshly mixed, warm Fahraeus medium into sterile 250 mL flasks. Wait for it to solidify.
2. Take a germinated seed with a sterile dissecting and stick the radicle in the center of the medium inside the flask (Fig. 6a, b).
3. Close the flask with the sterile foam stopper.

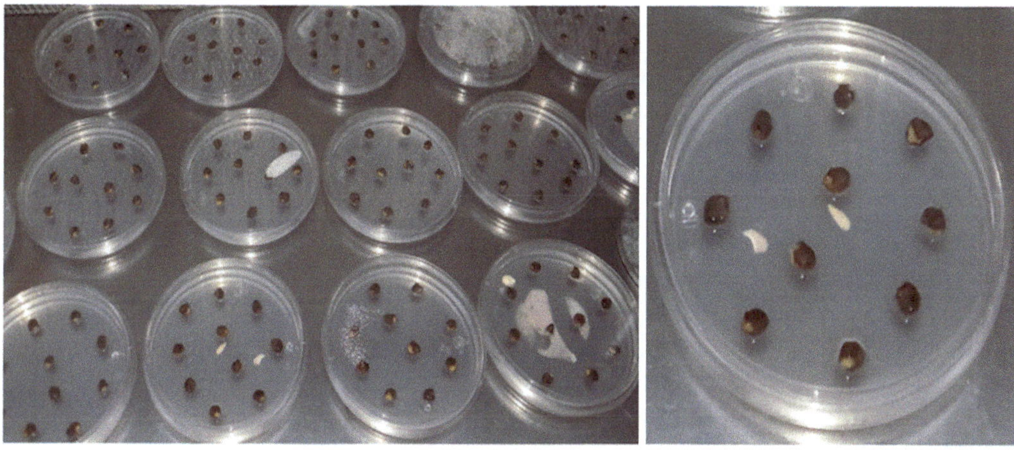

Fig. 4 Distribution of disinfected scarified seeds on Fahraeus medium (semi-solid) plates. Up to ten seeds can be placed to allow for optimal water absorption and radicle emergence

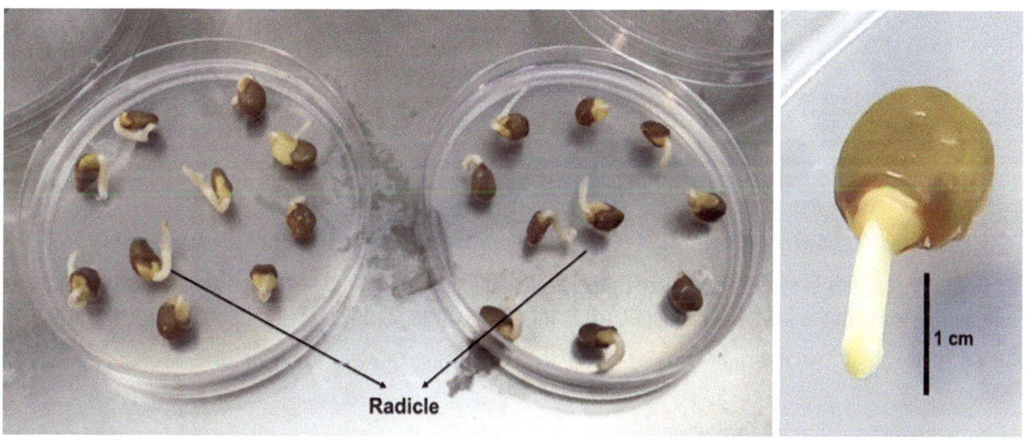

Fig. 5 Germination of sweet acacia seeds. Seed radicle emergence is observed after hydration on Fahraeus medium at 28°C under dark conditions

4. Incubate at 28°C in complete darkness for 72 h in a plant growth chamber (*see* **Note 6**).
5. Once etiolation has occurred, allow the foam stopper to push the stem against the flask mouth. This will allow the stalk to come out of the flask without damaging the cotyledons (Fig. 6c).
6. Incubate at 28°C in a light:dark cycle (16:8 h) in a plant growth chamber.
7. Monitor plant growth for 30 days and evaluate the morpho-physiological parameters of interest (Fig. 6d).

Fig. 6 In vitro culture of sweet acacia. (**a**) Radicle of seedlings. (**b**) Planted seedlings in Fahraeus medium. (**c**) Stem supported with the foam flask stopper. (**d**) Plant growth after 30 days in a light:dark cycle (16:8 h)

4 Notes

1. The sodium hypochlorite solution can be prepared with commercial bleach considering the purity of the product.
2. Mechanically scarify the seeds enough so that you can see the difference between the color of the seed coat and the endosperm.
3. This step can be repeated once more if the chlorine odor is still perceived.
4. Only in the case of micronutrients, a single stock can be made by mixing all the reagents. All solutions should be sterilized by autoclaving and then kept in bottles at 4°C protected from light until use.

5. The CaCl2 and Fe-citrate stock solutions should be added to the rest of the components after autoclaving; otherwise, a white precipitate will form.
6. Be sure to lift the stopper as the stem grows to allow the stem to come out of the flask.

References

1. Wilson P (2006) New combinations in the genus Vachellia (Fabaceae: Mimosoideae) from Australia. Telopea (Syd) 11:233–244
2. Clarke HD, Seigler DS, Ebinger JE (1989) Acacia farnesiana (Fabaceae: Mimosoideae) and Related Species from Mexico, the Southwestern U.S., and the Caribbean. Syst Bot 14:549
3. Herrera-Arreola G, Herrera Y, Reyes-Reyes BG, Dendooven L (2007) Mesquite (*Prosopis juliflora* (Sw.) DC.), huisache (*Acacia farnesiana* (L.) Willd.) and catclaw (*Mimosa biuncifera* Benth.) and their effect on dynamics of carbon and nitrogen in soils of the semi-arid highlands of Durango Mexico. J Arid Environ 69:583–598
4. García-Winder LR, Goñi-Cedeño S, Olguín-Lara PA, Díaz-Salgado G, Arriaga-Jordán CM (2009) Huizache (Acacia farnesiana) whole pods (flesh and seeds) as an alternative feed for sheep in Mexico. Trop Anim Health Prod 41:1615–1621
5. Vibrans H (2009) Malezas de México. http://www.conabio.gob.mx/malezasdemexico/mimosaceae/acacia-farnesiana/fichas/ficha.htm
6. Erkovan HI, Clarke P, Whalley R (2016) A review on general description of *Vachellia farnesiana* (L.) Wight & Arn. J Fac Agric 47:71–76. https://doi.org/10.17097/ZFD.11126
7. Cavalcante A, Cavalcante AMB, Cox RD (2016) *Acacia farnesiana* (L.) Willd. – a potentially invasive alien species? Int J Ecol Environ Sci 42:209–215
8. Richardson DM, Rejmánek M (2011) Trees and shrubs as invasive alien species – a global review. Diversity Distributions 17:788–809. https://doi.org/10.1111/j.1472-4642.2011.00782.x
9. Huizache (*Acacia farnesiana*)· iNaturalist Mexico. https://mexico.inaturalist.org/taxa/47446-Acacia-farnesiana
10. *Vachellia farnesiana* L. Wight & Arn. GRIN-Global. https://npgsweb.ars-grin.gov/gringlobal/taxon/taxonomydetail?id=450067
11. Chauhan BS, Campbell S, Galea VJ (2021) Seed germination biology of sweet acacia (*Vachellia farnesiana*) and response of its seedlings to herbicides. Weed Sci 69:681–686
12. Smith TP, Wilson SB, Marble SC, Xu J (2022) Propagation for commercial production of sweet acacia (*Vachellia farnesiana*): a native plant with ornamental potential. Native Plants J 23:337–348
13. Deka L et al (2024) Exploration of the antioxidant and anti-inflammatory potential of *Vachellia farnesiana* bark by in vitro method. J Pharmacogn Phytochem 13:702–707
14. Hernández-Mata SJ, Garza-Ocañas F, Villalón-Mendoza H, Goche-Telles JR, Carrillo-Parra A (2023) Bioenergetic potential of the Huizache *Vachellia farnesiana* (L.) Willd. In: Garza-Ocañas F (ed) Sustainable management of natural resources: diversity, ecology, taxonomy and sociology. Springer International Publishing, Cham, 49–59 https://doi.org/10.1007/978-3-031-33394-1_4
15. Ceccon E, Almazo-Rogel A, Martínez-Romero E, Toledo I (2012) The effect of inoculation of an indigenous bacteria on the early growth of Acacia farnesiana in a degraded area. CERNE 18:49–57
16. Maphuhla NG, Oyedeji OO (2024) Evaluation of Acacia karroo's potential aspect in the phytoremediation of soil pollution. Sustainability 16:3315
17. Urías-García C et al (2023) Micropropagation of Vachellia spp. as an alternative for the reforestation of desert areas in Mexico. Cult Trop 44:1–9.
18. Mata Balderas JM et al (2022) Monitoring of a reforestation survival with native species of the Tamaulipas thorn scrub. Rev Mex Cienc For 13:28–52
19. Garcidueñas-Piña C et al (2022) *Prosopis laevigata* and *Acacia farnesiana* in association with Rhizobium sp. remove copper from soil. S Afr J Bot 151:417–424
20. Romero MF et al (2021) Metagenomics of mine tailing rhizospheric communities and its selection for plant establishment towards bioremediation. Microbiol Res 247:126732
21. Salazar-Ramírez G et al (2020) Characterization of fungal endophytes isolated from the metal hyperaccumulator plant *Vachellia farnesiana* growing in mine tailings. Microorganisms 8:226. 2020 Vol. 8, Page 226

22. Arvizu-Valenzuela LV et al (2020) Barriers for plant establishment in the abandoned tailings of Nacozari, Sonora, Mexico: the influence of compost addition on seedling performance and tailing properties. Environ Sci Pollut Res 27:39635–39650
23. Morales-Domínguez JF, de León DCS-D, Garcidueñas-Piña C, Pérez-Molphe-Balch E (2019) Germination, in vitro propagation and soil acclimatization of *Acacia farnesiana* and *Prosopis laevigata*. S Afr J Bot 124:345–349
24. Chauhan BS, Campbell S, Galea VJ (2021) Seed germination biology of sweet acacia (*Vachellia farnesiana*) and response of its seedlings to herbicides. Weed Sci 69:1–6
25. Xu JJ, Wilson SB, Vendrame WA, Beleski DG (2023) Micropropagation of sweet acacia (*Vachellia farnesiana*), an underutilized ornamental tree. In Vitro Cell Dev Biol Plant 59:74–82
26. Fahraeus G (1957) The infection of clover root hairs by nodule bacteria studied by a simple glass slide technique. J Gen Microbiol 16:374–381
27. Aguirre-Noyola JL, Rosenblueth M, Santiago-Martínez MG, Martínez-Romero E (2021) Transcriptomic responses of rhizobium phaseoli to root exudates reflect its capacity to colonize maize and common bean in an intercropping system. Front Microbiol 12:740818

Check for updates

Chapter 7

Germination of *Calibanus hookeri* (Lem.) Trel. Seeds Through Imbibition in Distilled Water Plus Physical Scarification

Juan M. Pichardo-González, Martín Quintana-Camargo, Esmeralda J. Cruz-Gutiérrez, Carlos I. Cruz-Cárdenas, Francisco F. Calvillo-Aguilar and María G. Pérez-Reynozo

Abstract

Calibanus hookeri (Lem.) Trel., known in Mexico as "tinaja," is a succulent plant highly valued as an ornamental plant, endemic to the states of Hidalgo, Tamaulipas, and San Luis Potosí. It is classified as threatened by NOM-059-SEMARNAT-2010, as its natural populations have been reduced due to excessive exploitation and habitat destruction. The environment in which the species grows experiences prolonged periods of drought and extreme temperatures, resulting in only 39% of the seedlings that emerge surviving. Therefore, it is urgent to undertake accessions for its ex situ conservation in germplasm banks. In this sense, in the seed laboratory of the Seed Collection of the National Center for Genetic Resources of INIFAP, a research work was carried out that resulted in the generation of this protocol, which aims to increase the percentage of germination in *Calibanus hookeri* seeds in the evaluation of the germination of the seeds of this species through pre-germinative treatments of seed imbibition plus physical scarification.

Keywords Seed imbibition, Physical scarification, Seed germination

1 Introduction

Calibanus hookeri (Lem.) Trel., known in Mexico as "tinaja," is a caudiform plant from the deserts of northern Mexico, endemic to the states of Hidalgo, Tamaulipas, and San Luis Potosí [1]. It is succulent highly valued as an ornamental plant [2]. It belongs to the Asparagaceae family and is registered in the NOM-059-SEMARNAT-2010, cataloged as a threatened plant. Its natural populations have been significantly reduced due to excessive exploitation and destruction of its habitat [3]. In addition to the above, the environment in which the species develops experiences prolonged periods of drought and extreme temperatures; this causes only 39% of the seedlings that sprout to survive [2]. Therefore, it is urgent to ensure its ex situ conservation.

Marco A. Ramírez-Mosqueda et al. (eds.), *Conservation of Plant Genetic Resources*, Methods in Molecular Biology, vol. 3011,
https://doi.org/10.1007/978-1-0716-5130-8_7,

One way to contribute to the conservation of this species is through the storage and conservation of seeds in orthodox seed germplasm banks; these contribute to the development of protocols for evaluating seed quality and improving their conservation. Studies on the quality and conservation of *C. hookeri* seeds are very scarce. Therefore, research on these topics was carried out in the seed laboratory of the Seed Collection of the National Center for Genetic Resources of INIFAP, and this protocol for the evaluation of seed quality was generated with the aim of contributing to its better conservation as a plant genetic resource. By reproducing this protocol, the germination percentage of *Calibanus hookeri* seeds can be increased in the evaluation of the germination of seeds of this species through pre-germination treatments of seed imbibition with distilled water plus physical scarification of the seeds (seed without fruit structures [without capsule]) [4].

2 Materials

2.1 Plant Material, Materials, and Equipment

1. "Tinaja" seeds (*Calibanus hookeri* [Lem.] Trel.). The quantity of seeds will depend on the quantity available.
2. Distilled water (1 L or more depending on the quantity of seeds to be treated). Pour the water into 500 mL beakers.
3. Acrylic germination boxes measuring 16 cm in length, 11 cm in width and 11 cm in height or plastic containers or plastic bags for sowing the seeds.
4. Tweezers.
5. Scalpel with knob.
6. Substrates: peat moss, agrolite, and vermiculite.
7. Seed germination chamber or seed germination room with controlled conditions.
8. Trowel.

3 Methods

3.1 Imbibition of "tinaja" (*Calibanus hookeri*) Capsules for Scarification

1. Place the *Calibanus hookeri* seeds in a container with a capacity of 250 mL or 500 mL or 1000 mL depending on the quantity available.
2. Pour distilled water into the container with the capsules until they are covered with enough distilled water to soak them.
3. Leave the *Calibanus hookeri* capsules to soak for 24 h at room temperature (Fig. 1).

Fig. 1 Seeds of *Calibanus hookeri* (Lem.) Trel. soaked in distilled water

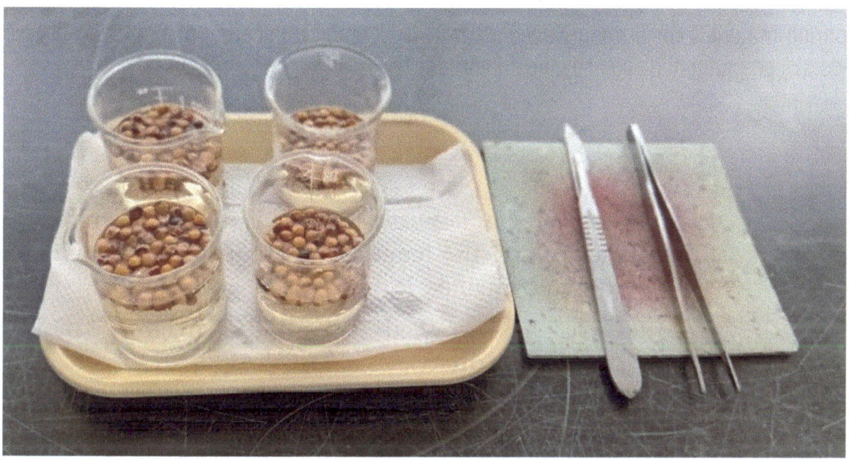

Fig. 2 Physical scarification of capsules of *Calibanus hookeri* (Lem.) Trel

3.2 Physical Scarification of Calibanus hookeri Capsules

1. Take a *Calibanus hookeri* capsule and, using tweezers and a scalpel, remove the fruit structures (the pericarp) until the bare seeds are left (Fig. 2) (*see* **Note 1**).
2. Once the seeds are extracted from the fruit, they are ready for growing (*see* **Note 2**).

3.3 Preparation of the Substrate for Sowing Calibanus hookeri Seeds in the Evaluation of Germination

1. In a container or tray, pour three parts peat moss, one part agrolite, and one part vermiculite (peat moss, agrolite, and vermiculite, 3:1:1 ratio) (Fig. 3).
2. Mix the substrates.
3. Add tap water and continue mixing the moistened substrates (*see* **Note 3**).

3.4 Sowing Calibanus hookeri Seeds in the Prepared Substrate

1. In acrylic germination boxes measuring 16 cm long, 11 cm wide, and 11 cm high, empty the moistened substrate mix into the container, leaving a free space of 4 cm at the top (Fig. 3) (*see* **Note 4**).

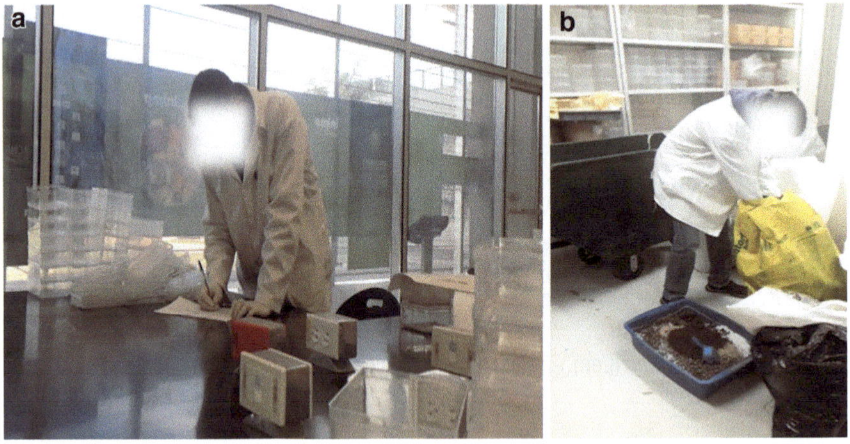

Fig. 3 Preparing materials for sowing *Calibanus hookeri* (Lem.) Trel seeds. (**a**) Preparing acrylic boxes for sowing seeds; (**b**) preparing the seed sowing substrate

2. Sow the *Calibanus hookeri* seeds in the substrate, distributing them along the length and width of the germination box or chosen container, leaving space between them.
3. Cover the sown seeds with a 1–2.5 cm layer of moistened substrate (*see* **Note 5**).

3.5 Incubation of *Calibanus hookeri* Seeds in a Germination Chamber or in a Germination Room with Controlled Conditions

1. Place the germination boxes or containers with the sown seeds in a germination room or germination chamber programmed at a temperature between 25°C and 30°C, with a relative humidity of 70% and a photoperiod of 16 h of light and 8 h of darkness.
2. Four replicates of 100 seeds (400 seeds total) can be considered for germination evaluation if the sample contains sufficient seeds [5] (*see* **Note 6**).
3. Irrigate periodically when the substrate is observed to be lacking moisture.
4. Count the seedlings that have emerged in the substrate, water them, and monitor their growth until the number of seedlings emerges no longer changes.
5. The final result of the germination test is obtained by averaging the four replicates of normal seedlings, abnormal seedlings, dead seeds, and non-germinating seeds, which is expressed as a percentage [5].

4 Notes

1. Because the *Calibanus hookeri* fruit is a capsule, it sometimes contains a single seed or two seeds in a single fruit [2].
2. It is suggested to keep the seeds hydrated while they are being extracted from the capsule and while they are being planted. Continue adding water until the substrate mix reaches "field capacity," meaning the mix is moist but not over-watered. At this point, the substrate will be ready for planting.
3. The substrate must be moistened to actual field capacity (when squeezed by hand, one to two drops of water come out, but without dripping) to avoid waterlogging or water deficit during germination.
4. When using other containers such as trays, preferably use 10–12 cm high containers. When using plastic bags, such as those used in nursery gardens, fold the bags to reduce the amount of substrate used.
5. When using acrylic germination boxes, during the first few days of germination assessment, the boxes can be covered with their lids. When the first seedling emerges, remove the lid. Also, if the appearance of microorganisms such as fungi is observed, apply a fungicide (such as Captan®) according to the manufacturer's recommendation.
6. The seed laboratory of the Seed Collection of the National Center for Genetic Resources of INIFAP has a policy that when the quantity of seeds in the sample or accession is low, germination evaluation is performed using four replicates of 25 seeds (100 seeds total).

References

1. Hernández-Sandoval L (2020) Flora del Bajío y de regiones adyacentes. Instituto de Ecología A.C. Centro Regional del Bajío Pátzcuaro, Michoacán, México, Fascículo 213, p. 40
2. Nuñez-Coronado C, García de los Santos G, Fernández-Pavía YL (2024) Semillas de *Calibanus hookeri* caracterización morfológica, germinación y viabilidad. Agro-Divulgación 4(1):55–59
3. Núñez CC, González RH, Fernández PYL (2021) Micropropagación de *Calibanus hookeri* (Lem.) Trel (1911). Una Especie Amenazada Rev Colomb Biotecnol 23(1):46–54
4. Pichardo GJM, Rivera TE, Cruz CCI, Quintana CM, Cruz GEJ, Calvillo AFF (2021) Evaluación de tratamientos pregerminativos en semillas de Calibanus *hookeri* (Lem.) Trel. *In*: Memoria de la XI Reunión Nacional de Investigación Forestal. Ciudad de México. 10–12 noviembre de 2021. pp 244–246
5. International Seed Testing Association (ISTA) (2017) International rules for seed testing. Rules 2017

Chapter 8

Osmotic Conditioning of Amaranth Seeds to Increase Germination

Juan M. Pichardo-González, Martín Quintana-Camargo, Carlos I. Cruz-Cárdenas, Francisco F. Calvillo-Aguilar, Esmeralda J. Cruz-Gutiérrez, María G. Pérez-Reynozo and Edgar Torres-García

Abstract

Amaranth can help solve food, malnutrition, and health problems in the population because it provides energy, proteins, vitamins, and minerals, and is also a natural source of lysine, an amino acid rare in other cereals. However, when the seed is stored, it deteriorates relatively rapidly, which is primarily reflected in low germination. Osmotic conditioning, or "priming," is a treatment that can improve the germination capacity of seeds once they have deteriorated. To generate the present procol, an osmotic conditioning experiment was conducted with the aim of increasing the germination percentage in amaranth seeds from accessions conserved at the National Center for Genetic Resources (CNRG) of INIFAP. A study conducted by the working team of the Seed Collection of the CNRG showed that of a total of 40 amaranth seed accessions received for conservation and safekeeping in said collection, only 16 samples had a germination percentage greater than or equal to 85%. Therefore, a research work was carried out on osmotic conditioning with seeds of accessions that had low germination, which led to the obtaining of this protocol.

Keywords *Amaranthus* sp., Priming, Seed deterioration, Seed germination

1 Introduction

The demand for amaranth is currently on the rise, as it can help solve food, malnutrition, and health problems in the population, since it provides energy, proteins, vitamins and minerals and is also a natural source of lysine, an amino acid rare in other cereals [1]. When amaranth seed is stored for sowing or as a genetic resource without adequate storage conditions, it deteriorates rapidly, which is mainly reflected in low germination [2]. There are some treatments such as osmotic conditioning or priming to improve the germination capacity of seeds, once they have deteriorated [3, 4].

Marco A. Ramírez-Mosqueda et al. (eds.), *Conservation of Plant Genetic Resources*, Methods in Molecular Biology, vol. 3011, https://doi.org/10.1007/978-1-0716-5130-8_8,

Osmotic conditioning or priming consists of the hydration of seeds under controlled conditions in a solution of a certain concentration using KNO_3, KH_2PO_4, K_3PO_4, $MgSO_4$, NaCl, among other compounds and for a set period of time, which seeks to condition the seeds by delaying their physiological deterioration [5, 6]. With osmotic conditioning, seeds retain viability and accelerate the germination process, as well as root length and vigor, since it repairs damage associated with deterioration [3].

A study conducted by the work team of the Seed Collection of the CNRG showed that of a total of 40 amaranth seed accessions that were received for conservation and safekeeping in said collection, only 16 samples had a germination percentage greater than or equal to 85%. Therefore, a research work on osmotic conditioning with seeds of accessions that had low germination was developed [2], which led to the obtaining of this protocol. With the application of this protocol of osmotic conditioning or priming of amaranth seeds, it will be possible to increase germination in amaranth seeds, especially in those "old" or deteriorated seeds.

2 Materials

2.1 Plant Material

1. Amaranth seeds (*Amaranthus* sp.). The quantity of seeds will depend on the quantity available.

2.2 Distilled Water and Potassium Nitrate (KNO_3)

1. Distilled water (1 L or more depending on the quantity of seeds to be treated). Pour the water into 500 mL beakers.
2. Potassium nitrate (KNO_3) reagent grade (*see* **Note 1**).

2.3 Equipment

1. Analytical balance.
2. Stirring plate.
3. Magnetic stirrer.
4. Glassware: 80 mL beakers, 500 mL beaker, glass rod.
5. Drying oven.
6. Metal boxes with lids for oven-drying seeds.
7. Seed germination chamber or seed germination room with controlled conditions.
8. Clear acrylic germination boxes.
9. Paper towels.
10. Filter paper.
11. Small bottle with distilled water.
12. Tweezers.
13. Small bottle of 70% alcohol.
14. Marker pen.

3 Methods

3.1 Preparation of Potassium Nitrate (KNO_3) Solution

1. On an analytical balance, weigh the reagent grade potassium nitrate (KNO_3) according to the quantity of solution to be prepared at a concentration of 2% (Table 1).
2. In a beaker, empty the volume of distilled water according to the solution needed to prepare (250 mL, 500 mL, 1000 mL, or more).
3. Empty the amount of KNO_3 according to the volume to be prepared into the beaker with distilled water.
4. Place the beaker containing the KNO_3 solution on a shaking plate and, using a magnetic stirrer, stir the distilled water with the KNO_3 to create the solution (*see* **Note 2**).

3.2 Imbibition of Amaranth Seeds in the KNO_3 Solution and Drying of Amaranth Seeds

1. Once the KNO_3 solution is prepared, the seeds are immersed in the beaker containing said solution and soaked for 2 h (Fig. 1).
2. Immediately after the 2 h imbibition period, all seeds are removed from the KNO_3 solution and dried superficially using paper towels.

Table 1
Quantity of KNO_3 to be weighed according to the volume of distilled water to be prepared to a solution at a concentration of 2%

Volume of distilled water to be prepared for a concentration of 2% of KNO_3	Amount of KNO_3 to weigh (g)
250 mL	0.50
500 mL	1.0
1000 mL	2.0

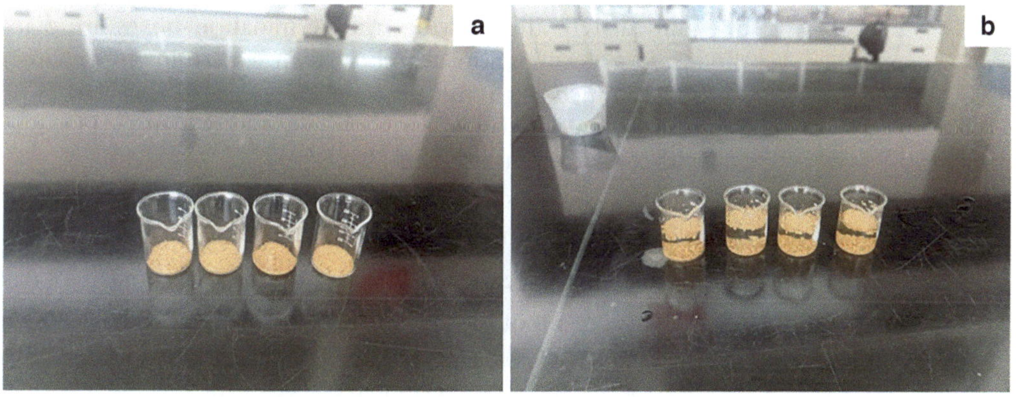

Fig. 1 *Amaranthus* sp. seeds. (**a**) In beakers before placing them in the KNO_3 solution; (**b**) soaked in the 2% KNO_3 solution

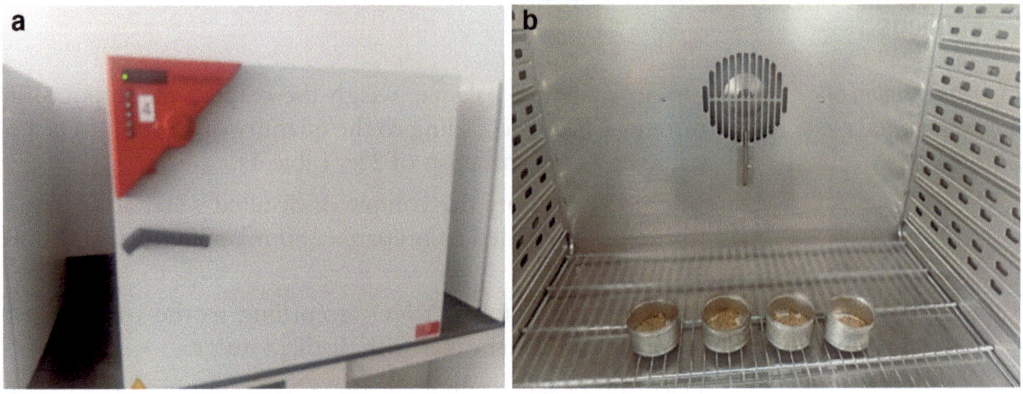

Fig. 2 Drying of *Amaranthus* sp. seeds. (**a**) Drying oven; (**b**) seeds dried in a drying oven programmed at 25°C

3. These seeds are then placed in metal boxes to dry the seeds.
4. The boxes with the amaranth seeds are placed inside a drying oven and left to dry for 24 h at a constant temperature of 25°C (Fig. 2).

3.3 Evaluation of Amaranth Seeds Germination Using the Top of Paper (TP) Method

1. Once the seeds are dried, they are evaluated for physiological quality with a germination test using the top of paper method [7] (Fig. 3).
2. In a transparent acrylic germination box of 16 cm long, 11 cm wide, and 11 cm high, a folded paper towel with the dimensions of the bottom of the box and a paper filter cut to the dimensions of the bottom of the box are placed (*see* **Note 3**).
3. The filter paper and the paper towel must be humid with distilled water using a pestle, and immediately afterward the seeds of amaranth are sown, distributed over the length and width of the filter paper (*see* **Note 4**).
4. Once the seeds are sowed, the acrylic box is covered and with it the box remains ready to incubate in the germination conditions.
5. Immediately after sowing, the germination boxes are placed inside a germination chamber or in a germination chamber programmed at a temperature of 25 ± 2°C (ISTA, 2017) [7], with a photoperiod of 16 h of light and 8 h of darkness and with a relative humidity of 70%.
6. Manually irrigate the seeds throughout the duration of the test using a small bottle with distilled water (*see* **Note 5**).
7. Perform the first count at 4–5 days and the final count at 14 days.
8. The final result of the germination test is obtained by averaging the four repetitions of normal seedlings, abnormal seedlings, dead seeds, and non-germinated seeds, which is expressed as a percentage [7].

Fig. 3 Establishment of the *Amaranthus* sp. seed germination test in the laboratory in transparent acrylic boxes using the top of paper method

4 Notes

1. The concentration of this reagent will be 2%, so the amount used will depend on the number of seeds subjected to osmotic conditioning. The amount of KNO_3 to be used will be removed from its container and weighed on an analytical balance.
2. This stirring will last until it is observed that the KNO_3 is completely dissolved in the distilled water; once this is observed, the solution is ready.
3. The paper towel and the filter paper will serve as a substrate and support for the seedlings in the germination test.
4. ISTA recommends using four repetitions of 100 seeds (400 seeds in total) for the germination test [7]. However, the Manual of Seed Handling in Genebanks by Bioversity International [8] mentions that to determine the viability of the beginning of storage, it is recommended to perform a germination test with a fixed sample size of 200 seeds.
5. Watering should be done so that the seeds do not lack moisture, taking care not to over-water to avoid damage to the seeds due to excess moisture.

References

1. INDESO L (2014) Manual para la producción de amaranto cultivo, cosecha y post cosecha. Oaxaca, México, Consultado el 17 de agosto de 2019
2. Pichardo-González JM, Quintana-Camargo M, Narváez-Esparza MA, Barrera-Ortiz ED (2019) Efecto del acondicionamiento osmótico con soluciones de KNO3 en la germinación de semillas de Amaranthus *hypochondriacus* L. Rev de Invención Técnica 3(11):1–4
3. Khan AA, Tao K, Knypl S, Borkowska B (1978) Osmotic conditioning of seeds: physiological and biochemical changes. Acta Hort 83:267–278
4. Marín SJ, Mejía J, Hernández A, Peña A, Carballo A (2007) Acondicionamiento osmótico de semillas de tomate de cáscara. Agricultura Técnica en México 33:115–123
5. Black M, Bewley JD (2000) Seed technology and its biological basis. Sheffield Academic Press Ltd, Sheffield,
6. Moosavi A, Tavakkol R, Afshari F, Aynehband A (2009) Effects of priming on germination characteristics of four amaranth cultivars. J Food Agric Environ 7(3–4):353–358
7. International Seed Testing Association (ISTA) (2017) International rules for seed testing. Rules 2017
8. Rao NK, Hanson J, Dulloo ME, Ghosh K, Nowell D, Larinde M (2006) Manual of seed handling in genebanks. Handbooks for Genebanks No. 8. Bioversity International, Rome

Check for updates

Chapter 9

In Vitro Propagation of *Limonium cordatum* (L.) Mill. for Short-Term Conservation

Marco Savona and Carlo Mascarello

Abstract

Limonium cordatum (L.) Mill. (Plumbaginaceae) is a halophilous plant. Due to the intense anthropic pressure that influenced its habitat, it is classified as Vulnerable (VU) according to the conservation status of all the Italian endemic vascular plants. Since micropropagation could ensure massive plants multiplication, in order to preserve and restore gene pool of rare and endangered species, this method was successfully applied in *L. cordatum*. The uses of two different light intensities and two levels of temperature were tested to promote better plant quality. BA at 0.2 mg/L ensured the highest multiplication rate (12 new shoots/explant/month). Single-rooted explants are the best material to obtain plants suitable for acclimatization activities.

Keywords Micropropagation, BA (6-benzylaminopurine), Quality plant, Italian vulnerable species, Cluster, Acclimatization, Halophilous, In situ conservation

1 Introduction

Limonium cordatum (L.) Mill. (Plumbaginaceae) is a halophylous plant that grows near the sea in the Northern Mediterranean area; it is sporadically present from Savona (Italy) to Nice (France), in an area of 7.5 square km. It's a small but vigorous plant (15–30 cm), bearing many rosettes of greyish pubescent leaves. Its leaves are linear-spatulate and their surface is quite rough due to calcareous secretions. Flowers appear from July to August; the corymbose inflorescence with erect branches presents many violet flowers visible from a great distance. The plant produces very small fruits in September. This plant is exclusively a rocky plant and needs a calcareous substrate: this is an ecological limitation of the species. Plant propagation is done by seeds. Due to the intense anthropic pressure (tourism and industrial activities) that influenced this mediterranean area, this species is classified as Vulnerable (VU) according to the conservation

Marco A. Ramírez-Mosqueda et al. (eds.), *Conservation of Plant Genetic Resources*, Methods in Molecular Biology, vol. 3011, https://doi.org/10.1007/978-1-0716-5130-8_9,

status of all the Italian endemic vascular plants assessed by IUCN categories and criteria [1].

Micropropagation is currently applied for massive plants multiplication to preserve and restore gene pool of rare and endangered species. This technique allows to obtain a large number of healthy plants in a small space, independently of natural seasons and climate. In vitro propagation is therefore considered a suitable technique for flora conservation and a protection strategy [2]. Few reports are present in literature on the application of micropropagation of several *Limonium* species: in vitro protocol was applied for threatened *L. cavanillesii* [3], for five species of the genus were studied by [4] Martin and Perez in 1995 and *L. perezii* was micropropagated for ornamental purpose [5]. Recently, it has been reported that the use of silver nanoparticles on *L. sinuatum* could promise high economic efficiency in micropropagation of this species [6]. The role of biological active compounds and alternative methods for micropropagation of *Limonium insigne* was reported by [7].

In the present protocol, micropropagation technique was applied in *L. cordatum* with the aim to obtain healthy, robust, and high-quality plants to be used for ex vitro reintroduction programs.

2 Materials

2.1 Plant Material

1. Immature inflorescence stems (20–30 cm long) [8].

2.2 Surface Sterilization of Source Material

1. Washing solution: fresh water added with some drops of liquid soap.
2. Sterilization solution: aqueous sodium hypochlorite solution NaOCl prepared by mixing one part of commercial bleach (5% free chlorine) and four parts of deionized water (1% free chlorine) plus two drops of surfactant (Tween 20®).
3. Sterile deionized water in glass vessels with caps.
4. Magnetic stirrer, 1000 mL flask or beaker.
5. Tissue culture facilities: instruments (scalpel, forceps, spirit burner to flame sterilize instruments), laminar flow hood, culture room.

2.3 Culture Media

1. Glass jars with transparent caps.
2. Media based on the formulation of Murashige and Skoog (MS) [10] for shoot propagation from apical and axillary buds of young branches and root induction (4.3 g/L), MS vitamins, agar (8 g/L), sucrose (30 g/L), 6-benzylaminopurine (BA) (0.2 mg/L), indole-3-acetic acid (IAA) (0.5 mg/L) [9].
3. pH meter and autoclave.

4. Sterile Petri dishes (9 cm diameter).
5. Parafilm.
6. Duran glass flasks or beakers, 1000 mL capacity.
7. Tissue culture facilities: instruments (scalpel, forceps, spirit burner to flame sterilize instruments), laminar flow hood, culture room.

2.4 Acclimatization of Plants to Ex Vitro Conditions

1. Greenhouse with a mist system.
2. Plastic alveolar pots.
3. Plastic pot of 10 cm in diameter.
4. Potting medium for acclimatization consists of peat and perlite (40:60, v/v).
5. Commercial potting medium for transplanting consists of peat with pumice.

3 Methods

3.1 Preparation and Sterilization of Culture Media

1. Dissolve MS powder (43 g), containing a micro- and macroelements complex, in 9 L of deionized water to prepare a 10× stock solution; while stirring the water, add the powder and stir until complete dissolution; bring the solution to a final volume (10 L) by adding water; use 1 L of this solution (4.302 mg) for each liter of culture medium.
2. Dissolve MS vitamin powder, containing 25.8 g mixed vitamins, to prepare 250 mL of 1000× stock solution in deionized water and stir until completely dissolved, eventually warming the solution up to 30°C. Use 1 mL vitamin stock solution (103.1 mg) for each liter of culture medium.
3. Add any desired stable supplement (commercial sucrose, 30 g/L) weighing the powder and dissolve by stirring.
4. Plant growth regulators stocks: to prepare a 1 mg/mL of stock solution, weigh 100 mg of the plant growth regulator 6-benzylaminopurine (BA), indole-3-acetic acid (IAA), and add in a volumetric flask 20–30 mL ethanol (99.8%). Gently stir until completely dissolved and slowly bring the volume to 100 mL with distilled water at room temperature. Store the stock solution as recommended (4°C in light for BA; 4°C in dark for IAA).
5. After adding all the components, while stirring, adjust the pH of the medium (5.7) using 1 N NaOH or 1 M HCl.

6. Add 8 g/L technical agar, heat until clarity of the solution, stirring the medium on an electric plate or heating in a microwave.
7. Dispense 62.5 mL medium into 16 culture vessels (200 mL).
8. Sterilize the medium in autoclave at 121°C, 1 atm, for 20 min. Allow medium to cool and solidify prior plant inoculation.

3.2 Inflorescence Sterilization

1. Excise immature inflorescence stems from selected in vivo mother plants (Fig. 1).
2. Rinse plant material in the washing solution, with stirring for 10 min.
3. Dip the inflorescence in the sterilization solution and stir for 20 min.
4. Rinse the material twice with sterile distilled water for 10 min. The previous procedure and the following steps should be performed in sterile conditions under laminar flow hood (*see* **Note 1**).

3.3 Explants Culture

1. Transplant the single aseptic explants into glass vessels containing culture media. The jars must be maintained in a growth chamber with 33 PPFD (μE $m^{-2}s^{-1}$) and 23°C (*see* **Note 2**, Figs. 2 and 3).

Fig. 1 Immature inflorescence stems of *Limonium cordatum*, cultured in vitro, after sterilization treatment

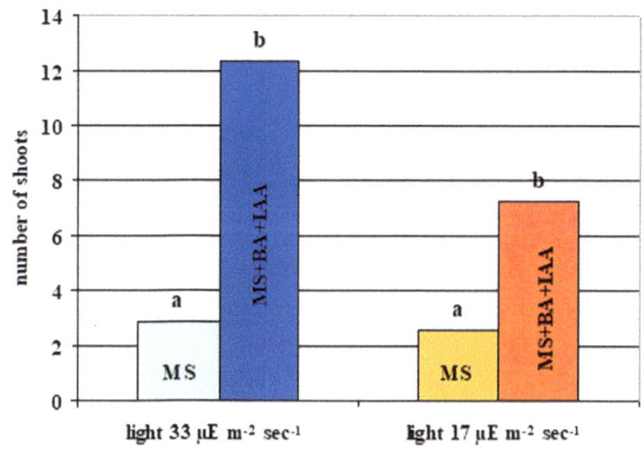

Fig. 2 Multiplication rate (number of shoots/explant/month) under two levels of light intensity and achieved on two different media compositions (with or without PGRs)

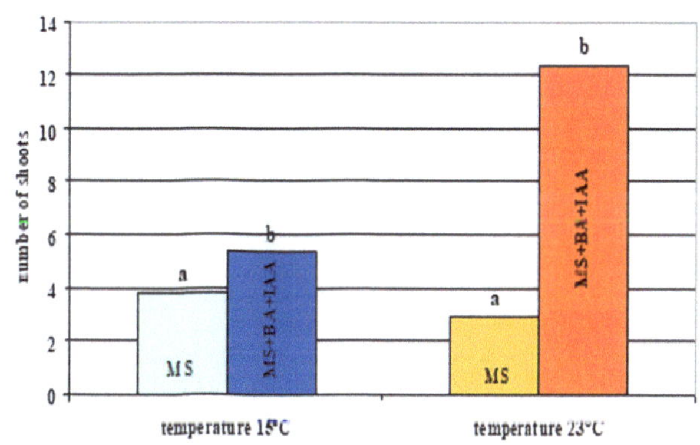

Fig. 3 Multiplication rate (number of shoots/explant/month) under two levels of temperature and achieved on two different media compositions (with or without PGRs)

2. Single regenerated vegetative shoots are separated and transferred to fresh culture media. The explants were subcultured every 40 days till the production of enough material to perform further experiments (*see* **Note 3**).
3. For in vitro rooting phase, it is not necessary to add any PGR to the medium and it is mandatory to apply reduced light intensity to improve single plantlets height (Figs. 4 and 5).

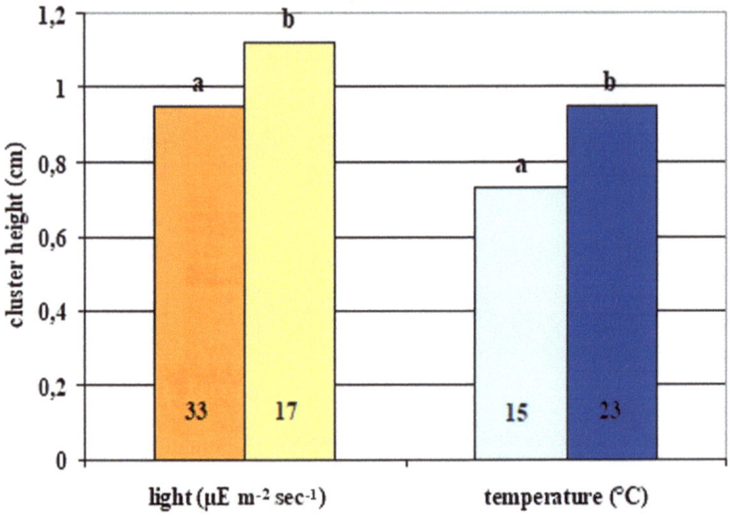

Fig. 4 Effect of light intensity and temperature on cluster height

Fig. 5 In vitro single plantlets in multiplication phase

3.4 Plants Acclimatization

1. Single in vitro rooted plants can be acclimatized in spring after 6 months of in vitro cultivation (Figs. 6 and 7).
2. Transfer the rooted plants in the substrate in an unheated greenhouse under a mist system (10 s every 30 min) for the first 15 days, then gradually decrease the spray frequency until new leaves appear (60 days).

Fig. 6 In vitro rooted explants ready to acclimatization

Fig. 7 Young plants acclimatized ready to be transplanted

Fig. 8 Flowered potted plants, ready to be in situ reintroduced

3. When the plants start to grow, transplant them in pots containing a commercial substrate with peat and pumice (Fig. 8); after 2 months, the plants are able to produce regular inflorescences and are ready to be transferred in open air (*see* **Note 4**).

4 Notes

1. The proposed sterilization protocol allowed 100% sterile explants.
2. The addition of plant growth regulators (PGRs) to the basal MS medium increased shoots proliferation at both light intensities, and the highest value was achieved with the higher light intensity when explants were cultured on medium supplemented with PGRs (more than 12 new neo-formed shoots/explant). The addition of growth regulators to the MS basal medium guarantees a higher multiplication rate. At the lowest value of temperature (15°C), the number of shoots per explant is statistically different but not so relevant compared to MS. At 23°C, it is possible to appreciate the highest value (almost double compared to 15°C).
3. The best results to improve cluster height can be obtained by applying the lowest light intensity (17 μE $m^{-2}s^{-1}$) or the highest temperature (23°C). The best acclimatization performance

is guaranteed with single shoots with a specific height and not by compact cluster.

4. The acclimatization protocol described above allows to obtain 97% of plants ready for reintroduction in the place where *L. cordatum* was present.

Acknowledgments

The authors woud like to thank Dr. Barbara Ruffoni, Dr. Elena Zappa, Professor Luigi Minuto, and Professor Gabriele Casazza for their contribution to the research.

References

1. Rossi G et al (2020) Lista rossa della flora italiana: 2. Endemiti e altre piante minacciate
2. Chokheli VA, Dmitriev PA, Rajput VD, Bakulin SD, Azarov AS, Varduni TV, Stepanenko VV, Tarigholizadeh S, Singh RK, Verma KK, Minkina TM (2020) Recent development in micropropagation techniques for rare plant species. Plants 8–9(12):1733. https://doi.org/10.3390/plants9121733
3. Amo-Marco JB, Ibanez MR (1998) Micropropagation of *Limonium cavanillesii* Erben, a threatened statice, from inflorescence stems. Plant Growth Regul 24(1):49–54
4. Martín C, Pérez C (1995) Micropropagation of five endemic species of Limonium from the Iberian Peninsula. J Hortic Sci 70(1):97–103
5. Ruffoni B, Savona M, Mascarello C, Pamato M, Farina E, Sanremo I et al (2002) In vitro culture and neomorphogenesis from immature floral scapes of *Limonium perezii*. Italus Hortus (Italy) 9(5):57–61
6. Cuong DM, Mai NTN, Tung HT et al (2023) Positive effect of silver nanoparticles in micropropagation of *Limonium sinuatum* (L.) Mill. 'White'. Plant Cell Tissue Organ Cult 155:417–432. https://doi.org/10.1007/s11240-023-02488-5
7. Ortuño A, Díaz L, Pérez I, Sánchez F, Del Río JA (2017) Biological active compounds from *Limonium insigne* and alternative methods for its micropropagation. Sci Hortic (Amsterdam) 230:78–85
8. Savona M, Mascarello C, Mantovani E, Minuto L, Casazza L, Ruffoni B (2009) Strategy to improve the quality, acclimatization and ex vitro re-introduction of micropropagated plants of *Limonium cordatum* (L.) Mill., a Mediterranean endemic. Acta Hort 812:527–531
9. Casazza G, Savona M, Carli S, Minuto L, Profumo P (2002) Micropropagation of *Limonium cordatum* (L.) Mill. for conservation purposes. J Hortic Sci Biotechnol 77(5):541–545
10. Murashige T, Skoog F (1962) A revised medium for rapid growth and bioassays with tobacco tissue cultures. Physiol Plant 15:473–497

Chapter 10

Uses and Conservation of Chaya in the Southeast of Mexico

Pedro Osorio-Montalvo, Germán Nic-Matos, Iván Córdova-Lara and Luis Sáenz-Carbonell

Abstract

Chaya (*Cnidoscolus aconitifolius*) is a native Mesoamerican plant that's been tapped for its nutritional and medicinal properties for centuries. Preserving this plant is crucial, given its significance in ensuring food security and supporting traditional medicine. Chaya is packed with protein, vitamins, and minerals, making it an excellent nutrient source for rural communities. Research has also highlighted its medicinal benefits, including lowering blood sugar levels and improving digestive health. This drought-resistant plant thrives in poor soil, making it perfect for sustainable farming in regions with harsh climates. By conserving chaya, we can also protect biodiversity and local ecosystems. However, chaya faces threats from deforestation, soil erosion, and the loss of traditional knowledge about its use and management. Implementing conservation strategies and promoting sustainable cultivation practices is key to ensuring its availability for future generations.

Keywords *Cnidoscolus aconitifolius*, Food security, Medicinal benefits

1 Introduction

Genetic resources are genetic materials from plants, fungi, animals, and microorganisms that contain some type of functional unit, such as nutrients, proteins, or compounds with medicinal properties, used directly or indirectly in agriculture or food production. These resources are valuable because they contribute to human health, food security, have social significance, and create new opportunities for product development [1].

One valuable genetic resource is the chaya plant, which has been used since pre-Hispanic times in regions influenced by Mayan culture, including southeastern Mexico, Belize, and parts of Central America. The wild chaya has stingy leaves that make it difficult to cultivate and manipulate. The Mayan people domesticated chaya to produce plants with larger and softer leaves [2]. Historical accounts from the Spanish mention that chaya was highly valued by Mayan leaders and is even referenced in the book "Chilam Balam of Chumayel" [3].

Marco A. Ramírez-Mosqueda et al. (eds.), *Conservation of Plant Genetic Resources*, Methods in Molecular Biology, vol. 3011, https://doi.org/10.1007/978-1-0716-5130-8_10,

Chaya belongs to the genus *Cnidoscolus* in the family Euphorbiaceae. It is a perennial shrub that can grow up to 6 m tall, with dichotomous branching and large leaves measuring approximately 30 cm, each with three or more lobes (Fig. 1a, b). The flowers are white and monoecious (Fig. 1c), with fertility stages that are temporally separated, resulting in rare seed production [3]. The genus *Cnidoscolus* includes 50 species found throughout the Americas, with 25 species endemics to Mexico [4]. In the Yucatan Peninsula, three species of the genus *Cnidoscolus* are known to exist: (Mill) I.M. Johnst, *C. souzae* McVaught (Fig. 1d), and *C. urens* (L.). The first two species are distributed in the Yucatan Peninsula, Belize, and Guatemala, while the last species can be found as far south as Argentina [4].

The chaya plant was classified by Breckon in 1975 as a representative of its genus *Cnidoscolus aconitifolius* spp. *aconitifolius*. In practical terms, we can say that the chaya plant is any species of the genus *Cnidoscolus*, which is part of the family Euphorbiaceae. Four varieties have been recorded ("chayamansa," "redonda," "estrella," and "picuda"). "Chayamansa" is the best-known and one of the most used for food [3]. There are very few studies on the genetic diversity of chaya; one of these found very low levels of diversity in the Yucatan Peninsula, which clearly reflects an effect of clonal propagation of the species [4].

Fig. 1 A panel of six photographs showing chaya plants and vegetation (**a**), plant of chaya growing along the road (**b**), a close-up of a plant stem with small white flowers (**c**), plants of *Cnidoscolus souzae*, another species of chaya (**d**), plants in a commercial plantation (**e**) and enlargement of a chaya plant growing in a rocky area (**f**)

2 Uses of Chaya in the Yucatan Peninsula

Chaya has been used since pre-Hispanic times in the areas influenced by the Mayans. However, for a long time, its use was only practiced by their descendants. Nowadays, its medicinal and organoleptic properties are being rediscovered.

The main use of chaya is as a vegetable as part of various dishes. It is often compared to spinach, but its nutritional content is higher than the latter [5]. Approximately 18 dishes in regional cuisine include this plant. Some notable dishes are brazo de reina (commonly eaten during Lent), fried chaya with eggs, chaya sancochada, chaya tamales, and chaya beverages [6]. Additionally, due to the plant's nutritional properties, the leaves are used to feed backyard animals, mainly poultry and pigs [6]. Currently, new generations are embracing chaya leaves as food. Efforts have been made to incorporate them into products such as yogurt, jams, and sausages. The use of chaya as a complement in chorizo has been well received by residents in certain areas of the state of Quintana Roo, Mexico [6].

3 Medicinal Properties of Chaya

Chaya is one of the most widely used plants in southern Mexico and in the Mayan region of Guatemala and Belize, holding great cultural and gastronomic value in this geographic region as the main source of leafy vegetables for the population [3, 7]. The part of the chaya plant that is most utilized is the leaves, as they contain the highest number of proteins, vitamins, amino acids, fatty acids, and active principles such as polyphenols [8] and flavonoids [9, 10]. Chaya is exceptionally high in fiber (31.165%), calcium (50 mg/g), iron (10 mg/g), potassium (20 mg/g), and vitamin C (892.025 mg/100 g) [11]. When compared to spinach, chaya surpasses it in nutrient content, with 78% more protein, 111% more fiber, 100% more iron, and 242% more vitamin C [12].

Chaya offers a wide range of health benefits, including antidiabetic properties [11], cardioprotective effects [13], antioxidant, antimicrobial, anti-inflammatory, hematopoietic, and anticancer properties [14]. Additionally, chaya is commonly used to address rheumatism, gastrointestinal disorders [15], and inflammatory diseases [8, 16]. The use of chaya leaves in traditional medicine for various pathologies is well-documented, with beliefs in its hepatoprotective properties [17]. It has also been reported that rats fed with chaya extracts had effect on the blood levels of sex hormones produced in the pituitary-gonadal axis [18] and chaya leaf extracts have shown indomethacin-like activity in chronic models [16, 19].

There is sufficient information to affirm that chaya leaves contain different phenolic groups, such as coumarins, phenols, tannins, anthraquinones, phlorotannins and flavonoids [12]. Among the flavonoids contained in chaya, quercetin and kaempferol stand out, which together confer a high nutritional and therapeutic value as anti-inflammatory agents [9]. Flavonoids possess anti-inflammatory activity because they contribute to the process of repair and prevention of injuries [10]. Many cell types of the immune system participate in this process, such as neutrophils or polymorphonuclear leukocytes, mast cells, macrophages and lymphocytes [20] which travel through the bloodstream to the site of action and together activate a complex and efficient mechanism of action, repairing the tissue and healing the injury [21]. Flavonoids inhibit the enzyme that synthesizes arachidonic acid, a precursor of prostacyclin, prostaglandins, and thromboxane, which are highly active vasodilators in the body. Additionally, flavonoids can regulate the release of certain mediators such as histamine and tryptase, molecules that promote the cessation of the inflammatory process [22].

4 The Chaya Plant in the Mayan *Milpa*

The Mesoamerican region is considered one of the world's major centers of plant origin and domestication, having contributed several important crops to modern agriculture, including maize, beans, squash, cotton, tomatoes, cacao, avocados, and agave [23]. *La Milpa* (swidden corn fifields) is the best known agroecological system in the area and has been well described in the academic literature [24–26] that has probably persisted for more than 5,000 years. It is not known exactly where or when it emerged, but it is very likely that the *milpa* was the cultivation system in which the plant species that form it were first dispersed throughout Mesoamerica, and then to the rest of North America and finally to South America [27]. They arrived in the Maya region at an early stage when the population was already cultivating roots, tubers, fruits and common jack-bean, among others.

In addition to these well-known crops, Mesoamerican peoples cultivated and domesticated many other useful plants that remain relatively unknown outside this region. One of these is chaya, cultivated mainly in *milpas* and home gardens for its nutritious leaves. Chaya is known in the Yucatan Peninsula as sweet chaya or "chaay" in the Mayan-Yucatecan dialect. It is appreciated for its sweet flavor and easy cooking, and is even used as animal fodder. Chaya is the only plant that can be harvested year-round. Nowadays, Mayan communities use traditional techniques combined with agroecological management, which has improved the quality and quantity of chaya production. This sustainable production benefits Mayan families, who use chaya as a food with high nutritional value with additional multiple uses. The surplus can be sold locally to support family income and contribute to achieving food sovereignty in the region [27].

5 Conservation of Chaya

In light of the various benefits of this crop, it is crucial to establish cultivation and propagation methods to increase its production for food or as raw material to obtain different secondary metabolites [28]. Generally, chaya is able to survive adverse environmental conditions such as drought, high temperatures, and low light conditions; however, the best soil for its growth is well-drained loam soil [29]. Among the propagation methods, traditional and biotechnological methods (such as in vitro culture) can be found. One of the main traditional methods that has been efficiently tested for this crop is the use of cuttings with rooting solutions for the formation of healthy seedlings. It has been observed that the combined use of soil, organic fertilizer, and sand in a ratio of 3:2:1, along with cuttings containing three to four nodes [30], is sufficient to propagate healthy chaya seedlings ready for planting in open fields (Fig. 1e, f).

Regarding to the biotechnological methods efficiently tested for chaya propagation, one approach involves inducing embryogenic callus from leaf explants. Studies have shown that using 6-Benzylaminopurine (6-BAP) and 2,4-dichlorophenoxyacetic acid (2,4-D) at concentrations of 5 mg/L and 2.5/5 mg/L, respectively, in MS medium (Murashige and Skoog) can promote callus formation with 100% effectiveness [19]. This propagation method is viable for in vitro cultivation of chaya and for obtaining extracts or compounds with potential pharmacological applications.

Another method for chaya propagation and conservation is suspension cell culture. This involves proliferating individual cells while they are suspended in a liquid medium under constant agitation. This method has been used to multiply and conserve chaya cultures, as well as to obtain bioactive secondary metabolites with anti-inflammatory effects such as lupeol acetate [31]. Studies have shown that chaya suspension cell extracts contain higher levels of lupeol acetate compared to extracts obtained from in vivo plants [28]. This method represents a sustainable alternative for the production of natural resources and the conservation of the chaya plant.

6 Diseases That Affect the Chaya Plant

Currently, there are limited reports on diseases and pests affecting the chaya crop. Some studies have identified fungi [32] and bacteria [33] on chaya plants. According to these reports, the plants studied did not exhibit symptoms at the time of collection, suggesting that these fungi do not cause significant damage to the chaya crop. The fungi detected include *Colletotrichum siamense*, *Penicillium griseofulvum*, *Pezizomycotina* sp., *Penicillium citrinum*, *Aspergillus niger*, and *Lasiodiplodia theobromae* [32]. Additionally, the pathogen *Puccinia* ssp. (rust), which damages the leaves of the crop, was found (Mayorga-Mayorga, 2017). Regarding bacteria, two strains of *Pseudomonas putida* and one strain of *Pseudomonas* sp. were identified [33].

Viruses are also known to affect the growth, development, and productivity of chaya plants. In 2023, Montero-Astúa et al. reported the presence of cassava common mosaic virus (CsCMV) in chaya plants in Costa Rica for the first time [34]. This virus is transmitted asexually and can infect cassava and other plant species in countries such as South America, Brazil, Colombia, Paraguay, Peru, and Argentina [35, 36]. To date, no vector for this virus has been identified [37]. Subsequently, this pathogen was reported in Florida (USA) and Yucatan (Mexico) [38, 39].

In the case of pests, a study conducted in Guatemala reported the presence of the lace bug (*Corythucha* ssp). This insect, belonging to the order Hemiptera, feeds on the sap juices of the plant and lays its eggs on the underside of chaya leaves to avoid being eaten by predators. The symptoms caused by this insect on chaya leaves include a whitish discoloration on the upper side of the leaves and dark spots (adult excrement) on the back of the leaf [40]. Additionally, the mite *Tetranychus urticae* was found, which causes severe lesions on the back of chaya leaves. Although not fatal, a massive infection of this mite can hinder the photosynthesis process of chaya, leading to a deficiency in its growth and development.

Recently, in Yucatan, Mexico, two species of chaya (*Cnidoscolus aconitifolius* spp. *aconitifolius* and *C. souzae*) were found to be infected with phytoplasmas belonging to subgroup 16SrIII-L (Data in the process of publication). These phytoplasmas were initially detected in Colombia in the 1970s in the Yucca plant (*Manihot esculenta*) [41]. Cassava is a plant native to South America and is a crucial crop for the food security of millions of people in the tropics. It is also a significant global source of industrial starch [42].

7 Discussion and Perspectives

As mentioned above, chaya originated from Mesoamerica and was domesticated by the Mayan people to utilize its leaves as a food source. This plant is well adapted to the soil of the Yucatan Peninsula, which is shallow, poor in nutrients, and highly permeable. Despite these conditions, it is easy to cultivate with minimal labor and can be harvested year-round. Chaya is part of the ancient agroecological system known as *milpa*, which allowed the ancient Mesoamerican people to survive.

The use of chaya as an edible plant with high nutritional content, along with its various medicinal properties such as lowering cholesterol and blood sugar levels, high antioxidant activity, and reducing inflammation, has increased its popularity among non-native people. Currently, its leaves, leaf powder, and leaf extract have gained significant commercial interest. This increased interest has led to a rise in consumption and monetary value of the plant, both within and outside the Mayan region of Mexico, Guatemala, and Belize. Nowadays, chaya is a promising crop for both food production and medicinal use. However, chaya cultivation faces important threats such as overharvesting and loss of land. It is important to develop sustainable agroecological practices to conserve this ancient crop for the benefit of future generations.

Acknowledgments

This work was partially supported by CONAHCYT Project number 320993.

References

1. Secretaria de Agricultura y desarrollo rural (2022) Recursos genéticos, ¿qué son y para qué sirven? https://www.gob.mx/agricultura/articulos/recursos-geneticos-que-son-y-para-que-sirven
2. Munguía-Rosas MA (2022) Domestication reduces phenotypic plasticity in chaya (*Cnidoscolus aconitifolius* (Mill.) I.M. Johnst). Botanical. Sciences 100(1):93–106. https://doi.org/10.17129/botsci.2879
3. Ross-Ibarra J, Molina-Cruz A (2002) The ethnobotany of chaya (*Cnidoscolus aconitifolius* ssp. aconitifolius Breckon): a nutritious maya vegetable. Econ Bot 56(4):350–365
4. Chin-Chan T, Ortiz-García MM, Ruiz-Gil PJ, Martínez-Castillo J (2021) Genetic diversity of Chaya (*Cnidoscolus aconitifolius* (Mill.) I. M. Johnst. ssp. aconitifolius) in Yucatan, Mexico, its putative domestication center. Polibotánica 51:185–201
5. Kuri-García A, Chávez-Servín JL, Guzmán-Maldonado S (2017) Phenolic profile and antioxidant capacity of *Cnidoscolus chayamansa* and *Cnidoscolus aconitifolius*: a review. J Med Plants Res 11(45):713–727
6. Méndez-Aguilar MJ, Cálix de Dios H, Xolalpa-Aroche A, Brito Estrella EE, Alvarado Dzul SH, Luckson J (2021) Chaya [Cnidoscolus aconitifolius (Mill.) I. M. Johnst]: aportaciones sobre su uso como alimento y manejo en comunidades de Quintana Roo En: Experiencias productivas en los sistemas agroalimentarios de la Zona Maya de Quintana Roo, Primera edición 2021. Universidad Intercultural Maya de Quintana Roo, p.
7. Booth S, Bressani R, Johns T (1992) Nutrient content of selected indigenous leafy vegetables consumed by the Kekchi people of Alta Verapaz, Guatemala. J Food Compos Anal 5(1):25–34
8. Pérez-González MZ, Gutiérrez-Rebolledo GA, Yépez-Mulia L, Rojas-Tomé IS, Luna-Herrera J, Jiménez-Arellanes MA (2017) Antiprotozoal, antimycobacterial, and anti-inflammatory evaluation of *Cnidoscolus chayamansa* (Mc Vaugh) extract and the isolated compounds. Biomed Pharmacother 89:89–97
9. Kuti JO, Konuru HB (2004) Antioxidant capacity and phenolic content in leaf extracts of tree spinach (*Cnidoscolus* spp.). J Agric Food Chem 52(1):117–121
10. Equihua VM, Martinez DLD, Galicia DSG, Marrufo RAS, España ADH, Kumul RGC (2024) *Cnidoscolus chayamansa* (chaya): una planta medicinal con propiedades terapéuticas y antiinflamatorias. Desde el Herbario CICY 16:30–34
11. Pillai KK, Chidambaranathan N, Halith MM, Jayaprakash S, Narayanan N (2012) Antihyperglycemic effect of alcoholic extracts of *Cnidoscolus chayamansa* in experimental diabetes and their effects on key metabolic enzymes involved in carbohydrate metabolism. Int J Res Pharm Chem 2(1):179–187
12. Kuri-García A, Chávez-Servín JL, Guzmán-Maldonado S (2017) Phenolic profile and antioxidant capacity of Cnidoscolus chayamansa and *Cnidoscolus aconitifolius*: a review. J Med Plants Res 11(45):713–727
13. Miranda-Velasquez L, Oranday-Cardenas A, Lozano-Garza H, Rivas-Morales C, Chamorro-Cevallos G, Cruz-Vega DE (2010) Hypocholesterolemic activity from the leaf extracts of *Cnidoscolus chayamansa*. Plant Foods Hum Nutr 65:392–395
14. Panghal A, Shaji AO, Nain K, Garg MK, Chhikara N (2021) Cnidoscolus aconitifolius: nutritional, phytochemical composition

and health benefits – a review. Bioact Compd Health Dis 4(11):260–286
15. Mena Linares Y, González MDM, Valido DA, Escobar RR, Pizarro EA, Castillo AO (2017) Actividad gastroprotectora y toxicidad aguda del extracto de hojas de *Cnidoscolus chayamansa* Mc Vaugh. Medicentro Electrónica 21(1):11–21
16. García-Rodríguez RV, Gutiérrez-Rebolledo GA, Méndez-Bolaina E, Sánchez-Medina A, Maldonado-Saavedra O, Domínguez-Ortiz MÁ et al (2014) *Cnidoscolus chayamansa* Mc Vaugh, an important antioxidant, anti-inflammatory and cardioprotective plant used in Mexico. J Ethnopharmacol 151(2):937–943
17. Jiménez-Arellanes MA, García-Martínez I, Rojas-Tomé S (2014) Potencial biológico de especies medicinales del género Cnidoscolus (Euphorbiacea). Rev Mex Cienc Farm 45(4):1–6
18. Lucky SC, Festus OA (2014) Effects of aqueous leaf extract of Chaya (*Cnidoscolus aconitifolius*) on pituitary-gonadal axis hormones of male Wistar rats. J Exp Clin Anat 13(2):34–39
19. Pérez-González MZ, Siordia-Reyes AG, Damián-Nava P, Hernández-Ortega S, Macías-Rubalcava ML, Jiménez-Arellanes MA (2018) Hepatoprotective and anti-inflammatory activities of the *Cnidoscolus chayamansa* (Mc Vaugh) leaf extract in chronic models. Evid Based Complement Alternat Med 2018:3896517. https://doi.org/10.1155/2018/3896517
20. Sotelo MGR, Cortés EG, Llanos AIC, Guzmán ABP, del Toro GV (2019) Desarrollo de una forma farmacéutica semisólida de Ibervillea sonorae con actividad anti-inflamatoria. Instituto Politécnico Nacional, Unidad Profesional Interdisciplinaria de Biotecnología, p.
21. Jeschke MG, van Baar ME, Choudhry MA, Chung KK, Gibran NS, Logsetty S (2020) Burn injury. Nature reviews Disease primers 6(1):11
22. Agundis MAA, Morales MAR, Alcaraz ERM, Álvarez MAD, Macías CLM, Betancourt CA et al (2018) Formulación de un emulgel con quercetina, con propiedades antiinflamatorias. Jóvenes en la Ciencia 4(1):1146–1149
23. McClung de Tapia E (1992) The origins of agriculture in Mesoamerica and Central America. In: Cowan CW, Benco NL, Watson PJ (Eds.) The origins of agriculture: an international perspective. The University of Alabama Press, 143–171
24. Gómez-Pompa A, Flores SJ, Sosa V (1987) The 'pet kot': a man-made tropical forest of the Maya. Interciencia 12(1):10–15
25. Nigh R (2008) Trees, fire and farmers: making woods and soil in the Maya forest. J Ethnobiol 28(2):231–243
26. Diemont SA, Martin JF (2009) Lacandon Maya ecosystem management: sustainable design for subsistence and environmental restoration. Ecol Appl 19(1):254–266
27. Méndez RM (2015) La milpa maya yucateca en el siglo XVI: evidencias etnohistóricas y conjeturas. Etnobiología 13(1):1–25
28. Pérez-González MZ, Nieto-Trujillo A, Gutiérrez-Rebolledo GA, García-Martínez I, Estrada-Zúñiga ME, Bernabé-Antonio A et al (2019) Lupeol acetate production and antioxidant activity of a cell suspension culture from Cnidoscolus chayamansa leaves. S Afr J Bot 125:30–38
29. Grubben G, Denton OA (eds) (2004) Plant Resources of Tropical Africa 2. Vegetables. PROTA Foundation. Backhuys Publishers, Wageningen, Netherlands, p. 668
30. Kitila C, Mohammed W, Omotosho OO, Olopade JO (2022) Influence of rooting media, number of nodes and seedling growing methods on rooting, seedling establishment and early growth of Chaya (*Cnidoscolus aconitifolius* McVaugh) stem cuttings at Dire Dawa, Eastern Ethiopia. Arch Anat Physiol 7(1):012–025
31. Hassan LG, Liman MG, Msheila HE, Ogbiko C, Babagana A, Andrew O (2018) Lupeol acetate isolated from n-hexane extract of *Tapinanthus globiferus* leaf. ChemSearch. Chem Search J 9(1):83–88
32. Ogbuji N, Ataga A, Nwachukwu E (2024) Molecular characterization of fungi associated with *Cnidoscolus aconitifolius* (Mill.) I. M. Johnston. J Agric 6:1–12
33. Lyanyi NG, Ataga AE, Nwachukwu EO (2019) Biochemical and molecular characterization of bacteria associated with *Cnidoscolus aconitifolius* (Mill.) IM Johnston. IOSR J Pharm 14(4):78–85
34. Montero-Astúa M, Sandoval-Carvajal I, Moreira-Carmona L, Villalobos-Muller W, Garita-Salazar L, Carvajal-Rojas S (2023) Phytopathology and cultural behaviors: putative introduction of Chaya-strain of Cassava common mosaic virus to Costa Rica. Mexican. J Phytopathol 41(4):7
35. Di Feo L, Zanini A, Rodríguez Pardina P, Cuervo M, Carvajal-Yepes M, Cuellar WJ (2015) First report of Cassava common mosaic virus and Cassava frogskin-associated virus infecting Cassava in Argentina. Plant Dis 99:733–733. https://doi.org/10.1094/PDIS-10-14-1088-PDN
36. Fernández E, Espinoza I, Lozano I, Bolaños C, Carvajal-Yepes M, Cuellar WJ (2017) First Report of Cassava Common Mosaic Disease and Cassava common mosaic virus Infecting Cassava (*Manihot esculenta*) in Peru. Plant Dis 101:1066–1066. https://doi.org/10.1094/PDIS-10-16-1540-PDN
37. Costa AS, Kitajima EW (1972) Cassava common mosaic virus. CMI/ABB Description of Plant

Viruses No. 90. *Commonwealth Agricultural Bureaux and the Association of Applied Biologists.* Wm. Cuiross and Son Ltd., UK, 4
38. Zettler FW, Elliott MS (1986) An antigenically distinct strain of cassava common mosaic virus infecting *Cnidoscolus aconitifolius.* Phytopathology 76:632–638
39. Elliott MS, Zettler FW (1987) Cassava common mosaic virus infections of chaya (*Cnidoscolus aconitifolius*) in Yucatan, Mexico. Plant Dis 71:353–356. https://www.apsnet.org/publications/PlantDisease/BackIssues/Documents/1987Articles/PlantDisease71n04_353.pdf
40. Mayorga Mayorga VH (2017) Plagas y enfermedades del follaje en plantas de chaya (*Cnidoscolus aconitifolius* (Mill.) IM Johnst) en dos ambientes del sur del país. Universidad del Valle de Guatemala. (Ph.D. thesis). https://repositorio.uvg.edu.gt/xmlui/handle/123456789/5141
41. Pineda B, Jayasinghe U, Lozano JC (1983) La enfermedad 'cuero de Sapo' en yuca (Manihot esculenta Crantz). ASIAVA 4:10–12
42. Ceballos H (2012) Chapter 1: Cassava in Colombia and the world: new prospects for a millennial crop. In: Ospina B, Ceballos H (Eds.) Cassava in the third millenium: modern production, processing, use, and marketing systems. International Center for Tropical Agriculture (CIAT); Latin American and Caribbean Consortium to Support Cassava Research and Development (CLAYUCA). Technical Centre for Agricultural and Rural Cooperation (CTA), Cali, Colombia, 1–11

Check for updates

Chapter 11

Plant Tissue Culture: A Short-Term Conservation Tool

José Roberto Bautista-Aguilar, Marco A. Ramírez-Mosqueda, José Humberto Caamal-Velázquez, José Luis Aguirre-Noyola and Carlos Alberto Cruz-Cruz

Abstract

Plant tissue culture (PTC) is a biotechnology tool with multiple applications in plants, among which highlights the enhancement in obtaining plant propagules through in vitro propagation or micropropagation. However, the production of a large number of plants in a short period of time and reduced space is used in the field of research to provide a greater number of samples (cells, tissues, shoots, plants, etc.) used in various tests. In vitro propagation by PTC plays an important role in the application of techniques such as in vitro preservation or cryopreservation. In this context, a large number of shoots and plant tissue (particularly apical meristems) is required, which must be obtained aseptically and under controlled conditions. Thus, micropropagation is an essential tool in the first steps of ex situ conservation strategies through in vitro conservation and cryopreservation. Likewise, this scientific discipline is essential in different biological assays where plant material is required under laboratory conditions, such as physiology, phytopathology, and molecular biology experiments, among others. In this chapter, we will discuss how in vitro propagation is considered as a short-term conservation tool and how it is linked to strategies for the conservation of plant genetic resources.

Keywords Micropropagation, Plant genetic resource, In vitro conservation, Cryopreservation

1 Introduction

The conservation of plant genetic resources involves the use of biotechnological tools that allow short-, medium-, and long-term conservation [1]. Medium-term conservation by in vitro propagation and long-term conservation by cryopreservation have been mainly applied to various plant species [2, 3]. However, short-term conservation by in vitro propagation has been underestimated due to the high cost of continuous subculturing [4].

Conservation in the broadest sense refers to the care and maintenance of a natural resource to ensure its survival [5]. Two types of conservation have been defined: in situ, which consists of preserving the resource in its natural habitat, and ex situ, which consists

Marco A. Ramírez-Mosqueda et al. (eds.), *Conservation of Plant Genetic Resources*, Methods in Molecular Biology, vol. 3011, https://doi.org/10.1007/978-1-0716-5130-8_11,

of preserving the resource outside its natural habitat [6, 7]. Within these, strategies can be divided according to the time frame in which the resource is protected [8]. However, a comprehensive strategy for the plant genetic resource conservation would involve a sequential process of conservation in the short, medium, and long term. This review discusses plant tissue culture as a short-term conservation strategy and its importance as a first step in the development of medium- and long-term techniques.

2 Plant Tissue Culture

Plant tissue culture (PTC) allows us to establish, multiply, conserve and manipulate any type of plant material under artificial and controlled conditions in an aseptic environment. The plant material established in vitro is called explant and can be from a cell, tissue, organ to a whole plant [9]. These techniques contribute to solving both basic and applied problems in the biological sciences, since they offer a series of model systems, ideal for anatomical, physiological, biochemical and genetic research [10]. They also have a practical application in the in vitro cloning of any plant material. Currently, different types of PTC are performed: cell culture, organ culture, tissue or organ segment culture, whole plant culture and co-culture. According to Pasternak [11], the success or failure of PTC is based on three basic principles: explant selection, medium selection, and incubation conditions. The response of an explant to in vitro culture depends on the interaction between the explant, the medium, and the culture conditions. This response may be a process of morphogenesis or simply vegetative development. However, if the in vitro conditions are not adequate, explants may remain dormant or die [11].

In vitro cultured tissue is kept dormant or alive by the addition of components (macro and micronutrients, carbon source, and organic compounds) that it requires to survive. This feature makes PTC an ex situ conservation technique, although it is considered a short-term strategy due to the constant subculturing of the plant material. A subculture is defined as the refreshing or changing of the culture medium, thus avoiding the loss of its components [12]. Hamad and Taha [13] mention that it is possible to maintain pineapple (*Ananas comosus* L. Merr.) plant tissue for a maximum of 300 days through four subcultures every 75 days. For cannabis (*Cannabis sativa* L.), it was possible to obtain four subcultures of a period of 45, 60, 60, and 70 days each, which indicates an indirect preservation of approximately 235 days [14]. Nevertheless, Martinez-Estrada et al. [15] found that in sugarcane (*Saccharum* spp. hybrids), nine subcultures of 30 days each favored the appearance of high percentages of polymorphism (somaclonal variations). This is an undesirable consequence in conservation strategies. Somaclonal variation can be defined as genetic and epigenetic changes that occur in cells, tissues or whole plants cultured under in vitro conditions [16]. For this reason, short-term conservation by PTC has not been possible in some plant species that require a large number of subcultures with high periodicity.

3 Applications of Plant Tissue Culture

PTC brings different benefits to modern plant biotechnology, including plant health by obtaining pathogen-free plants, the production of a large number of plant propagules through micropropagation, the regeneration of genetically modified plants, the obtainment of secondary plant metabolites, and the in vitro conservation of plant germplasm [12]. For cryopreservation of plant germplasm, these techniques are mainly used to obtain the plant material needed for micropropagation trials. In addition, PTC allows long-term regeneration of cryopreserved plant germplasm [17]. The applications of PTC are broad and complex. However, this plant biotechnology tool has enabled the manipulation and preservation of plant material (cells, tissues or whole plants) in an aseptic manner under controlled environmental conditions [11]. This has led to advances in several fields of science, including ex situ conservation in the long and medium term.

4 Micropropagation as a Short-Term Conservation Method

Plant micropropagation, or in vitro propagation, consists of obtaining a large number of copies (clones) of plants through plant tissue culture, a process that can be repeated indefinitely [18] (Fig. 1). In this sense, a mother plant is selected that must have excellent phytosanitary and agronomic characteristics in order to clone it into daughter plants [19]. The resulting plants contain the same genetic information as the mother plant and are used as propagules to establish commercial plantations [20] (Fig. 1). The large number of plants and the genetic uniformity of the plant material obtained make micropropagation a conservation tool [21] (Fig. 2). However, the multiplication of plant material is so rapid that it is necessary to perform several subcultures periodically to maintain it, so it is considered short-term. Nevertheless, genetic and epigenetic abnormalities sometimes occur in this type of in vitro propagated plants [22]. The somaclonal variation increases with the number of subcultures performed on the plant material [15].

Commercial scale-up of plant propagation requires plant tissue culture techniques that exploit the totipotency and multipotency of plant cells to generate new individuals from a small explant [23, 24]. Some innovations, however, have been explored in the field of mass

Fig. 1 Vanilla micropropagation process. (**a**) In vitro multiplication (a process that can be repeated countless times). (**b**) Acclimatization of plants under greenhouse conditions (final stage of the plant micropropagation process)

Fig. 2 Short-term conservation of plant germplasm through plant tissue culture. (**a**) ex situ conservation of sugarcane; (**b**) ex situ conservation of damiana; (**c**) ex situ conservation of strawberry

plant micropropagation, such as temporary immersion systems (TIS) [23]. Using these TIS, a larger number of plants can be obtained from the mother plant in less time and in a smaller space, combining the advantages of liquid culture media (greater availability of nutrients) and obtaining plants that are physiologically better adapted to gas exchange [25]. Currently, there are numerous studies establishing protocols for commercial-scale mass micropropagation of various plant species by TIS. Therefore, the use of these systems could be considered the greatest technological advance in plant tissue culture in the last decades [26].

In scientific research, in vitro propagation ensures the availability of a large amount of plant material (samples) to carry out different experiments, as well as the uniformity of samples and abiotic controlled conditions [27]. There are multiple scientific works that previously test the phytostimulant or phytotoxic effect of chemical compounds (fertilizer, biostimulant, herbicide, etc.) before its use in greenhouse or field [28–30].

5 Role of PTC During In Vitro Preservation

Medium-term ex situ plant preservation involves plant tissue culture techniques using modified culture media that slow down growth [2]. However, in order to establish in vitro preservation protocols for minimal growth, it is first necessary to have a large amount of homogeneous plant material. For this reason, after in vitro establishment, in vitro propagation of plant germplasm is essential to have the necessary material to establish medium- and long-term conservation protocols [31]. There are previous studies in which in vitro conservation protocols have been generated, which describe that it is essential to increase the number of clonal plant material in order to have as many samples as possible during the development of these protocols. Juan-Vicedo et al. [32] mentioned that for the medium-term conservation of *Tetraclinis articulata*, a previous phase of multiplication of the established plant material was necessary. In order to conserve germplasm in vitro, Sharma et al. [33] multiplied *Bacopa monnieri* material by micropropagation.

Another benefit of plant tissue culture in medium-term conservation is the regeneration of the preserved plant material, considering that in vitro conservation refers to storage periods of 6 to 12 months without subcultures [34]. The key to the success of an in vitro preservation protocol is the ability to regenerate the preserved material into complete plants after the storage period, as well as the ability to multiply them in vitro [35]. Cruz-Cruz et al. [36] mentioned that in vitro regeneration of the orchid *Stanhopea tigrina* was achieved after an in vitro conservation period of 6 months. On the other hand, the need to induce in vitro regeneration of *Notylia barkeri* orchid material preserved in minimal growth was emphasized by García-Merino et al. [37]. One issue that needs to be considered in genetic resource banks is the periodic evaluation of the regeneration and propagation of plant germplasm through PTC.

6 Role of PTC in Cryopreservation

Plant cryopreservation techniques require a large number of plant materials to be frozen, primarily apical meristems. However, most plants have only one apical meristem, which dramatically increases the number of plants required [38]. Therefore, techniques are needed to obtain a large number of plants in a short time and in a small space. To establish a protocol for cryopreservation of *Malus* cultivars,

Halmagyi et al. [39] mentioned that they used a previous phase of in vitro propagation to obtain the necessary plant material for their experimental design. Malhotra et al. [40] mention that in the cryopreservation of *Humulus lupulus*, the multiplication phase of the material is fundamental before experimenting with the technique of drop vitrification. The cryopreservation of *Lamprocapnos spectabilis* required 840 apical meristems, which were obtained by in vitro propagation [38].

As in vitro preservation, cryopreservation requires a regeneration phase of the plant material. In this sense, plant meristems are usually cryopreserved and pre-cultured in a modified medium with different osmoregulators that dehydrate the tissue cells. They are then cryoprotected and frozen in liquid nitrogen. Subsequently, their viability (survival) is assessed by the percentage of meristems that develop into plants when cultured in vitro in a nutrient-rich medium under controlled environmental conditions [41].

Since cells and tissues are sometimes dehydrated, frozen, and then recovered (reheated), an in vitro regeneration and propagation protocol is needed to convert the plant material into complete individuals [41, 42]. In other words, if there were no way to regenerate this germplasm through in vitro culture, cryopreservation (long-term conservation) would not make sense because it would not be possible to restore the conserved material to its initial state (plant).

7 Conclusions and Perspectives

The use of plant tissue culture is essential in ex situ conservation techniques such as in vitro conservation and cryopreservation. Its applications range from the establishment and in vitro propagation of plant material to providing tools to verify the viability of conserved material after a period of storage. It also allows short-term conservation through micropropagation, which is based on obtaining many plants in a short time and space.

Furthermore, the PTC allows studies and experiments with plants under controlled abiotic conditions, as well as the production of plant propagules on a commercial scale using temporary immersion systems. The PTC is a biotechnological tool developed more than a century ago for the short-, medium-, and long-term conservation of plant genetic resources.

References

1. Salgotra RK, Chauhan BS (2023) Genetic diversity, conservation, and utilization of plant genetic resources. Genes 14(1):174
2. Benelli C, Tarraf W, Izgu T, De Carlo A (2022) *In vitro* conservation through slow growth storage technique of fruit species: an overview of the last 10 years. Plants 11(23):3188
3. Jaiswal AN, Vagga A, Jaiswal A Jr, Vagga AA Sr (2022) Cryopreservation: a review article. Cureus 14(11):e31564

4. Tripathi MK, Tiwari S, Tripathi N, Tiwari G, Bhatt D, Vibhute M, Gupta N, Mishra N, Parihar P, Singh P, Sharma A, Ahuja A, Tiwari S (2021) Plant tissue culture techniques for conservation of biodiversity of some plants appropriate propagation in degraded and temperate areas. In: Current topics in agricultural sciences, vol 4. BP International Publisher, Bhanjipur, India, pp 30–60
5. Jhariya MK, Meena RS, Banerjee A, Meena SN (eds) (2021) Natural resources conservation and advances for sustainability. Elsevier, San Diego
6. Wang W, Li J (2021) In-situ conservation of biodiversity in China: advances and prospects. Biodivers Sci 29(2):133
7. Naruka KPS, Reddy PB (2023) Assessing the efficacy of in-situ vs. ex-situ biodiversity conservation measures. In: Srivastava B, Reddy PB (eds) Recent advances in biodiversity research. IMRF International Publishing House, Krishna, India, pp 67–73.
8. Priyanka V, Kumar R, Dhaliwal I, Kaushik P (2021) Germplasm conservation: instrumental in agricultural biodiversity—a review. Sustainability 13(12):6743
9. Reddy J (2024) Plant tissue culture. CRC Press
10. Custódio L, Charles G, Magné C, Barba-Espín G, Piqueras A, Hernández JA et al (2022) Application of *in vitro* plant tissue culture techniques to halophyte species: a review. Plants 12(1):126
11. Pasternak TP, Steinmacher D (2024) Plant growth regulation in cell and tissue culture *in vitro*. Plants 13(2):327
12. Phillips GC, Garda M (2019) Plant tissue culture media and practices: an overview. In Vitro Cell Dev Biol Plant 55::242–257
13. Hamad AM, Taha RM (2008) Effect of sequential subcultures on *in vitro* proliferation capacity and shoot formations pattern of pineapple (*Ananas comosus* L. Merr.) over different incubation periods. Sci Hortic (Amsterdam) 117(4):329–334
14. Hesami M, Adamek K, Pepe M, Jones AMP (2023) Effect of explant source on phenotypic changes of *in vitro* grown cannabis plantlets over multiple subcultures. Biology 12(3):443
15. Martínez-Estrada E, Caamal-Velázquez JH, Salinas-Ruíz J, Bello-Bello JJ (2017) Assessment of somaclonal variation during sugarcane micropropagation in temporary immersion bioreactors by intersimple sequence repeat (ISSR) markers. In Vitro Cell Dev Biol Plant 53:553–560
16. Duta-Cornescu G, Constantin N, Pojoga DM, Nicuta D, Simon-Gruita A (2023) Somaclonal variation—advantage or disadvantage in micropropagation of the medicinal plants. Int J Mol Sci 24(1):838
17. Agrawal A, Srivastava V (2019) Plant tissue culture interventions for management of plant genetic resources. Training manual on plant genetic resources management and utilization. NAHEP-CAAST Sponsored Short Term Training Programme, p 76
18. Lédo AS, Vendrame WA (2021) Coconut micropropagation and cryopreservation. Hortic Rev (Am Soc Hortic Sci) 48:307–337
19. Singh A (2015) Micropropagation of plants. Plant biology and biotechnology: volume II: plant genomics and biotechnology. Springer, pp 329–346
20. Podwyszyńska M, Orlikowska T, Trojak-Goluch A, Wojtania A (2022) Application and improvement of *in vitro* culture systems for commercial production of ornamental, fruit, and industrial plants in Poland. Acta Soc Bot Pol 91(1):914
21. Oseni OM, Pande V, Nailwal TK (2018) A review on plant tissue culture, a technique for propagation and conservation of endangered plant species. Int J Curr Microbiol Appl Sci 7(7):3778–3786
22. Ferreira MDS, Rocha ADJ, Nascimento FDS, Oliveira WDDS, Soares JMDS, Rebouças TA et al (2023) The role of somaclonal variation in plant genetic improvement: a systematic review. Agronomy 13(3):730
23. Bello-Bello JJ, Mancilla-Álvarez E, Spinoso-Castillo JL (2025) Scaling-up procedures and factors for mass micropropagation using temporary immersion systems. In Vitro Cell Dev Biol Plant 61:321–332
24. Almeida MD, Graner ÉM, Brondani GE, Oliveira LSD, Artioli FA, Almeida LVD et al (2015) Plant morphogenesis: theorical bases. Advances in Forestry. Science 2(1):13–22
25. Ramírez-Mosqueda MA, Bello-Bello JJ (2021) SETIS™ bioreactor increases *in vitro* multiplication and shoot length in vanilla (*Vanilla planifolia* Jacks. Ex Andrews). Acta Physiol Plant 43(4):52
26. Mirzabe AH, Hajiahmad A, Fadavi A, Rafiee S (2022) Temporary immersion systems (TISs): a comprehensive review. J Biotechnol 357:56–83
27. Chukwu SC, Awala SK, Angombe S, Valombola JS, Nanhapo PI, Mberama C et al (2025) Recent progress in tissue culture techniques and biotechnological innovations for banana production (Musa spp.): a review. Discover Plants 2(1):13
28. Boa Ventura TF, Espírito Santo LA, Alves Silva B, Neto BS, de Moraes CM, Bittencourt SNDS et al (2025) Liquid extract of Kappaphycus alvarezii enhances growth and metabolic responses in in vitro cultivated explants of this same species. J Appl Phycol 37:2147–2156
29. Chaouch R, Kthiri Z, Soufi S, Jabeur MB, Bettaieb T (2023) Assessing the biostimulant

effect of micro-algae and thyme essential oil during in-vitro and ex-vitro rooting of strawberry. S Afr J Bot 162:120–128
30. Ambros E, Kotsupiy O, Karpova E, Panova U, Chernonosov A, Trofimova E, Goldenberg B (2023) A biostimulant based on silicon chelates enhances growth and modulates physiological responses of in-vitro-derived strawberry plants to in vivo conditions. Plants 12(24):4193
31. Radomir AM, Stan R, Florea A, Ciobotea CM, Bănuţă FM, Negru M et al (2023) Overview of the success of *in vitro* culture for *ex situ* conservation and sustainable utilization of endemic and subendemic native plants of Romania. Sustainability 15(3):2581
32. Juan-Vicedo J, Serrano-Martínez F, Cano-Castillo M, Casas JL (2022) *In vitro* propagation, genetic assessment, and medium-term conservation of the coastal endangered species *Tetraclinis articulata* (Vahl) Masters (Cupressaceae) from adult trees. Plants 11(2):187
33. Sharma N, Singh R, Pandey R (2016) In vitro propagation and conservation of Bacopa monnieri L. Protocols for in vitro cultures and secondary metabolite analysis of aromatic and medicinal plants. 2nd edn. Springer, pp 153–171
34. Gaidamashvili M, Benelli C (2021) Threatened woody plants of Georgia and micropropagation as a tool for *in vitro* conservation. Agronomy 11(6):1082
35. Ramírez-Mosqueda MA, Cruz-Cruz CA, Atlahua-Temoxtle J, Bello-Bello JJ (2019) *In vitro* conservation and regeneration of *Laelia anceps* Lindl. S Afr J Bot 121:219–223
36. Cruz-Cruz CA, González-Arnao MT, Bautista-Aguilar JR, Ramírez-Mosqueda MA (2022) *In vitro* short-term storage of Stanhopea tigrina Bateman ex Lind. S Afr J Bot 151:334–338
37. García-Merino GF, Mosqueda MAR, Mata-Alejandro H, López-Larios AV, López-Aguilar DR (2024) Conservación in vitro de Notylia barkeri Lindl: Rescate de Notylia barkeri. Revista Bio Ciencias 11:e1633
38. Kulus D, Tymoszuk A, Kulpińska A, Wojnarowicz J, Szałaj U (2024) Nanoparticle-mediated enhancement of plant cryopreservation: cultivar-specific insights into morphogenesis and biochemical responses in *Lamprocapnos spectabilis* (L.) Fukuhara 'Gold Heart' and 'Valentine'. PLoS One 19(5):e0304586
39. Halmagyi A, Deliu C, Isac V (2010) Cryopreservation of Malus cultivars: comparison of two droplet protocols. Sci Hortic (Amsterdam) 124(3):387–392
40. Malhotra EV, Mali SC, Sharma S, Bansal S (2024) A droplet vitrification cryopreservation protocol for conservation of hops (*Humulus lupulus*) genetic resources. Cryobiology 115:104887
41. Popova E, Kulichenko I, Kim HH (2023) Critical role of regrowth conditions in post-cryopreservation of *in vitro* plant germplasm. Biology 12(4):542
42. Wang MR, Zhang Z, Zámečník J, Bilavčík A, Blystad DR, Haugslien S, Wang QC (2020) Droplet-vitrification for shoot tip cryopreservation of shallot (*Allium cepa* var. aggregatum): effects of PVS3 and PVS2 on shoot regrowth. Plant Cell Tissue Organ Cult 140:185–195

Chapter 12

Ex Situ Conservation of *Opuntia* spp. Germplasm in a Complementary Agriculture Model

Fernanda Díaz-Sánchez, Jorge Cadena-Iñiguez, Jorge David Cadena-Zamudio, Víctor M. Ruiz-Vera, Daniel Alejandro Cadena-Zamudio, Luis A. Barrera-Guzmán, Brenda I. Trejo-Téllez and María Isabel Iñiguez-Luna

Abstract

Mexico, as a megadiverse country, harbors a large part of its biodiversity in socially owned lands, such as ejidos and agrarian communities. These areas are fundamental for the conservation of local varieties that ensure food availability, an example of which is the Genetic Resources for Food and Agriculture (GRFA), such as *Opuntia* spp., *Agave* spp., *Yucca* spp., and other native plants. Despite this biological wealth, rural communities face challenges such as poverty and the abandonment of agricultural production. However, this also represents an opportunity to implement conservation and production schemes with accessible technologies. The conservation of these resources is crucial for food security and nutrition, especially in a context of climate change and other challenges. The Convention on Biological Diversity and the Nagoya Protocol are important instruments for biodiversity conservation and equity in access to and use of genetic resources. There is a need to reconcile germplasm conservation with food production, and strategies such as the establishment of low-cost conservation-production modules can be beneficial to rural communities whose livelihoods often depend on agriculture.

Keywords Genetic resources, Agrobiodiversity, *Opuntia* spp., Conservation, Germplasm

1 Introduction

The Genetic Resources for Food and Agriculture (GRFA) native to Mexico, such as *Opuntia* spp., Agave spp., Yucca spp., succulents, and cacti, among others, are primarily located on socially owned lands, such as ejidos and agrarian communities [1]. These resources are essential for agrobiodiversity, providing key options for agriculture, food, and industry [2]. However, the rural communities that host these resources face high levels of poverty, migration, and abandonment of productive activities, which hinders the use and exploitation of local GRFA [3]. Nevertheless, these challenges also present opportunities

Marco A. Ramírez-Mosqueda et al. (eds.), *Conservation of Plant Genetic Resources*, Methods in Molecular Biology, vol. 3011, https://doi.org/10.1007/978-1-0716-5130-8_12,

to implement conservation and production schemes using accessible, low-cost technologies adapted to rural conditions [4].

The conservation of genetic resources is essential for ensuring food security and nutrition, as they play a strategic role in agricultural production systems [5]. Ex situ conservation of germplasm, such as in the case of *Opuntia* spp., presents a viable and low-cost option for rural farmers by allowing the association of different biological variants of nutritional and commercial importance within conservation-production modules. This approach is particularly relevant in the face of the challenges posed by climate change, which severely impacts traditional agricultural production systems [6].

Mexico, recognized for its vast biodiversity, ranks fourth among megadiverse countries [7]. Within this richness, agrobiodiversity, which includes locally cultivated varieties, is key to food stability, particularly under extreme climatic conditions or limited water availability [8]. A significant portion of this biodiversity is located on socially owned lands, which account for 66% of common-use lands in Mexico [9, 10]. The resources available on these lands are diverse, ranging from grasses and construction materials to forests and minerals, offering great opportunities for sustainable conservation and utilization [11].

The conservation of landrace varieties, a product of biological and cultural heritage, is crucial to protect these resources from biopiracy and exploitation by highly industrialized countries seeking access to their genetic properties without implementing proper conservation models on socially owned lands [12]. In response, the Second Global Plan of Action of the FAO, emphasizes the importance of improving and conserving GRFA, highlighting the need for alternative models that integrate germplasm conservation with agricultural production to address the challenges of climate change and ensure food security [13].

The Convention on Biological Diversity and the Nagoya Protocol reinforce the importance of conserving genetic resources for the benefit of local and indigenous communities, promoting fair and equitable access to these resources while ensuring their conservation and sustainable use [14]. In this context, the aim of this study is to establish an ex situ germplasm conservation model for *Opuntia* spp., using low-cost technologies that benefit both biodiversity conservation and agricultural production in rural areas.

2 Materials

2.1 Plant Material

1. Seven biological variants of *Opuntia ficus-indica* were selected for human or animal consumption (Table 1).

2.2 Reagents and Supplies

1. Stainless-steel pruning shears, scalpels, and tweezers (autoclavable).
2. 1% (v/v) sodium hypochlorite solution.

Table 1
Biological variants and planting densities used in the complementary agriculture model

Species	Variety	Use	Density (plants/ha^{-1})
ficus-indica Mill.	Copena	Vegetable	20,952
Opuntia ficus-indica Mill.	Pelón blanco	Forage	20,952
Opuntia ficus-indica Mill.	Verdura	Vegetable	20,952
Opuntia ficus-indica Mill.	Pelón rojo	Forage	20,952
Opuntia ficus-indica Mill.	Rojo liso	Vegetable	20,952
Opuntia ficus-indica Mill.	Atlixco	Vegetable	20,952
Opuntia ficus-indica Mill.	Villanueva	Vegetable	20,952
Ficus carica L.	Muratalla	Fruit	6945
Zea mays L.	Criollo local	Grain	20,000

3. 70% (v/v) ethanol.
4. Portable first-aid kit and personal protective equipment (PPE: nitrile gloves, safety glasses, dust mask).
5. BanGerMex passport-data forms (paper or digital tablet).
6. GPS receiver (±3 m accuracy).
7. QGIS 3.x (open-source GIS).
8. Data-logging weather station (air temperature, rainfall, RH).
9. Drip-irrigation tubing and emitters (2 L h^{-1}).
10. Organic compost (1–2 kg m^{-2}) and green manure seed mix.
11. Shade mesh (35% porosity).

2.3 Equipment

1. 4 × 8 m plots delimited with wooden stakes and sisal cord.
2. Soil auger (Ø 5 cm) for texture and pH sampling.
3. Knapsack sprayer (15 L) for foliar nutrition/biocontrol.
4. Digital callipers for cladode thickness.
5. 0.5 m PVC quadrat for density validation.

3 Methods

3.1 Site Selection and Climatic Monitoring

1. Identify a rural community that presents a semi-arid climate suitable for *Opuntia* spp. (mean annual temperature ≥18°C; rainfall <550 mm).
2. Install automatic dataloggers to record mean temperature, absolute maximum/minimum temperature, and rainfall.

3. Expand the collection range by including at least two neighboring communities with similar edaphoclimatic conditions to capture regional diversity.
4. Compile climatic summaries every quarter to adjust agronomic management if extremes are detected.

3.2 Field Collection and Accession Documentation

1. At each locality, select three healthy adult plants representative of local variability (*see* **Note 1**).
2. Detach one mature cladode per plant with tools disinfected with 70% ethanol (*see* **Note 2**).
3. Record passport data (GPS, altitude, slope, soil texture, collector ID) in the BanGerMex format (*see* **Note 3**).
4. Place cladodes in labeled mesh bags and transport to the conservation site within 24 h in a cooled container.
5. Obtain Prior Informed Consent (PIC) from the community assembly and issue mutually agreed terms where material transfer is envisaged (*see* **Note 4**).
6. Confirm that all personnel are trained in BanGerMex documentation standards and SADER-SNICS regulations (*see* **Note 5**).

3.3 High-Density Establishment of the AgriMod Module 7

1. Deep-plough to 30–40 cm and incorporate 1–2 kg/m^{-2} compost (*see* **Note 6**).
2. Mark beds 1.5 m wide with 0.3 m furrows; open 45° planting slits.
3. Transplant cladodes vertically at the densities in Table 1 (20,000–21,000 plants ha^{-1}) (*see* **Note 7**).
4. Irrigate with drip lines delivering 6 mm every 5 days during the first 3 months; thereafter, adjust to rainfall and evapotranspiration (*see* **Note 8**).
5. Erect a 35% shade mesh on the western edge to reduce afternoon thermal stress.
6. Label each accession with a waterproof aluminum tag bearing its unique BanGerMex code.
7. Distribute each variant in three replicated plots arranged in concentric rings; outer rings act as windbreaks. The spatial design, rainwater-harvesting trenches, and service corridors are depicted (Fig. 1).

3.4 Routine Agronomic Management

1. Fertigate monthly with 100 kg/ha^{-1} N–P–K (5-2-4) split in equal doses.
2. Apply a foliar Ca–B complex at bud emergence to reduce cladode burn.
3. Scout weekly for pests and diseases; remove affected tissue and disinfect tools.

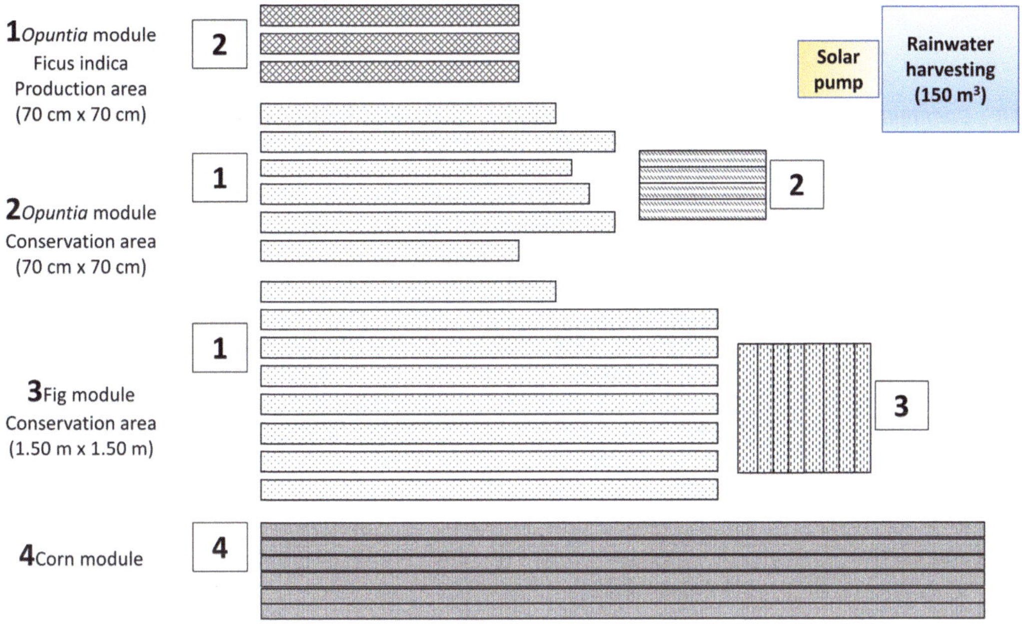

Fig. 1 Distribution of agricultural production areas, conservation, and rainwater-harvesting elements for irrigation

4. Diversify the system with other semi-desert crops (e.g., *Agave*, *Yucca*) to improve ecosystem services and income (*see* **Note 9**).

3.5 Continuous Harvest

1. Harvest 8–10 months after planting.
2. Cut cladodes ≥8 cm every 7–10 days, adjusting to growth rate (*see* **Note 10**).
3. Sort cladodes by caliber and intended market; store at 10°C, 90% RH (relative humidity) for up to 7 days before dispatch.

3.6 Performance Monitoring and Adaptive Management

1. Record each harvest (number of cladodes ha^{-1}) and prevailing climatic conditions (*see* **Note 11**).
2. Analyze yield data monthly; adjust irrigation frequency and fertilization rates accordingly.
3. Compare the productivity of variants; upscale high-performing genotypes or replace poorly adapted ones.

3.7 Germplasm Maintenance and Ex Situ Conservation

1. Conduct weekly visual inspections for stress, malnutrition, or disease (*see* **Note 11**).
2. Perform viability and germination tests every 6 months on a subsample of cladodes.
3. Maintain detailed records of conservation status, morphological traits, and any interventions.

4. Propagate replacement material asexually to maintain genetic integrity.

3.8 Community Knowledge Survey and GIS Mapping

1. Design a short questionnaire (≤12 items) covering culinary, medicinal, and forage uses.
2. Interview a representative group of local residents; obtain written informed consent (*see* **Note 4**).
3. Merge survey data with BanGerMex records; export as CSV.
4. Import the CSV into QGIS, create shapefiles per accession and generate high-resolution distribution maps (TIFF, 600 dpi) for archival.
5. Validate GPS coordinates and metadata integrity before final map generation (*see* **Note 12**).

3.9 Bioprospecting and Potential Use

1. Establish collaborations with public or private research groups to screen cladodes and fruits for bioactive metabolites (*see* **Note 13**).
2. Perform untargeted metabolomics followed by targeted assays for antioxidants or nutraceuticals.
3. Set up field trials (water-stress gradients, soil types) to evaluate how the environment influences metabolite concentration.
4. Encourage active participation of local communities in bioprospecting benefits (*see* **Note 14**).
5. Share benefit-sharing agreements with communities as per the Nagoya Protocol.

4 Notes

1. Sampling strategy. Three plants per site capture intra-population diversity with manageable effort.
2. Tool sterilization. Seventy percent ethanol dip or brief flaming prevents fungal contamination.
3. BanGerMex format. Nationally standardized system under SADER-SNICS for traceability of every accession.
4. Informed consent. Required for surveys and any use of genetic resources under the Nagoya Protocol.
5. BanGerMex training. All collectors must be trained in BanGerMex documentation and SADER-SNICS regulations to guarantee accurate passport data.
6. Furrow Depth. Deep tillage improves root establishment in low-organic semi-arid soils.

7. Planting density. 20,000–21,000 plants ha^{-1} balances yield and competition.
8. Drip irrigation. Reduces water use and disease incidence.
9. System diversification. Incorporating other semi-desert species (e.g., *Agave*, *Yucca*) enhances resilience and broadens income sources.
10. Weekly harvests. Regular cutting maintains market supply and stimulates new cladode growth.
11. Germplasm checks and climate logging. Systematic inspections and linked climate data detect early stress and genetic drift.
12. Coordinate validation. Verify GPS accuracy before integrating data into GIS layers to avoid mapping errors.
13. Germplasm-bank links. Engage national and international germplasm banks to exchange knowledge and strengthen conservation networks.
14. Community participation. Facilitating local involvement in bioprospecting and conservation ensures equitable benefit sharing and long-term adoption.

References

1. Cadena Iñiguez J, González Santos R, Cuevas Sánchez J, Riviello Flores MDLL, Ruiz Posadas LDM (2020) La Conservación in situ de la Biodiversidad Agrícola, y la generación de proyectos en ejidos y comunidades de México
2. Mastretta-Yanes A, Bellon MR, Acevedo F, Burgeff C, Piñero D, Sarukhán J (2019) Un programa para México de conservación y uso de la diversidad genética de las plantas domesticadas y sus parientes silvestres. Rev Fitotecnia Mexicana 42(4):321–334
3. Cogle LC, Cualchi DVT, Morocho CMP, Torres DXT, de Aparicio CXP (2021) La migración de zonas rurales a zonas urbanas en el Ecuador. Recimundo 5(1):14–21
4. Márquez-Rosano C, Sánchez-Reyes GA, Vásquez-Maldonado B, Legorreta-Díaz MdC (2022) Contribuyen las UMA a la conservación de la biodiversidad en el sureste del país?, En Gómez-Martínez, Emanuel, Biodiversidad y políticas públicas en México Chapingo, Texcoco, Estado de México (México). Universidad Autónoma Chapingo
5. Alba TA (2021) Contribución de los recursos genéticos a los sistemas alimentarios. J Boliviano de Cienc 17(Especial):98–100
6. Lluch-Cota SE, Zapata JAV, Delgado CN (2022) Agricultura, agua y cambio climático en zonas áridas de México. Recursos Naturales Y Soc 8(2):35–48
7. Rodríguez AG, Chumacero LMG (2022) Biodiversidad y artesanías de México. Investigación y Ciencia: de la Univ Autónoma de Aguascalientes 87:9
8. Bélanger J, Pilling D (2019) The state of the world's biodiversity for food and agriculture. FAO
9. de México ACC (2021) Comisión Nacional para el Conocimiento y Uso de la Biodiversidad (CONABIO)
10. DE C (1999) Comisión de Recursos Genéticos para la Alimentación y la Agricultura. Agricultura 25(29)
11. del Campo MAPM (2004) Las tierras de uso común en ejidos certificados1
12. Bonilla BEH, Reynoso AMR, Cortés VR, Trujillo SJS, Hernández MD (2020) Análisis económico de productores y comercializadores de nopal en el Valle de Teotihuacán. RICEA Rev Iberoamericana de Contaduría, Economía y Administración 9(17):72–108
13. Rodriguez Valverde EA (2024) La contribución de la cooperación internacional en materia de conservación de agrobiodiversidad para el cumplimiento del ODS 2 en el Perú en el periodo del 2018–2022
14. Genéticos R (2011) 1. Protocolo de Nagoya sobre Acceso a los Recursos Genéticos y Participación Justa y Equitativa en los Beneficios que se Deriven de su Utilización

Chapter 13

Ex Situ Conservation of Epiphytic Orchids: A Bibliometric Review

Arturo Hernández-García and Obdulia Baltazar-Bernal

Abstract

Epiphytic orchids, representing 70% of the Orchidaceae family, face critical challenges due to habitat loss, climate change, and overexploitation. Ex situ conservation aims to preserve these species outside their natural habitats while maintaining their ecological characteristics for potential reintroduction. This review examines various ex situ strategies, including seed banking, living collections, tissue culture, and mycorrhizal conservation, as well as acclimatization techniques and reintroduction efforts. Botanical gardens, private collections, and orchid seed banks play essential roles in safeguarding genetic diversity and supporting reintroduction projects. The success of these efforts hinges on a multidisciplinary approach that incorporates scientific research, community involvement, and habitat restoration to ensure the long-term survival of epiphytic orchids. The findings emphasize the need for integrated conservation strategies to bridge the gap between orchid ecology studies and practical conservation actions.

Keywords Orchidaceae, Orchid acclimatization, Orchid hardening, Phorophyte, Orchid reintroduction, Orchid adaptation, Orchid preservation

1 Introduction

According to recent sources, there are currently 31,482 accepted orchid species in the world, distributed on 703 genera [1], around 70% of these species have an epiphytic habit [2]. It has been calculated that Orchids originated 112 million years ago in Australia and epiphyte orchids originated in Southeast Asia among the ancient epidendroid orchids [3]. Epiphyte orchids germinate and live attached to other plants; they have evolved to survive without soil nutrients [4, 5].

To endure the tree crown conditions, epiphytic orchids developed a wide range of traits such as thicker roots protected with a velamen composed of dead, dry, and spongy tissue, which helps them retain moisture [6]; although tender roots make photosynthesis. They have also developed thicker and larger pseudobulbs which help them to store more water and nutrients. Also, their leaves have a

Marco A. Ramírez-Mosqueda et al. (eds.), *Conservation of Plant Genetic Resources*, Methods in Molecular Biology, vol. 3011, https://doi.org/10.1007/978-1-0716-5130-8_13,

leathery texture with thick cell walls that help them endure drought periods [5]. Another helpful adaptation is the partial or total shift to Crassulacean Acid Metabolism (CAM) from C3 Metabolism, which allowed epiphytic orchids to colonize more dryer environments [7]. This shift to CAM photosynthesis occurred separately on different orchid genus around late Miocene and helped orchids to optimize the water use on carbon fixation [8]. These adaptations, alongside other factors like the evolution of pollinia, pollinator specialization, and miniature seeds, may be the drivers of the immense orchid diversity we have in the present [2, 5].

The relation between epiphytic orchids and their phorophytes may be stronger than it seems, orchids not only receive a living space from their host tree, other parameters like temperature, moisture, and light may vary depending on the host species and age [9]. Since their seeds lack an endosperm, epiphytic orchids need assistance from mycorrhiza to germinate, mycorrhizal communities present on tree bark have different arrangements depending on the host tree species [10]. In addition to this, orchids present variations in their capacity to survive on different conditions [11], and therefore they present affinity to certain host species [12]. All of this makes orchids highly dependent on their phorophytes and more vulnerable to habitat loss [13]. Pollination is another crucial factor in orchid conservation since many orchid species rely on a few specific pollinators to reproduce. This causes orchid populations to be highly susceptible to pollinator availability [14].

Although this relationship between orchids and their phorophytes has been beneficial for their diversity, recently, threats like deforestation, habitat loss, and anthropogenic pressure have reduced their populations drastically [15]. In this context, orchid conservation efforts should focus on preserving and restoring their natural habitats, especially native tree species and forest at mid elevations (around 1500 m.a.s.l.) [16]. These conservation efforts to slow the orchid population decline rates need a closer connection with scientific research, since most research focuses on orchid biology and ecology instead of direct conservation efforts [17]. Conserving the genetic information of orchid species may have medical importance due to the complex diversity of secondary metabolites that orchids possess [18]. The main themes and research areas in ex situ conservation of epiphytic orchids are diverse and interconnected, encompassing topics such as biodiversity, mycorrhizal relationships, germination, and species conservation (Fig. 1). This co-occurrence network provides a visual representation of the frequency and association strength between key concepts, helping to identify focal areas and gaps in current research. Node size represents the frequency of each keyword's appearance in the literature, while edge thickness indicates the strength of co-occurrence relationships. By understanding these connections, conservation strategies can be better informed and prioritized.

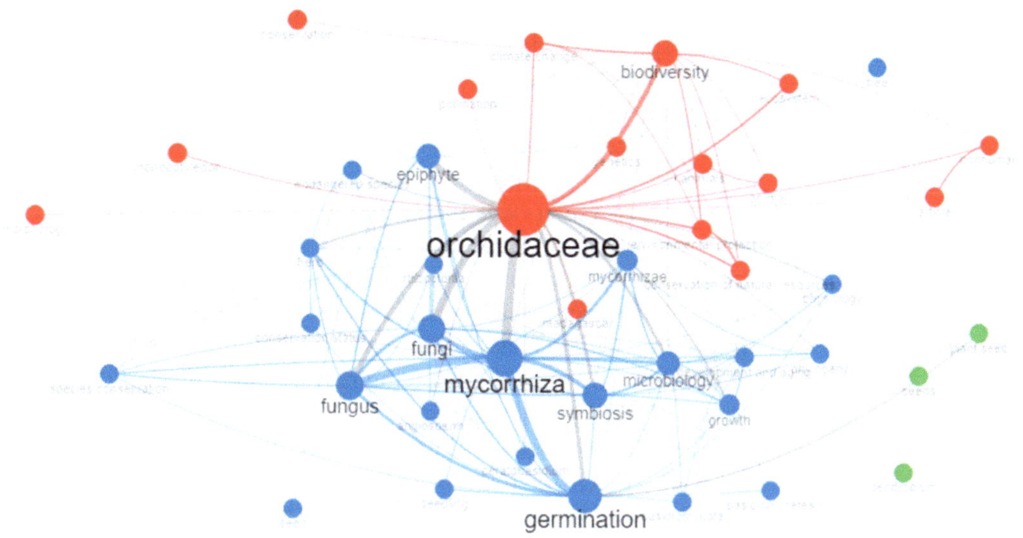

Fig. 1 Co-occurrence network of keywords related to the ex situ conservation of epiphytic orchids. Red, blue, and green nodes group themes by research focus, such as conservation, ecological interactions, and propagation techniques. Figure made with Bibliometrix software [19]

2 Research Methodology

This review was conducted to synthesize current approaches and methodologies in the ex situ conservation of epiphytic orchids, drawing from peer-reviewed articles, conservation reports, institutional publications, and relevant case studies. The study focused on identifying and analyzing the different strategies used in ex situ conservation, including seed banking, tissue culture, mycorrhizal conservation, acclimatization, and reintroduction efforts, with specific emphasis on evaluating the effectiveness of these methods and identifying key challenges and gaps in the existing research.

A comprehensive literature search was performed using academic databases such as JSTOR, Scopus, Google Scholar, and Web of Science, alongside specific journals focused on botany, conservation biology, and orchid research. Studies and reports were included if they detailed specific ex situ conservation methods, provided data on the success rates of reintroduction efforts, or discussed ecological interactions critical to orchid survival. Exclusion criteria removed articles that focused solely on in situ conservation, studies without empirical data, and articles lacking peer review or scientific rigor.

Data from selected articles were systematically extracted and categorized into key areas: seed banking, living collections, tissue culture, mycorrhizal associations, acclimatization, and reintroduction protocols. Extraction focused on quantifiable measures, such as survival rates, acclimatization success, and reintroduction outcomes.

Studies were further categorized based on regional differences, orchid species studied, and conservation settings (e.g., botanical gardens, private collections, institutional efforts). Findings were then synthesized to identify patterns, recurring challenges, and emerging trends in orchid conservation. Additionally, the role of institutions, community involvement, and funding in the success of conservation efforts was analyzed to provide a comprehensive view of factors influencing ex situ conservation outcomes. Figure 2 shows the countries where research on ex situ conservation of epiphytic orchids has been conducted. This map highlights contributions from countries such as the United States, China, and Brazil, which play significant roles in advancing research and methodologies for orchid preservation.

Limitations in available data were also considered, such as the scarcity of long-term studies on reintroduced orchid populations and the lack of standardized methodologies for measuring conservation success. Recommendations for future research and policy were developed based on identified gaps, emphasizing interdisciplinary approaches and collaboration. This structured approach enabled a systematic review of current ex situ conservation practices, offering insights into the effectiveness of various methodologies and highlighting areas where further research could enhance the survival and adaptability of epiphytic orchids outside their natural habitats.

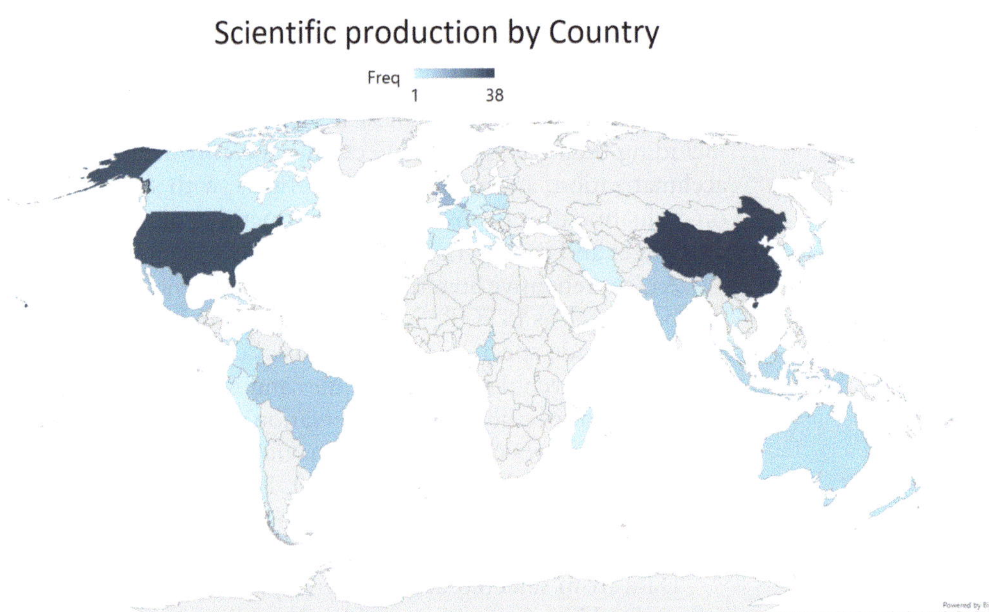

Fig. 2 Global distribution of scientific research on ex situ conservation of epiphytic orchids, illustrating the scientific output by country. Darker shades indicate higher scientific production, reflecting regional focus and resource availability in orchid conservation studies. Figure made with Bibliometrix software [92]

3 Epiphytic Orchids Life Cycle

Epiphytic orchids are herbaceous perennial plants with evergreen or deciduous leaves [19]. The epiphytic orchid seeds will germinate in the tree crown on mature branches, populated with mosses [20] and mycorrhizal fungi; these fungi are essential for tiny seeds to germinate [21]. Seedlings of *Pomatocalpa maculosum* (Lindl.) J.J.Sm. were found on trunks with cryptogam cover, rough surface bark, and humidity availability [22]; similar characteristics were found in *Lepanthes caritensis* seedlings [23]. In orchids, research has been done on the phenology of pseudobulb formation [24]. There are very few studies in orchid juvenility, but in 192 species, the studies are focused on the reproductive phase [25]. Phenological orchids studies are focused on vegetative or reproductive phases. The former includes dormancy and leaf development, leaf falling and juvenility, and the latter includes flowering, fructification, and seed dispersal [26, 27].

Some species have shorter juvenility periods, such as *Erycina pusilla*, which blooms within a year under natural conditions [28] and they can live up to two to three years [29]; the largest orchid in the world, *Grammatophyllum speciosum*, flowers after 32 years [30, 31]. Thus, the average vegetative period lasts for 1–13 years [27]. In nature three stages are defined: (i) seedlings are small plants with a pair of little leaves, (ii) juveniles are individuals with one or more leaf on the stem without flowering, and (iii) adults embraced individuals that are flowering or have been flowered in the past [32]. Orchid flowering also occurs in response to seasonal climate changes, for instance, Johansson [33] described two flowering peaks for orchids during the dry season in eastern Africa; however, Sahagún-Godínez [34] reported that epiphytic orchids in western Mexico have a flowering peak at the beginning of the rainy season and another during the dry season. Similar flowering patterns among genera of epiphytes orchids were reported [33, 35, 36].

About the number of pollinators, out of 456 orchid species, 67% were pollinated by a single pollinator, 14% were pollinated by two, and 17% by more species [37]. In *Specklinia digitale*, once the flower is pollinated, the fruit develops and matures in less than a month; in contrast, *Trichocentrum stramineum* reaches maturation in 14 months [38]. In epiphytic families, 80% of the seeds are dispersed by the wind [39]; however, on epiphytic orchids species 100% of the seeds are wind dispersed. Although one fruit holds and disperses tree millions of seeds; symbiotic germination and survival rates of orchid plants in natural environments are less than 1% [40].

4 Ex Situ Orchid Conservation

The main goal of ex situ orchid conservation is to preserve orchid species out of their natural habitats in response to habitat loss, climate change, and overexploitation [41]. This also means preserving the morphological, biochemical, and ecological interactions of species to ensure a correct re-establishment to in situ conditions [42]. A multifaced approach must be taken to address the different threats and challenges orchids face [14]. Ex situ conservation strategies include seed banking, living collections, tissue culture, mycorrhizal fungi conservation, and reintroduction programs, these strategies must be complemented with research, monitoring, public engagement, and education [43].

4.1 Living Orchid Collections

Orchid populations are affected by many threats, most of them caused by human activities. A wide variety of actions are being implemented to safeguard and recover wild orchid populations; seed banks, cryopreservation and live collections are some examples of this [41]. Ex situ orchid conservation sites have different origins, founding, objectives, and size. In countries with higher orchid diversity the most common sites are private collections and public parks with an important abundance of natural occurring populations [44, 45]. On the other hand, in countries with a lower diversity of native orchids but higher resources and technology botanic gardens and institutional collections are most found [46, 47]. Some types of living collections are botanical gardens [47], institute collections, urban orchids in public spaces, and private collections.

4.2 Botanical Gardens and Public Collections

Botanical gardens are large living plant collections that shelter an important number of plant taxa such as cacti, carnivore plants, trees, palms, and orchids, and have the important mission of conserving and maintaining ex situ collections of plants for scientific, educational, display, and conservation purposes [48]. As natural orchid populations are plummeting down efforts for their conservation are fruitless, in contrast, botanic gardens can help restoring these populations through breeding and reintroduction programs while providing shelter for endangered orchids [47]. If these orchid collections are well documented and cared for, they can become important sources of genetic material for ex situ conservation programs and further reintroductions to their natural habitat [49]. The collection of living specimens for their further conservation and studying has been important for the inclusion of orchids on endangered species lists [50].

Botanic gardens must seek to become science-based orchid conservation centers of international importance, for this goal they must expand their national and international networking, produce scientific publications and collaborate with other orchid conservation institutions [51]. Also, botanic gardens have the important mission of educating and involucrate society in the efforts made to conserve orchids and their natural habitat [52].

Although countries like Colombia and Ecuador are home to the largest orchid species diversity, the largest orchid collections belong to North American and European countries like United States, United Kingdom, Sweden, Canada, and Germany; also, 38% of botanic gardens have the main goal of displaying the specimens, leaving aside the conservation, education, and scientific research purposes. For example, the United States and Canada have around 210 orchid species, while the Atlanta Botanical Garden has the largest Orchid collection with more than 1600 species [45]. In contrast, Colombia has more than 4000 registered orchid species, while the Bogotá Botanical Garden José Celestino Mutis collection has only 339 orchid species [44].

Institutions play an especially significant role in orchid conservation, through ex situ conservation efforts like maintaining living collections, which are helpful for research and reproduction of endangered orchid species, spreading awareness and educating about orchid conservation. All of this while maintaining collaboration networks with other institutions and implementing conservation strategies [53]. Some examples of this are the botanic garden of the Biological Institute and the orchidarium "Miguel Angel Soto Arenas" of the National Autonomous University of Mexico (UNAM), which focus on the research and promotion of orchids as environmental education tools that can help people and children to understand the ecologic importance of native orchids [54].

4.3 Private Collections and Urban Orchids

Among orchid enthusiasts, there are certain collections that stand out for their diversity and diligent care of their specimens [55]. The access to these collections is important for research on taxonomy and biology of rare orchid species, as well as an opportunity to spread awareness about native orchids [56]. Even small private collections can be a reservoir of native wild orchids and other plants; also, these collectors can engage in orchid conservation efforts like reproduction, habitat conservation, and reintroduction [57]. A well-established and coordinated network of private reserves and collections can make remarkable efforts of conservation of native and endemic orchids, and at the same time provide an income source for owners [58].

Orchids can also inhabit parks and public spaces that host an important number of urban orchid species [59]. Orchid diversity may vary depending on site conditions [60] and they can serve as indicators of a healthy urban ecosystem and as a genetic reservoir [61]. Several authors have studied orchid diversity in urban areas [62], touristic parks [59], and schools [60].

4.4 Orchid Seed Banks

Orchid seed banks store seeds from distinct species for its conservation and future use, they collect, clean, and save orchids in controlled conditions assuring their viability to germinate in the long term [63]. Since epiphyte orchids possess truly orthodox seeds, they can maintain viability up to 8 to 14 years when stored dry at 5°C to 8°C; and they can be stored indefinitely using cryopreservation [64]. For this reason, cryopreservation of orchid seeds at ultra-low temperatures can be suitable for germplasm conservation [65]. Seed preservation is necessary for long-term conservation of epiphytic orchids, studies focused on seed germination media and seed longevity are needed [66]. Orchid Seeds Store for Sustainable Use (OSSSU) is a global network of orchid seed banks; it has the participation of 31 institutions in 22 countries and it stores seeds of more than 250 orchid species [67].

4.5 Tissue Culture

Orchids can be germinated in laboratory conditions using asymbiotic or symbiotic techniques [68]. Orchid species respond differently to growth media and culture conditions, in vitro study's main goal is to determine the best culture media for each species [69]. Some studies report a greater germination when supplementing growth media with organic supplements [70].

Orchids can also be propagated by explant multiplication or Protocorm Like Bodies (PLB) culture; although these techniques are useful for multiplication, the resulting orchids are not as heterogeneous as seed germinated orchids [71]. Orchid tissues can also be conserved on in vitro conditions; this technique relies on slow growth culture media to conserve orchid plantlets for further use [72]. Tissue cryopreservation is also a viable option to preserve orchid genome, especially of rare and endangered species [68].

4.6 Mycorrhiza

Mycorrhizal fungi are crucial for orchid conservation; they not only play a vital role on germination, but also can help orchids to better resist pathogens and adapt to their host tree microclimate [73, 74]. Some orchids present affinity for a certain mycorrhizal species [75]. The use of orchid mycorrhizal fungi on orchids improves plantlet survival during acclimatization and reintroduction phases [76]. Studying the relationship between fungi and orchids can help us improve conservation efforts for native orchids [74].

4.7 Acclimatization

Another key factor to reintroduction success is acclimatization. During this phase, the survival can vary from species to species, reaching between 35% and 95% of survival rate [77, 78]. Orchids are placed on trays or pots and are covered with domes to control humidity; temperature and light intensity also play a significant role during this stage [74, 75]. The most used substrates are Sphagnum moss, coconut choir, pinewood bark chips, vegetable coal, and mineral substrates [76, 78]. This phase is important for orchid survival out of the greenhouse, and a longer acclimatization period will increase the survival of plantlets on field conditions on the long term [79].

5 Reintroduction Efforts

The best site to conserve orchids is in their natural habitat, and ex situ conservation aims to preserve orchid germplasm for their future use in reintroduction efforts [80]. The main objective of orchid reintroduction is to restore orchid populations until there are enough specimens to attract pollinators and their reproduction cycle can continue autonomously [81]. Some key factors for a successful orchid reintroduction are the site conditions, which need to be similar to those where the plant naturally distributes [78]. The presence of native host trees that can host a high diversity of epiphytes, specific pollinators, and mycorrhizal species will facilitate the plantlet adaptation [82]. Correct habitat management can ensure the presence of these key factors to orchid adaptation; also, a well-focused field study can help us identify reintroduction sites [14].

Native epiphytic orchids are vulnerable to multiple threats, some examples are deforestation, climate change, illegal collection, forest fragmentations, pollinator decline, wildfires, and droughts. Reintroduction helps restore orchid species that have become critically endangered or extinct in their habitats [43]. Most orchid species have low growth rates, which causes reintroduction efforts to last for extended periods when constant monitoring is necessary [82].

Regarding plant characteristics, the initial size of the reintroduced plants is also a key factor to survival since larger plants can hold more water inside them [83]. The use of mycorrhizal inoculation of plantlets can increase the survival of acclimatized and re-introduced orchids [80]. Additionally, it has been registered that mycorrhizal fungus colonizes orchid plantlets after 240 days of reintroduction [84]. Constant monitoring is important to avoid orchids to be collected by poachers and to keep control of orchids response to reintroduction management [79].

Few authors have studied the reintroduction process from in vitro culture to field establishment (Table 1). The most successful reintroductions needed the participation of researchers, authorities and residents to relocate 13,000 orchid plantlets to their natural habitat; they reported a 97% of survival after 1 year [79]. Other authors [78] reported a survival rate of 69.7% of an epiphytic orchid after 2 years of being reintroduced on a natural reserve. Unfortunately, most reintroduction studies only evaluate the first months of orchid growth on in situ conditions [81]. There is a need for studies on the long-term survival of reintroduced orchids, as well as on the effectiveness of reintroduction efforts and methods [82].

Table 1
Summary of reintroduction strategies, survival rates, and key factors influencing reintroduction of epiphytic orchids

Country	South Korea	Brazil	Brazil
Reference	[79]	[85]	[86]
Main factors influencing reintroduction	Extended acclimatization period, site selection, host compatibility, and monitoring.	Site selection and initial plant condition	Host tree selection and watering
Reintroduction survival	97% after 24 months	60.23% after 24 months	N/E
Reintroduction method	Groups of ten plants were tied together, their roots were covered with sphagnum moss. The tree species were *Pinus densiflora*, *Camelia japonica*, *Quercus salicina*, *Quercus acuta*, *Machilus thunbergii*, *Castanopsis cuspidata* var. *Thunbergii*, and *Mallotus japonicus*	Plantlets attached to pine bark fragments were tied to phorophytes with at least 10 cm diameter at breast height (DBH) and hosted other epiphytes	Plantlets were attached to trees known for their ability to support *Schomburgkia crispa* Lindl. Plantlets were watered after reintroduction
Acclimatization time and survival (%)	97% after 24 months	N/E	N/E
Acclimatization methods	Plantlets were grown on trays containing sphagnum moss, the greenhouse was kept at a humidity of 80%–90% at a temperature of 22 ± 2°C	Plantlets were tied to pine bark substrate after 9 months of sowing, seedlings were watered periodically and fertilized every 2 months	N/E
Species	*Dendrobium moniliforme* (L.) Sw	*Cattleya intermedia* Graham	*Schomburgkia crispa* Lindl.
Country	Brazil	Brazil	China
Reference	[87]	[88]	[78]
Main factors influencing reintroduction	Host tree selection, light exposure, and fertilization	Initial acclimatization and phorophyte compatibility	Selection of suitable reintroduction sites, host tree compatibility, watering

Table 1
continued

Reintroduction survival	*Vitex megapotamica* showed between 0% and 90% and *Myrcia palustris* showed a 100% survival rate after 6 months	Non-fertilized plantlets reached 80% survival and fertilized plants reached 50% survival after 360 days	Highest survival was recorded on Yuanjiang nature reserve with 84.67% after 2 years
Reintroduction method	Plantlets were tied to adult trees of *Vitex megapotamica* and *Myrcia palustris* using natural fiber cords. They were placed between 0.8 and 2 m. Plants were sprayed with diluted Hoagland solution every two weeks during 6 months after reintroduction	Plantlets were reintroduced to *Hymenaea courbaril* using a mix of buriti palm fiber and coconut chips as substrate. Plants were divided into two groups, one receiving controlled release fertilizer (NPK 15-15-15) and the other was not fertilized	Plantlets were either potted on a sand, peat and fir bark mixture or mounted on *Pinus massoniana* tree trunks with their roots wrapped in sphagnum moss. This was done on three different sites, they were watered on days 1, 3, 7, 14, 21, and 28 after reintroduction
Acclimatization time and survival (%)	53% for sphagnum moss and 27% for pine bark after 4 months	Plants were acclimatized for 12 months; no survival rate was reported	The most effective substrate was sphagnum moss with 95% after 70 days
Acclimatization methods	Plantlets were placed on Sphagnum moss and pine bark substrates; they were watered periodically to maintain moisture and sprayed with diluted Hoagland solution every two weeks	Plantlets underwent a 12-month acclimatization period in a nursery with controlled shading	Plantlets were put onto five different substrates (Sphagnum moss, fir bark fragments, orchid sand and fir mixture, fir bark wrapped with sphagnum and sand, peat and fir mixture)
Species	*Cattleya intermedia* Graham	*Brassavola tuberculata* Hook	*Renanthera imschootiana* Rolfe
Country	India	Brazil	
Reference	[89]	[90]	
Main factors influencing reintroduction	Hardening phase, humidity control, substrate composition, and fertilization	Fungal association, microhabitat suitability, and controlled acclimatization	

(continued)

Table 1
continued

Reintroduction survival	Survival was 92.5% after 1 month, then it dropped to 12% after 9 months due to prolonged dry conditions	72% after 240 days	
Reintroduction method	Plantlets were attached to *Memecylon umbellatum* trees, orchid roots were covered with sphagnum moss for humidity retention, and then they were tied with a cotton string	Seedlings were reintroduced to an Atlantic Forest fragment; the seedlings were attached to host trees at heights of 1.6–2.0 m, using reed screens and sphagnum moss for moisture retention	
Acclimatization time and survival (%)	Plants were hardened for 4 weeks, then acclimatized for 5 weeks. No survival data was reported	Authors reported 100% but no evaluation time was reported.	
Acclimatization methods	Hardened plantlets were transferred to pots filled with a substrate mixture composed of coir, sand, brick fragments, charcoal fragments, dried moss, vermicompost, and tree bark. Plantlets were periodically sprayed with half-strength MS medium without sucrose	Seedlings were acclimatized in greenhouse conditions using a pine bark substrate	
Species	*Dendrobium aqueum* Lindley	*Hadrolaelia jongheana*	
Country	Mexico	Mexico	
Reference	[80]	[91]	
Main factors influencing reintroduction	The use of young plants led to a low survival	*Lycaste aromatica* plantlets survival increased in forest	*Jacquiniella teretifolia* survival was similar at all sites
Reintroduction survival	3.3%	Forest: 85%; Old coffee plantations: 65%; young coffee plantations: 35%	Forest: 57%; Old coffee plantations: 56%; young coffee plantations: 55%

(continued)

Table 1
continued

Reintroduction method	Seedlings were adapted on four substrates (Cedar bark (*Cedrela odorata*), Clay tile, *Agave fourcroydes* string	Seedlings were established on three habitat types: natural forests, old coffee plantations, and young coffee plantations. Each habitat had three repetitions and 16 trees were selected on each repetition. Plants were tied to upper and lower branches using a synthetic string	
Acclimatization time and survival (%)	25% after 4 months	Orchids were acclimatized for 3 months; survival percentage was not reported for this stage	
Acclimatization methods	Plantlets were put in trays on four substrates (Cedar bark (*Cedrela odorata*), dehydrated pataxte (*Theobroma bicolor*), Clay tile, and *Agave fourcroydes* string	Plantlets were put in propagation trays with a substrate mix (pine bark, oak charcoal, pumice) and placed in a shade-house	
Species	*Guarianthe skinneri* (Bateman) Dressler & W.E. Higgins	*Lycaste aromatica* Lindl.	*Jacquiniella teretifolia* (Sw.) Britton & P. Wilson

6 Conclusions

The conservation of epiphytic orchids is a complex and multifaceted endeavor that addresses the urgent need to protect one of the most diverse and vulnerable plant families on the planet. Epiphytic orchids, adapted to unique ecological niches and dependent on specific mycorrhizal relationships and pollinators, are particularly sensitive to environmental disturbances such as deforestation, climate change, and habitat fragmentation. Ex situ conservation offers a critical lifeline, enabling the preservation of genetic material, the study of adaptive traits, and the development of reintroduction protocols that are essential for maintaining biodiversity and ecological balance.

Effective ex situ conservation requires an integrated approach that combines scientific research with practical applications. Seed banking, tissue culture, and the establishment of living collections not only provide genetic reservoirs but also facilitate research on orchid biology and ecology, which is crucial for the success of reintroduction efforts. The role of botanical gardens, private collectors, and institutional partnerships is fundamental, as these entities provide both the resources and expertise needed to maintain viable orchid populations outside their natural environments.

Moreover, successful reintroduction efforts hinge on careful site selection, habitat restoration, and community engagement. Collaborative efforts that involve local communities, policymakers, scientists, and conservation organizations can create a sustainable framework for orchid conservation that benefits both biodiversity and local economies. Monitoring and long-term studies are essential to assess the survival and adaptation of reintroduced orchids, allowing for adjustments and improvements in conservation techniques.

The conservation of epiphytic orchids goes beyond preserving individual species; it represents a commitment to protecting ecosystems and the intricate relationships that sustain them. As conservation efforts progress, an emphasis on education, public awareness, and international collaboration will be pivotal in garnering the support needed to ensure the survival of these remarkable plants. Through a combination of scientific rigor, practical conservation strategies, and public engagement, ex situ orchid conservation can make a meaningful contribution to global biodiversity and inspire future generations to value and protect these extraordinary species.

References

1. WFO Plant List (2024) Orchidaceae Juss. Published on the Internet; https://wfoplantlist.org/taxon/wfo-7000000429-2024-06. Accessed 11 Oct 2024
2. Gravendeel B, Smithson A, Slik FJW, Schuiteman A (2004) Epiphytism and pollinator specialization: drivers for orchid diversity? Philos Trans R Soc B Biol Sci 359(1450):1523–1535. https://doi.org/10.1098/rstb.2004.1529
3. Givnish TJ, Spalink D, Ames M, Lyon SP, Hunter SJ, Zuluaga A, Doucette A, McDaniel J, Clements MA, Arroyo MTK, Endara L, Kriebel R, Williams NH, Cameron KM (2016) Orchid historical biogeography, diversification, Antarctica and the paradox of orchid dispersal. J Biogeogr 43(10):1905–1916. https://doi.org/10.1111/jbi.12854
4. Benzing DH (1990) Vascular epiphytes – general biology and related biota. Cambridge University Press, Cambridge
5. De LC, Biswas SS (2022) Adaptational mechanisms of epiphytic orchids: a review. Int J Biores Stress Manag 13(11):1312–1322. https://doi.org/10.23910/1.2022.3115a
6. Pridgeon AM (1987) The velamen and exodermis of orchid roots. In: Arditti J (ed) Orchid biology: reviews and perspectives. Cornell University Press, Ithaca, pp 150–185
7. Silvera K, Santiago LS, Winter K (2005) Distribution of crassulacean acid metabolism in orchids of Panama: evidence of selection for weak and strong modes. Funct Plant Biol 32:397–407. https://doi.org/10.1071/FP04231
8. Gamisch A, Winter K, Fischer G, Comes HP (2021) Evolution of crassulacean acid metabolism (CAM) as an escape from ecological niche conservatism in *Malagasy Bulbophyllum* (Orchidaceae). N Phytol 231:1236–1248. https://doi.org/10.1111/nph.17437
9. Rasmussen HN, Rasmussen FN (2018) The epiphytic habitat on a living host: reflections on the orchid–tree relationship. Bot J Linn Soc 186(4):456–472. https://doi.org/10.1093/botlinnean/box085
10. Wang X, Li Y, Song X, Meng Q, Zhu J, Zhao Y, Yu W (2017) Influence of host tree species on isolation and communities of mycorrhizal and endophytic fungi from roots of a tropical epiphytic orchid, *Dendrobium sinense* (Orchidaceae). Mycorrhiza 27(7):709–718. https://doi.org/10.1007/s00572-017-0787-7
11. Guevara-Perez CI, Delgado-Sánchez P, Torres-Castillo JA, Flores-Rivas JD, Mendieta-Leiva G, Rosa-Manzano ED (2019) Epiphytic orchids *Stanhopea tigrina* and *Prosthechea cochleata* are differentially affected by drought in a subtropical cloud forest. Photosynthetica 57(4):1053–1065. https://doi.org/10.32615/ps.2019.123
12. Hernández-Ríos E, Solano E, Ríos-Gómez R (2018) Host affinity and vertical distribution of epiphytic orchids in a montane cloud forest in southern Mexico. Bot Sci 96(2):200–217. https://doi.org/10.17129/botsci.1869
13. Parra-Sanchez E, Freckleton RP, Hethcoat MG, Ochoa-Quintero JM, Edwards DP

(2024) Transformation of natural habitat disrupts biogeographical patterns of orchid diversity. Biol Conserv 292:110538. https://doi.org/10.1016/j.biocon.2024.110538

14. Phillips RD, Reiter N, Peakall R (2020) Orchid conservation: from theory to practice. Ann Bot 126(3):345–362. https://doi.org/10.1093/aob/mcaa093
15. Adhikari YP, Fischer A, Pauleit S (2015) Sustainable conservation perspectives for epiphytic orchids in the Central Himalayas, Nepal. Appl Ecol Env Res 13(3):753–767. https://doi.org/10.15666/aeer/1303_753767
16. Timsina B, Münzbergová Z, Kindlmann P, Bhattarai BP, Shrestha B, Raskoti BB, Rokaya MB (2024) Associations between epiphytic orchids and their hosts and future perspectives of these in the context of global warming. Diversity 16(4):252. https://doi.org/10.3390/d16040252
17. Gale SW, Fischer GA, Cribb PJ, Fay MF (2018) Orchid conservation: bridging the gap between science and practice. Bot J Linn Soc 186(4):425–434. https://doi.org/10.1093/botlinnean/boy003
18. Pant B (2013) Medicinal orchids and their uses: tissue culture a potential alternative for conservation. Afr J Plant Sci 7:448–467. https://doi.org/10.5897/AJPS2013.1031
19. Trapnell DW, Hamrick JL (2006) Floral display and mating patterns within populations of the neotropical epiphytic orchid, Laeliarubescens (Orchidaceae). Am J Bot 93(7):1010–1018
20. Frei JK (1973) Effect of bark substrate on germination and early growth of *Encyclia tampensis* seeds. Am Orchid Soc Bull 42:701–708
21. Osorio-Gil EM, Forero-Montana J, Otero JT (2008) Variation in mycorrhizal infection of the epiphytic orchid *Ionopsis utricularioides* (Orchidaceae) on different substrata. Caribb J Sci 44:130–132
22. Watthana S (2004) Ecology and conservation biology of *Pomatocalpa naevata* J.J. Sm. (Orchidaceae). Nat Hist Bull Siam Soc 52:201–215
23. Tremblay RL, Zimmerman JK, Lebrón L, Bayman P, Sastre I, Axelrod F, Alers-García J (1998) Host specificity and low reproductive success in the rare endemic Puerto Rican orchid *Lepanthes caritensis*. Biol Conserv 85:297–304
24. Dressler, R, (1993). Phylogeny and classification of orchid family. Cambridge University Press, Cambridge
25. Ramírez-Martínez A, Mondragon D, Rivera-García R (2021) Vascular epiphytes: the ugly duckling of phenological studies. Acta Biol Colomb 26(2):247–261. https://doi.org/10.15446/abc.v26n2.83473
26. Tang J, Körner C, Muraoka H, Piao S, Shen M, Thackeray SJ, Yang X (2016) Emerging opportunities and challenges in phenology: a review. Ecosphere 7:8
27. Hew CS, Yong JW (2004) The physiology of tropical orchids in relation to the industry. World Scientific, Singapore
28. Chiu YT, Chang C (2018) *In vitro* flowering and breeding of *Erycina pusilla*. In: Lee YI, Yeung ECT (eds) Orchid propagation: from laboratories to greenhouses – methods and protocols. Humana Press, Totowa, NJ, 195–210 https://doi.org/10.1007/978-1-4939-7771-0_13edspp
29. Chase MW (1987) Obligate twig epiphytism in the Oncidiinae and other neotropical orchids. Selbyana 10:24–30
30. Otway B (2015) Queen of orchids, the largest orchid in the world in flower now at Kew! Pumpkin Beth. https://www.pumpkinbeth.com/2015/09/largest-orchid/. Accessed 26 Sep 2015
31. Rivera Gómez N, Tremblay RL, Meléndez-Ackerman E (2006) Distribution of life cycle stages in a lithophytic and epiphytic orchid. Folia Geobot 41:107–120
32. Johansson D (1974) Ecology of vascular epiphytes in West African rain forest. Acta Phytogeogr Suec 59:1–136
33. Sahagun-Godinez E (1996) Trends in the phenology of flowering in the Orchidaceae of Western Mexico. Biotropica 28(1):130–136. https://doi.org/10.2307/2388778
34. Wright SJ, Calderon O (1995) Phylogenetic patterns among tropical flowering phenologies. J Ecol 83(6):937–948. https://doi.org/10.2307/2261176
35. Texier N, Deblauwe V, Stévart T, Sonké B, Simo-Droissart M, Azandi L, Droissart V (2018) Spatio-temporal patterns of orchids flowering in Cameroonian rainforests. Int J Biometeorol 62(11):1931–1944. https://doi.org/10.1007/s00484-018-1594-3
36. Tremblay RL (1992) Trends in the pollination ecology of the Orchidaceae: evolution and systematics. Can J Bot 70(3):642–650. https://doi.org/10.1139/b92-083
37. Baltazar-Bernal O, De la Cruz-Martínez VM, Hernández-García A, Zavala-Ruiz J (2023) An exploratory study of orchidaceae species fruits in the central zone of Veracruz state, Mexico. Agrociencia 57(5):981–1007. https://doi.org/10.47163/agrociencia.v57i5.2860
38. Madison M (1977) Vascular epiphytes: their systematic occurrence and salient features. Selbyana 2(1):1–13
39. Teixeira da Silva JA, Tsavkelova EA, Ng TB, Parthibhan S, Dobránszki J, Cardoso JC,

Rao MV, Zeng S (2015) Asymbiotic *in vitro* seed propagation of *Dendrobium*. Plant Cell Rep 34:1685–1706. https://doi.org/10.1007/s00299-015-1829-2
40. Seaton PT, Kendon J, Pritchard H, Puspitaningtyas DM, Marks T (2013) Orchid conservation: the next ten years. Lankesteriana 13(1–2):93–101
41. Tremblay R, Alicea-Roman P, Anaya-Reyes A, Dulcrec-Rodas S, Medina-Tirado I (2022) Evidence of artificial selection: are orchids in cultivation an effective *ex situ* conservation strategy? Lankesteriana 22(3):263–284. https://doi.org/10.15517/lank.v22i3.53443
42. Seaton PT, Hu H, Perner H, Pritchard HW (2010) *Ex situ* conservation of orchids in a warming world. Bot Rev 76(2):193–203. https://doi.org/10.1007/s12229-010-9048-6
43. Ministerio de Ambiente y Desarrollo Sostenible (2015) Plan para el estudio y la conservación de las orquídeas en Colombia. Ministerio de Ambiente y Desarrollo Sostenible. Universidad Nacional de Colombia, Bogotá DC, Colombia, p 336
44. Vidal O, San Martín C, Mardones S, Bauk V, Vidal C (2012) The orchids of Torres del Paine Biosphere Reserve: the need for species monitoring and ecotourism planning for biodiversity conservation. Gayana Bot 69(1):136–146
45. Botanic Gardens Conservation International (BGCI) (2017) Orchids: 2017 Global *ex situ* Collections Assessment. Botanic Gardens Conservation International, U.S., pp 6. https://www.bgci.org/wp/wpcontent/uploads/2019/04/Orchids_2017_Global_Ex_Situ_Collections_Assessment.pdf. Accessed 07 Oct 2024
46. Swarts N, Dixon K (2009) Perspectives on orchid conservation in botanic gardens. Trends Plant Sci 14:590–598. https://doi.org/10.1016/j.tplants.2009.07.008
47. Girmay M (2023) Roles of Botanical Gardens for conservation and requirements for their establishment: review. Daagu Int J Basic Appl Res 5:182–192. https://doi.org/10.20372/dijbar.81657
48. Mirenda T (2011) Botanic gardens, education, and orchid conservation strategies: the need for a coordinated approach. Lankesteriana 11:217–227. https://doi.org/10.15517/lank.v11i3.18285
49. Pendersen H, Find J, Petersen G, Seberg O (2018) On the "Seidenfaden collection" and the multiple roles botanical gardens can play in orchid conservation. Lankesteriana 18(1):1–12. https://doi.org/10.15517/lank.v18i1.32587
50. Pendersen H (2010) Perspectives on orchid conservation in Queen Sirikit Botanic Garden: recommendations based on a SWOT analysis. Thai J Bot 2(Special Issue):1–13
51. Calderón S, Gasca T, Correa M, Calceido A (2023) Orchids of the Uniamazonia Botanical Garden (JBUDLA). Hortic Int J 7(2):45–50. https://doi.org/10.15406/hij.2023.07.00274
52. Cherevchenko T, Buyun L, Kovalska L, Long V (2007) *Ex situ* conservation of tropical orchids in Ukraine. Lankesteriana 7:129–133. https://doi.org/10.15517/lank.v7i1-2.18451
53. Téllez-Velasco A (2007) The important role that the botanical garden of the National Autonomous University of Mexico plays in the conservation of Mexican orchids. Lankesteriana 7(1-2):156–160
54. Kirby S (2003) Neotropical orchid eco-tourism: educational experience of an orchid neophyte at the Bosque De Paz Biological Preserve, central volcanic range, Costa Rica. Lankesteriana 7:121–124. https://doi.org/10.15517/lank.v3i2.23034
55. Muñoz M, Kirby S (2007) Orchid inventory and conservation project at Bosque de Paz Biological Reserve, upper Rio Toro valley, Alajuela, Costa Rica. Lankesteriana 7(1-2):60–65. https://doi.org/10.15517/lank.v7i1-2.18440
56. Rojas M, Morales C, Blanco M (2016) Conservación de orquídeas en una reserva privada de Palmares. Costa Rica Res J Costa Rican Distance Educ Univ 8(2):231–240
57. Orejuela J (2010) La conservación de orquídeas en Colombia y un caso en proceso en la cuenca del río Cali, municipio de Santiago de Cali, Valle del Cauca, Colombia. El Hombre y la Máquina, vol. 35. Universidad Autónoma de Occidente, Cali, Colombia, pp 53–66
58. Dewi NPYA, Angket S, Ledheng L, Sele Y, Welsiliana (2024) Diversity of orchid types in the Oeluan Tourism Forest, Bijeli Village, Noemuti District, North Central Timor Regency. Interdiscip Int J Conserv Cult 2(1):31–38. https://doi.org/10.25157/iijcc.v2i1.3906
59. Baltazar-Bernal O, Zavala-Ruiz J, Hernández-García A (2024) Orchid diversity (Orchidaceae) in two urban sites in the state of Veracruz, Mexico. Agrociencia. https://doi.org/10.47163/agrociencia.v58i5.3125
60. Newman BJ, Ladd P, Batty A, Dixon K (2007) Ecology of orchids in urban bushland reserves – can orchids be used as indicators of vegetation condition? Lankesteriana 7(1–2):313–315
61. Lussu M (2022) Urban orchids: an updated checklist of the orchid flora of Cagliari (Sardinia, Italy). Biogeographia J Integr Biogeogr 37:a015. https://doi.org/10.21426/B637156359

62. Seaton PT (2007) Establishing a global network of orchid seed banks. Lankesteriana 7(1-2):371–375
63. Machado NB, Custodio C (2005) Orchid conservation through seed banking: Ins and outs. Selbyana 25(2):1–10
64. Merritt DJ, Hay FR, Swarts ND, Sommerville KD, Dixon KW (2014) *Ex situ* conservation and cryopreservation of orchid germplasm. Int J Plant Sci 175(1):46–58. https://doi.org/10.1086/673370
65. Puspitaningtyas DM, Handini E (2021) Seed germination evaluation of *Phalaenopsis amabilis* in various media for long-term conservation. Biodiversitas 22(11):5231–5238. https://doi.org/10.13057/biodiv/d221162
66. Seaton PT, Pritchard HW (2011) Orchid seed stores for sustainable use: a model for future seed-banking activities. Lankesteriana 11(3):349–353
67. Bustam BM, Dixon K, Bunn E (2016) A cryopreservation protocol for *ex situ* conservation of terrestrial orchids using asymbiotic primary and secondary (adventitious) protocorms. In Vitro Cell Dev Biol Plant 52:185–195. https://doi.org/10.1007/s11627-015-9732-7
68. Jolman D, Batalla MI, Hungerford A, Norwood P, Tait N, Wallace LE (2022) The challenges of growing orchids from seeds for conservation: an assessment of asymbiotic techniques. Appl Plant Sci 10(5):e11496. https://doi.org/10.1002/aps3.11496
69. De Stefano D, Costa B, Downing J, Fallahi E, Khoddamzadeh A (2022) In-vitro micropropagation and acclimatization of an endangered native orchid using organic supplements. Am J Plant Sci 13:380–393. https://doi.org/10.4236/ajps.2022.133023
70. Lal N, Singh M (2021) Prospects of plant tissue culture in orchid propagation: a review. Indian J Biol 7:103–110. https://doi.org/10.21088/ijb.2394.1391.7220.15
71. Menezes-Sá TSA, Arrigoni-Blank MF, Da Costa A, Feitosa-Alcantara RB, Blank AF, Luz JM (2019) *In vitro* conservation and acclimatization of Epidendroideae (Orchidaceae) from Sergipe, Brazil. Biosci J 35(2):356–366. https://doi.org/10.14393/BJ-v35n2a2019-38776
72. Otero JT, Bayman P (2013) Tropical orchid mycorrhizae: a key to conservation and commercialization. Lankesteriana 13(1–2):57–63
73. Zettler WL, Poutler SB, McDonald KI (2007) Conservation-driven propagation of an epiphytic orchid (*Epidendrum nocturnum*) with a mycorrhizal fungus. HortScience 42(1):135–139
74. Rammitsu K, Yamamoto N, Chamara RMSR, Minobe M, Kinoshita A, Kotaka N, Ogura-Tsujita Y (2023) The epiphytic orchid *Vanda falcata* is predominantly associated with a single Tulasnellaceae fungus in adulthood, and Ceratobasidiaceae fungi strongly induce its seed germination *in vitro*. Plant Species Biol 38(6):306–318. https://doi.org/10.1111/1442-1984.12432
75. Zhao D, Selosse M, Wu L, Luo Y, Shao S, Ruan Y (2021) Orchid reintroduction based on seed germination-promoting mycorrhizal fungi derived from protocorms or seedlings. Front Plant Sci 12:701152. https://doi.org/10.3389/fpls.2021.701152
76. Decruse SW, Gangaprasad A (2018) Restoration of *Smithsonia maculata* (Dalz.) Saldanha, an endemic and vulnerable orchid of western Ghats through *in vitro* propagation. J Orchid Soc India 32:25–32
77. Wu K, Zeng S, Lin D, Teixeira da Silva JA, Bu Z et al (2014) In vitro propagation and reintroduction of the endangered Renanthera imschootiana Rolfe. PLoS One 9(10):e110033. https://doi.org/10.1371/journal.pone.0110033
78. Coopman J, Kane ME (2018) Greenhouse acclimatization methods for field establishment of *in vitro*-derived ghost orchid (Dendrophylax lindenii) plants. Native Plants J 19(2):100–108. https://www.jstor.org/stable/26524834
79. Kim Y, Kang K, Kim K (2016) Restoration of endangered orchid species *Dendrobium moniliforme* (L.) Sw. (Orchidaceae) in Korea. Korean J Pl Taxon 46(2):256–266. https://doi.org/10.11110/kjpt.2016.46.2.256
80. Emeterio-Lara A, Damon A (2024) Acclimatization with endophytic fungi and reintroduction of *Guarianthe skinneri* (Bateman) Dressler & W.E. Higgins, a threatened, native orchid of cultural value in southern Mexico. J Nat Conserv 78:1617–1381. https://doi.org/10.1016/j.jnc.2024.126573
81. Reiter N, Whitfield J, Pollard G, Bedggood W, Argall M, Dixon K, Davis B, Swarts N (2016) Orchid re-introductions: an evaluation of success and ecological considerations using key comparative studies from Australia. Plant Ecol 217:1–12. https://doi.org/10.1007/s11258-015-0561-x
82. Yam TW, Tay F, Ang P, Soh W (2011) Conservation and reintroduction of native orchids of Singapore – the next phase. Eur J Env Sci 1(2):38–47
83. Segovia-Rivas A, Meave JA, González EJ, Pérez-García EA (2018) Experimental reintroduction and host preference of the microendemic and endangered orchid *Barkeria whartoniana* in a Mexican Tropical Dry Forest. J Nat Conserv 43:156–164. https://doi.org/10.1016/j.jnc.2018.04.004
84. Vieira C, Bocayuva M, Veloso T, Moreira B, Freitas E, Bazzolli D, Kasuya MC (2020) Fungus

used for germination is supplanted after reintroduction of *Hadrolaelia jongheana* (Orchidaceae). Rev Agrar Acad 3:148–161. https://doi.org/10.32406/v3n62020/148-161/agrariacad
85. Endres-Júnior D, Sasamori MH, Silveira T, Schmitt JL, Droste A (2015) Reintroduction of *Cattleya intermedia* Graham (Orchidaceae) at the edge and interior of a seasonal semideciduous forest fragment in Southern Brazil. Rev Bras Bioc 13(1):33–40
86. Sorgato JC, Barros F (2020) Conservation of *Schomburgkia crispa* Lindl. (Orchidaceae) by reintroduction into a fragment of the Brazilian Cerrado. J Nat Conserv 53:125754. https://doi.org/10.1016/j.jnc.2019.125754
87. Dorneles LT, Trevelin V (2011) Aclimatização e reintrodução de *Cattleya intermedia* Graham ex Hook (Orchidaceae) obtidas por propagação *in vitro*. Iheringia, Sér Bot 66(2):167–174
88. Ribeiro IS, Ribeiro LM, Ramos JCM, Soares JS, Sorgato JC (2024) The Brazilian native orchid *Brassavola tuberculata* Hook.: ornamental potential and reintroduction. Ornamental Hortic 30:e242768. https://doi.org/10.1590/2447-536X.v30.e242768
89. Parthibhan S, Kumar TS, Rao MV (2014) Phenology and reintroduction strategies for *Dendrobium aqueum* Lindley – an endemic, near-threatened orchid. J Nat Conserv 22:274–278. https://doi.org/10.1016/j.jnc.2014.11.003
90. Vieira C, Bocayuva M, Veloso T, Moreira B, Freitas E, Bazzolli D, Kasuya MC (2020) Fungus used for germination is supplanted after reintroduction of *Hadrolaelia jongheana* (Orchidaceae). Rev Agrar Acad 3:148–161. https://doi.org/10.32406/v3n62020/148-161/agrariacad
91. Scheffknecht S, Winkler M, Hülber K, Mata Rosas M, Hietz P (2010) Seedling establishment of epiphytic orchids in forests and coffee plantations in Central Veracruz, Mexico. J Trop Ecol 26(1):93–102. https://doi.org/10.1017/S0266467409990332
92. Aria M, Cuccurullo C (2017) Bibliometrix: an R-tool for comprehensive science mapping analysis. J Informetr 11(4):959–975

Chapter 14

Medium-Term Preservation by the In Vitro Minimum Growth Method of *Cedrela odorata* L.

Esmeralda J. Cruz-Gutiérrez, Claudia Berenice Espitia-Flores, Humberto Antonio López-Delgado, Luis Alberto Gómez-Reyes, Juan M. Pichardo-González and Liliana Muñoz-Gutiérrez

Abstract

The species *Cedrela odorata* L. is characterized by having one of the precious woods with a high commercial value. It has been overexploited for timber production, which is why its population has declined. Consequently, it has been included in the Mexican Official Standard NOM-059-SEMARNAT-2010 on environmental protection and native species of flora and fauna as a species subject to special protection and listed in the Convention on International Trade in Endangered Species of wild fauna and flora. For this reason, the ex situ conservation of plant species listed in SEMARNAT's NOM 059 is of utmost importance, guaranteeing the survival of threatened and/or endangered species Botanic Gardens Conservation International (BGCI) (n.d.). In this context, the medium-term conservation method, in this case, minimum growth in vitro, is a complementary and efficient alternative to safeguard the germplasm of (Engelmann, 2011) species found in SEMARNAT's NOM 059. It manages to conserve plant material from nine months to one year in the same tube and the same culture medium.

Keywords Germplasm banks, Forestry, Ex situ conservation

1 Introduction

Ex situ conservation of plant genetic resources is an essential strategy to ensure the survival of threatened or endangered species, as well as species used for food and agricultural development, thus ensuring global biodiversity [1]. There are different techniques for ex situ conservation, one of which is in vitro minimal tissue growth. This technique allows for the conservation of plant material by extending subcultures, lowering conservation costs, and reducing the risks of contamination and somaclonal variation [2].

Furthermore, *Cedrela odorata* L. is one of the species characterized by its precious wood, which has been overexploited, leading to its inclusion in SEMARNAT's NOM 059 in the category: Subject

Marco A. Ramírez-Mosqueda et al. (eds.), *Conservation of Plant Genetic Resources*, Methods in Molecular Biology, https://doi.org/10.1007/978-1-0716-5130-8_14,

to Special Protection [3]. In Mexico, it is naturally distributed along the Gulf and Pacific coasts [4]. It is used for the production of fine furniture, musical instruments, and handicrafts, and has medicinal properties [5, 6]. Due to all these characteristics and the fact that it is included in SEMARNAT's NOM 059, it is extremely important to safeguard plant material in specialized germplasm banks in the area to ensure the survival of the species.

2 Materials

2.1 Plant Material

1. The germplasm used may be seeds or tissue from a healthy, vigorous, pest-free juvenile donor plant. For establishment, young shoots (with one or two buds) or seeds are used.

2.2 Disinfection of Surfaces and Equipment

1. Liquid detergent (*see* **Note 1**). 70% ethanol solution (*see* **Note 2**). 1.5% sodium hypochlorite solution (*see* **Note 2**). Sterile Distilled Water (*see* **Note 3**). Captan® solution ($3\,g{\cdot}L^{-1}$. *see* **Note 4**). pH adjustment solutions: 1 N sodium hydroxide (NaOH) and 0.1 N hydrochloric acid (HCl).

2.3 Culture Medium

1. The culture medium used was a Murashige and Skoog [7] and a Woody Plant Medium [8] (Table 1).
2. Medium additives (Table 1) used for the different stages of cultivation: Seed germination (GS), bud establishment (EY), multiplication (M), and minimum growth (CM).

2.4 Equipment

1. Culture room: Laminar airflow hood.
2. Surgical instruments: stainless steel forceps, surgical scalpels, removable sterile surgical blades, and sterile Petri dishes.

Table 1
Culture medium additives used for the different culture stages of *Cedrela odorata* L

Culture medium	Components
Seed establishment medium	Murashige and Skoog medium [7] with 30 g/L of sucrose, 8 $g{\cdot}L^{-1}$ of agar, and a pH of 5.7
Explant establishment medium	Woody Plant Medium [8] or Murashige and Skoog medium [7], added with 20 $g{\cdot}L^{-1}$ of sucrose, 1 g/L of activated carbon, 1 $mL{\cdot}L^{-1}$ of Plant Preservative Mixture™, 1 g/L of Polyvinylpyrrolidone, 8 $g{\cdot}L^{-1}$ of agar, and a pH of 5.7
Propagation medium	Murashige and Skoog medium [7], added with 30 g/L of sucrose, 0.5 $mg{\cdot}L^{-1}$ of Benzylaminopurine (BA), 8 g/L of agar, and a pH of 5.7
Minimum growth medium	Murashige and Skoog medium [7], with 15 g/L of sucrose, 15 $g{\cdot}L^{-1}$ of Mannitol®, 8 g/L of agar, and a pH of 5.7

3. Preparation of the culture medium: granatary and analytical scales for weighing, spatulas, magnetic stirrers, hot plate with stirring, potentiometer, microwave oven, graduated beaker, and autoclave.
4. Culture vessels: Beakers (capacity 1000 mL), reagent bottles, culture tubes, and Petri dishes.
5. Glassware cleaning: scrub the glassware with a commercial liquid detergent solution and rinse it with plenty of tap water.
6. Various materials: Film or plastic film for packaging food.

3 Methods

3.1 Disinfection of Plant Material

3.1.1 Seeds

1. The seeds are washed with liquid detergent (*see* **Note 1**) for a period of 10 min, repeating this process twice.
2. Left to soak for 16 h in double-distilled water.
3. Immersed in a Captan® 3 g/L solution for 50 min (*see* **Note 4**).
4. Placed in a 70% ethanol solution for a period of 45 s.
5. Introduced into a 30% sodium hypochlorite solution for a period of 13 min.
6. Wash the seeds three or four times with sterile distilled water to remove residues of disinfectant solutions (*see* **Note 3**).
7. Establishing Seeds in the growing medium (*see* **Note 5**).
8. The in vitro Seedlings, free of microorganisms, and more vigorous, were selected for the multiplication of the material.

3.1.2 Explants

1. The plant material is collected from a plant in a field free of pests and diseases, vigorous and nourished, making it the perfect donor plant.
2. The plant material is washed with liquid detergent (*see* **Note 1**) for 10 min.
3. Introduced into a Captan® 3 g/L solution for 45 min.
4. Immersed in a 70% ethanol solution for 30 s (*see* **Note 2**).
5. Immersed in a 30% sodium hypochlorite solution for 10 min (*see* **Note 2**).
6. Wash three or four times with sterile distilled water to remove residues of disinfectant solutions (*see* **Note 3**).
7. Establishment of plant material in the culture medium (*see* **Note 5**).
8. The in vitro plants, free of microorganisms and more vigorous, were selected for the multiplication of the material.

Fig. 1 (**A** and **B**) Seedlings of *Cedrela odorata* L., one year old, under minimum growth conditions. (**C**) Seedlings under in vitro conservation

3.2 Proliferation of Shoots

1. The shoots with a size of 15 mm are established in a semi-solid medium (*see* Table 1), with an incubation period of 30 days (*see* **Note 6**).
2. The shoots are incubated at a temperature of 24 ± 2°C and a photoperiod of 16 h of light and 8 h of darkness with a light intensity of 2000 lux.
3. After 30 days of incubation, new shoots are obtained per explant, which are separated and used in subsequent processes.

3.3 Minimum Growth

1. Shoots with a size of 15 mm in length and with at least one growth point (bud) are placed in a minimal growth medium (Table 1).
2. The shoots are incubated at a temperature of 18°C and 24°C, with a photoperiod of 16 h of light and 8 h of darkness with a light intensity of 2000 lux, for a period of 12 and 24 months. **Seedlings of *Cedrela odorata* L., 1-year-old, under minimum growth conditions are shown in** Fig. 1.

4 Notes

1. The detergent is a commercial brand for washing dishes.
2. Disinfectant solutions are prepared with distilled water.
3. The water is sterilized in an autoclave at 1.5 kg/cm^2 pressure at a temperature of 121°C for 15 min.
4. Captan® is a commercial fungicide.
5. The disinfection protocol using the 70% ethanol solution is carried out inside the laminar flow hood.
6. The tissues are dissected, with a size of 15 mm, within the laminar flow hood and established in the culture medium (Table 1).

Acknowledgments

This work was supported by the National Genetic Resources Center of National Institute of Forestry, Agriculture and Livestock Research.

References

1. FAO (2010) The second report on the state of the world's plant genetic resources for food and agriculture. Food and Agriculture Organization of the United Nations.
2. Villalobos VM, Engelmann F (2007) Cryopreservation of plant germplasm: a practical guide. Bioversity International.
3. SEMARNAT (2010) NORMA Oficial Mexicana NOM-059-Protección ambiental-Especies nativas de México de flora y fauna silvestres-Categorías de riesgo y especificaciones para su inclusión, exclusión o cambio-Lista de especies en riesgo. Secretaría del Medio Ambiente y Recursos Naturales. Diario Oficial de la Federación. Cd. Mx., México. http://www.profepa.gob.mx/innovaportal/file/435/1/NOM_059_SEMARNAT_2010.pdf. Accessed 27 May 2025
4. Pennington TD, Sarukhán J (2005) Árboles tropicales de México: manual para la identificación de las principales especies, 3rd edn. UNAM-Fondo de Cultura Económica, México, DF, México, p 523
5. Romo-Lozano JL, Vargas-Hernández JJ, López-Upton J, Ávila A ML (2017) Estimación del valor financiero de las existencias maderables de cedro rojo (*Cedrela odorata* L.) en México. Madera y Bosques 23(1):111–120. https://doi.org/10.21829/myb.2017.231473
6. Ruiz-Jiménez CA, De los Santos-Posadas HM, Parraguirre-Lezama JF, Saavedra-Millán FD (2018) Evaluación de la categoría de riesgo de extinción del cedro rojo (*Cedrela odorata*) en México. Revista Mexicana de Biodiversidad 89(3):938–949. https://doi.org/10.22201/ib.20078706e.2018.3.2192
7. Murashige T, Skoog F (1962) A revised medium for rapid growth and bio assays with tobacco tissue cultures. Physiol Plant 15(3):473–497
8. Lloyd G, McCown B (1981) Woody plant medium: a mineral nutrient formulation for microculture of woody plant species. HortScience 16.453

Check for updates

Chapter 15

In Vitro Preservation of Vanilla (*Vanila planifolia* Jacks. Ex Andrew) by Slow Growth

Mario Alejandro Hernández-Chontal, Lourdes Georgina Iglesias-Andreu, Zelzin Erendira Fernández-Villa, Gabriela Sánchez-Viveros and José Roberto Bautista-Aguilar

Abstract

Mexico is the center of origin and domestication of *Vanilla planifolia* Jacks. However, according to NOM-059-SEMARNAT-2010, it is listed as being at risk and subject to special protection. It is therefore important to undertake conservation strategies. Slow growth techniques are an option for short- and medium-term storage. They have as a starting point a good in vitro culture protocol, to which modifications are made to cause a decrease in the growth and metabolism of the plant and thus extend the culture period. This chapter describes the process of medium-term conservation of *V. planifolia* by the slow-growth technique.

Keywords Slow growth, Culture medium, Vanillin, Conservation protocol

1 Introduction

One of the techniques that has proven effective for ex situ germplasm conservation is the slow-growth in vitro conservation technique, as it allows the aseptic conservation of many individuals under controlled conditions for several months or even years in a reduced area [1]. To establish an effective protocol for in vitro preservation of slow-growing plants, it is necessary to reduce plant metabolism. For this purpose, low temperatures and light intensity have been used, as well as the addition of osmotic compounds such as mannitol or sucrose and different growth retardants in the culture medium [2].

V. planifolia is a species of great economic value for its pods, from which vanillin, a flavoring compound, is extracted, widely used in the pharmaceutical, food, and cosmetic industries [3]. However, wild populations of *V. planifolia* have been severely affected by excessive and illegal harvesting, to the point of bringing the species to the brink of extinction in the wild [4].

Marco A. Ramírez-Mosqueda et al. (eds.), *Conservation of Plant Genetic Resources*, Methods in Molecular Biology, vol. 3011, https://doi.org/10.1007/978-1-0716-5130-8_15,

Several studies have been conducted on the in vitro conservation of using osmotic agents such as mannitol at 10 g L^{-1} [5]. Excellent results have been found with the addition of growth inhibitors such as abscisic acid at 2 mg L^{-1} to MS medium [6, 7]. As well as employing blue LED light for slow growth conservation [8]. This chapter describes an effective protocol for the conservation of due to slow growth. This protocol can be effective for the conservation of different species.

2 Materials

2.1 Plant Material

Use shoots of of 0.5–0.8 cm in length previously established. The plants used should have the least number of subcultures possible (two to three subcultures) to avoid genetic variations.

2.2 Culture Media

1. Basal culture medium: Murashige and Skoog (MS) [9] (see Table 1).
2. Medium additives used: MS medium supplemented with abscisic acid (ABA) and sucrose (see Table 2).
3. pH adjustment solutions: 1 N sodium hydroxide (NaOH) and 1 N hydrochloric acid (HCl).

Table 1

Components for preparing 100 mL of MS culture medium solution

Nutrients	g/L
Solution 1	
KNO_3	19.0
NH_4NO_3	16.5
Solution 2	
$MgSO_4$ or $MgSO_4{\cdot}7H_2O$	1.8 or 3.7
$MnSO_4{\cdot}H_2O$	0.169
$ZnSO_4{\cdot}7H_2O$	0.086
$CuSO_4{\cdot}5H_2O$	0.00025
Solution 3	
$CoCl_2{\cdot}6H_2O$ or $CoCl_2{\cdot}2H_2O$	A grain
KI	0.0083
$CaCl_2$ or $CaCl_2{\cdot}2H_2O$	3.32 or 4.4
Solution 4	
KH_2PO_4	1.7

Table 1
continued

Nutrients	g/L
H_3BO_3	0.062
$Na_2Mo_4{\cdot}2H_2O$	0.0025
Solution 5	
$FeSO_4{\cdot}7H_2O$	0.27
Na_2EDTA or $Na_2EDTA{\cdot}2H_2O$	0.37 or 0.41
Solution 6	
Myo-inositol	1.0
Thiamine	0.001
Nicotinic acid	0.005
Pyridoxine	0.005
Glycine	0.02

Table 2
Substances that complement culture media

MS	100 %
Abscisic acid (ABA)	3 mg/L
Sucrose	30 g/L
Gelrite	2.5 g/L
pH	5.7

2.3 Lighting Conditions and Photoperiod

1. All cultures were maintained under a light irradiation of 50 ± 5 μmol/m^{-2} s^{-1}, provided by LED blue (460 nm) and a photoperiod (16 h light/8 h dark).

2.4 Equipment

1. Culture room: airflow cabinet.
2. Medium preparation: analytical balance, spatula, hot stirrer plate, pH meter, micropipettes, magnetic stirrer, refrigerator, and autoclave.
3. Glassware: bottles (200 mL capacity), beakers (500 mL and 1 L capacity), reagent bottles, and measuring cylinders.
4. Micropipette of 10–100 μL.
5. Culture vessels: tubes (22 × 220 mm).

3 Methods

3.1 Culture Media

3.1.1 Preparation of MS

1. We prepare concentrated solutions of MS medium adding 10 mL/L of each solution (1–6) (*see* **Notes 1 and 3**). The composition of the medium is shown in Table 1.

3.1.2 Plant Growth Regulators (PGR) Stock Solutions

1. Abscisic acid (ABA, 0.1 mg/mL): dissolve 10 mg of ABA in 1 mL 1 N NaOH and then add water to a volume of 100 mL.

3.1.3 Preparation Culture Media (Different Stages)

1. Prepare the culture medium from manufacturer specifications (*see* **Note 2**), add the supplementary components (see Table 2).
2. Distribute 10 mL of culture medium in tubes (22 × 220 mm).
3. Autoclave at 121°C with a pressure of 15 kPa for 20 min (all culture media).
4. After autoclaving, the glass tubes are kept in a clean room until use.

3.2 Explants

3.2.1 Plant Material

1. Shoots of 0.5–0.8 cm in length previously established of *V. planifolia*.

3.3 Culture (Incubation)

1. Shoot tips of 0.5–0.8 cm long.
2. Must be transferred to tubes (22 × 220 mm) containing 10 mL of incubation medium (see Table 2).
3. They should be incubated for 25 weeks at a temperature of 22°C in blue LED light for preservation (Fig. 1).

Fig. 1 *Vanila planifolia* conserved in slow growth

4 Notes

1. We use concentrated solutions for the preparation of the culture medium (see Table 1).
2. Semi-solid medium should be used by adding a gelling agent (Gelrite) (see Table 2).
3. For solution 6, heat separately, then mix and make up to volume (see Table 1).

References

1. Cruz-Cruz CA, Gonzalez-Arnao MT, Engelmann F (2013) Biotechnology and conservation of plant biodiversity. Resources 2:73–95. https://doi.org/10.3390/resources2020073
2. Rodrigues PHV, Arruda F, Forti VA (2018) Slow-grown in vitro conservation of *Heliconia champneiana* cv. Splash under different light spectra. Sci Agr 75:163–166. https://doi.org/10.1590/1678-992x-2016-0394
3. Buccellato F (2018) Vanilla in perfumery and beverage flavor. In: Havkin-Frenkel D, Belanger FC (Eds.) Handbook of vanilla science and technology. Wiley, New York, pp 367–373. https://doi.org/10.1002/9781119377320.ch19En
4. Soto-Arenas MA (1999) Filogeografía y recursos genéticos de las vainillas de México. Informe final SNIB-CONABIO proyecto No. J101. Instituto Chinoın, A.C., Herbario de la Asociación Mexicana de Orquideología, A.C. México, DF.
5. Divakaran M, Babu KN, Peter KV (2006) Conservation of Vanilla species, in vitro. Sci Hort 110:175–180. https://doi.org/10.1016/j.scienta.2006.07.003
6. Bello-Bello J, García-García G, Iglesias-Andreu L (2015) Conservacion de vainilla (*Vanilla planifolia* Jacks.) bajo condiciones de lento crecimiento in vitro. Rev Fitotecn Mex 38:165–171. http://www.scielo.org.mx/scielo.php?script=sci_arttext&pid=S0187-73802015000200006&lng=es&nrm=iso
7. Bautista-Aguilar JR, Iglesias-Andreu LG, Martínez-Castillo J, Ramírez-Mosqueda MA, Ortiz-García MM (2021) In vitro conservation and genetic stability in *Vanilla planifolia* Jacks. HortScience 56:1494–1498. https://doi.org/10.21273/HORTSCI16118-21
8. Bautista-Aguilar JR, Iglesias-Andreu LG, Flores-Estévez N, Sánchez-Coello NG (2024) Effect of led lighting on minimal growth for in vitro conservation of vanilla genotypes. Propag Ornamental Plants 24:13–19
9. Murashige T, Skoog F (1962) A revised medium for rapid growth and bio assays with tobacco tissue cultures. Physiol Plant 15:473–497. https://doi.org/10.1111/j.1399-3054.1962.tb08052.x

Check for updates

Chapter 16

Medium-Term Preservation by the In Vitro Minimum Growth Method of *Cordia alliodora* (Ruiz and Pavón) Oken

Claudia Berenice Espitia-Flores, Esmeralda J. Cruz-Gutiérrez, Marcelina Vélez-Torres, Luis Alberto Gómez-Reyes, Juan M. Pichardo-González and Jorge David Cadena-Zamudio

Abstract

The species *Cordia alliodora* (Ruiz and Pavón) Oken is found in the Neotropical region of great ecological and economic importance. Its wood is highly valued for its high quality, rapid growth, and adaptability to diverse ecosystems. In Mexico, it has a wide natural distribution, extending to Brazil and the Antilles. However, it also faces problems of deforestation, land-use change, and overexploitation. Ex situ conservation of this species is an alternative to ensure its survival, utilizing the resource without affecting ecosystems. There are several techniques for conserving plant genetic resources. One is in vitro minimal growth, which allows tissue to be preserved in the same tube and medium for up to a year without subculturing the material. Therefore, the objective of this work is to obtain a conservation protocol under minimal growth conditions, with the aim of conserving in a plant germplasm bank.

Keywords Forest conservation, Plant germplasm banks, In vitro culture

1 Introduction

Ex situ conservation techniques allow the preservation of plant material in germplasm banks, seed banks, or botanical gardens [1, 2]. In vitro germplasm banks are an option for medium-term plant conservation, utilizing strategies such as minimum growth [3]. To achieve this, modified culture media are used that restrict the normal growth of plants conserved under this condition. The proper maintenance and survival of the plant material must be ensured; the only requirement is that the plant size be reduced, but that the plant tissue retains all its metabolic functions [4, 5]. Therefore, it is necessary to seek modifications to the culture medium that reduce size (plant growth) without compromising survival.

The black laurel (*Cordia alliodora* (Ruiz and Pavón) Oken) is a tree with timber value in Central and South America. It is associated

Marco A. Ramírez-Mosqueda et al. (eds.), *Conservation of Plant Genetic Resources*, Methods in Molecular Biology, vol. 3011, https://doi.org/10.1007/978-1-0716-5130-8_16,

with crops such as cocoa, banana, and coffee in agroforestry systems [6]. Despite having abundant natural populations, there is a possibility of these decreasing due to issues of land use change [7]. In addition, it is important to conserve germplasm of this species due to its high timber value and possible future use. Therefore, the objective of this chapter was to establish an in vitro conservation protocol for *Cordia alliodora* (Ruiz and Pavón) Oken by minimum growth.

2 Materials

2.1 Plant Material

Cordia alliodora (Ruiz and Pavón) Oken is established from seeds, coming from young, healthy, vigorous, and pest-free plants.

2.2 Sanitizing Agents

1% Powder Util® Detergent (*see* **Note 1**).
70% ethanol solution (*see* **Note 2**).
30% sodium hypochlorite solution (*see* **Note 2**).
Sterile distilled water (*see* **Note 3**).
Captan® Solution (3 g/L).

2.3 Culture Medium

1. Medium Murashige and Skoog [8].
2. To prepare the culture medium for the different stages of cultivation, certain substances are added (Table 1): Seed germination (GS), multiplication (M), and minimum growth (CM).

2.4 Equipment

1. Culture room: Laminar airflow hood.
2. Surgical tools: stainless steel forceps, surgical scalpels, removable sterile surgical blades, and sterile Petri dishes.
3. Preparation of the culture medium: granular and analytical scales for weighing, spatulas, magnetic stirrers, hot plate with stirring, potentiometer, microwave oven, graduated beaker, and autoclave.

Table 1
Culture medium additives used for the different culture stages of *Cordia alliodora* (Ruiz and Pavón) Oken

Culture medium	Components
Seed establishment médium	Murashige and Skoog medium with 15 g/L of sucrose and 8 g/L of agar (pH 5.7)
Propagation medium	Murashige and Skoog medium, added with 15 g/L of sucrose, 0.3 mg/L of Benzylaminopurine (BAP), and 8 g/L of agar (pH 5.7)
Minimum growth medium	Murashige and Skoog medium at 50% of its salt concentration, with 15 g/L of sucrose, 15 g/L of Mannitol®, and 8 g/L of agar (pH 5.7)

4. Culture vessels: Beakers (capacity 1000 mL), reagent cups, culture tubes, and Petri dishes.

Miscellaneous materials: To clean glassware, it is suggested to scrub the glassware with a commercial liquid detergent solution and rinse it with plenty of tap water. To seal the culture tubes, use plastic wrap or film for packaging food.

3 Methods

3.1 Disinfection Plant Material

1. The seeds are washed for 10 min with a 1 % Util® powder detergent solution (*see* **Note 1**), rinsing them two or three times to remove any excess soap that may remain on the seeds, and this washing process is repeated twice.
2. Left to soak for 16 h in double-distilled water.
3. Immersed in a Captan® 3 g/L solution for 50 min (*see* **Note 4**).

3.2 In Laminar Flow Hood

1. Placed in a 70% ethanol solution for 45 s (*see* **Note 5**).
2. Introduced into a 30% sodium hypochlorite solution for 13 min (*see* **Note 2**).
3. Wash the seeds three or four times with sterile distilled water to remove residues of disinfectant solutions.
4. Establishing seeds in the growing medium.
5. In vitro seedlings, free of microorganisms and more vigorous, are selected for the multiplication of the material (*see* **Note 6**).

3.3 Proliferation of Shoots

1. Explants of 15 mm height are placed in a semi-solid medium (see Table 1), with an incubation period of 30 days.
2. The incubation conditions of the explants are: a temperature of 24 ± 2°C and a photoperiod of 16 h of light and 8 h of darkness with a light intensity of 2000 lux.
3. After the incubation period, the new shoots obtained by explant are separated and used in subsequent processes.

3.4 Minimum Growth

1. For this stage, 20 mm long explants with at least one growth point (bud) are used, and placed in a minimal growth medium (Table 1).
2. The explants are incubated at a temperature of 18°C, with a photoperiod of 16 h of light and 8 h of darkness with a light intensity of 2000 lux, for a period of 12 months(Fig. 1).

Fig. 1 (**A**) One-year-old *Cordia alliodora* (Ruiz and Pavón) Oken plant in minimal growth conditions, (**B**) *Cordia alliodora* (Ruiz and Pavón) Oken seedlings in conservation medium under minimal growth conditions

4 Notes

1. Powder detergent is a brand name for washes dishes.
2. Disinfectant solutions are prepared with distilled water.
3. The water is sterilized in an autoclave at 1.5 kg/cm^2 of pressure and at a temperature of 121°C for 15 min.
4. Captan® is a commercial fungicide.
5. The disinfection protocol using the 70% ethanol solution is carried out inside the laminar flow hood.
6. Tissues are dissected, 15 mm to 20 mm in size depending on the stage, within the laminar flow hood and established in the culture medium (Table 1).

References

1. Ogwu MC, Osawaru ME (2023) Disease outbreaks in ex-situ plant conservation and potential management strategies. In: Sustainable utilization and conservation of Africa's biological resources and environment. Springer Nature Singapore, Singapore, pp 497–515
2. Hurdu BI, Coste A, Halmagyi A, Szatmari PM, Farkas A, Puşcaş M, Butiuc-Keul A (2022) *Ex situ* conservation of plant diversity in Romania: a synthesis of threatened and endemic taxa. J Nat Conserv 68:126211
3. Yadav S, Priya K, Dhiman R, Godara S, Kandari P, Thakur A (2024) Long-and medium-term storage of germplasm for conservation of tree species. In: Biotechnological approaches for sustaining forest trees and their products. Springer Nature Singapore, Singapore, pp 241–275
4. Benelli C, Tarraf W, Izgu T, De Carlo A (2022) *In vitro* conservation through slow growth storage technique of fruit species: an overview of the last 10 years. Plants 11(23):3188
5. Chauhan R, Singh V, Quraishi A (2019) In vitro conservation through slow-growth storage. In: Faisal M, Alatar AA (eds) Synthetic seeds: germplasm regeneration, preservation and prospects. Springer, Cham, pp 397–416
6. Pineda-Herrera E, Manzano-Méndez F, Valdez-Hernández JI, Beltrán-Rodríguez LA

(2018) Crecimiento diamétrico de *Cordia alliodora* (Ruiz & Pav.) Oken en un sistema agroforestal de Oaxaca, México. Revista Forestal Mesoamericana Kurú 15(37):25–33

7. Peralta DEN, Verdesoto CAC, Pilozo DMS (2022) Determination of forest cover in 10 farms under traditional agroforestry systems in the Corotú area of El Empalme canton, Guayas. Centrosur Agraria 1(12):45–58.
8. Murashige T, Skoog F (1962) A revised medium for rapid growth and bio assays with tobacco tissue cultures. Physiol Plant 15(3):473–497. https://doi.org/10.1111/j.13993054.1962.tb08052.x

Medium-Term Preservation by the In Vitro Minimum Growth Method of *Gmelina arborea* Roxb

Claudia Berenice Espitia-Flores, Esmeralda J. Cruz-Gutiérrez, Manuel de Jesús Bermúdez-Guzmán, Luis Alberto Gómez-Reyes, Juan M. Pichardo-González and José Luis Aguirre-Noyola

Abstract

Gmelina arborea Roxb., commonly known as Melina, is a fast-growing, commercially important tropical forest species widely used in reforestation programs for timber, pulp and paper production, and other uses. Its adaptability to diverse site conditions and high productivity make it attractive for industrial plantations. Due to its importance, species conservation is essential. Plant tissue culture is a biotechnological tool for the propagation and conservation of forest species, allowing the production of plants with uniform genetic characteristics and high growth potential in a short period of time. One strategy is the minimum growth protocol for medium-term conservation (from 9 months to 2 years). These protocols aim to reduce the metabolic rate of explants, decreasing the frequency of manipulation due to the prolonged subculture time, minimizing the risk of contamination, and minimizing costs. This type of in vitro conservation ensures the availability of important genetic material for future use in reforestation or conservation programs.

Keywords Conservation, Forestry, Ex situ germplasm banks

1 Introduction

Currently, adverse environmental conditions caused by climate change have affected forest plantations worldwide [1]. In particular, abrupt temperature changes in various forested regions have caused a shift in tree growth patterns [2]. These timber resources are extremely important to humanity due to their widespread use in construction, fuel, furniture making, and sometimes for their medicinal properties [3].

Gmelina arborea Roxb is an important timber species due to the quality of its wood and rapid growth. It has also been reported that it is used ethnopharmacologically [4, 5]. However, studies conducted in various parts of the world highlight its susceptibility to rising temperatures. Furthermore, a negative correlation has been noted

Marco A. Ramírez-Mosqueda et al. (eds.), *Conservation of Plant Genetic Resources*, Methods in Molecular Biology, vol. 3011, https://doi.org/10.1007/978-1-0716-5130-8_17,

between rising temperatures and the growth of this species, which affects carbon sequestration [6]. Future global warming is expected to affect the carbon sequestration potential of tropical forests in the face of climate change. Due to the great complexity of avoiding global environmental changes, it is recommended to try to conserve forest resources through different ex situ conservation techniques, including in vitro germplasm banks [5]. Therefore, the objective of this chapter was to establish a protocol for the in vitro germplasm conservation of *Gmelina arborea* Roxb.

2 Materials

2.1 Plant Material, Sanitizing Agents, and Culture Medium

1. Melina seeds are collected from young, healthy, vigorous, and pest-free plants to be established in in vitro culture.
2. 1% powder detergent Util® solution (*see* **Note 1**), 70% ethanol solution (*see* **Note 2**), 30% sodium hypochlorite solution (*see* **Note 2**), sterile distilled water (*see* **Note 3**), and a Captan® solution (0.03 g/L).
3. Medium Murashige and Skoog [7].
4. In the different stages of in vitro culture for Melina, different substances were added to the culture medium (Table 1), seed germination (GS), multiplication (M), and minimum growth (CM).

2.2 Equipment

1. Culture room: Laminar airflow hood.
2. Surgical tools: stainless steel forceps, surgical scalpels, removable sterile surgical blades, and sterile Petri dishes.
3. Preparation of the culture medium: granular and analytical scales for weighing, spatulas, magnetic stirrers, hot plate with stirring, potentiometer, microwave oven, graduated beaker, and autoclave.

Table 1
Culture medium additives used for the different culture stages of *Gmelina arborea* Roxb

Culture medium	Components
Seed establishment medium	Murashige and Skoog medium [7] with 15 g/L of sucrose, 8 g/L of agar, and a pH of 5.7
Propagation medium	Murashige and Skoog medium [7], added with 15 g/L of sucrose, 3 mg/L of naphthaleneacetic acid (NAA), 8 g/L of agar, and a pH of 5.7
Minimum growth medium	Murashige and Skoog medium [7], with 15 g/L of sucrose, 8 g/L of agar, and a pH of 5.7

4. Culture vessels: Beakers (1000 mL capacity), reagent cups, culture tubes, and Petri dishes.
5. Miscellaneous materials: To clean glassware, it is suggested to scrub the glassware with a commercial liquid detergent solution and rinse it with plenty of tap water. To seal the culture tubes, use plastic wrap or film for packaging food.

3 Methods

3.1 Disinfection Plant Material

1. The seeds are washed twice for 10 min with a 1% powder detergent Util® solution (*see* **Note 1**), followed by two or three rinses.
2. Left to soak for 16 h in double-distilled water.
3. Immersed in a Captan® 0.03 g/L solution for 30 min (*see* **Note 4**).
4. Placed in a 70% ethanol solution for one minute (*see* **Note 5**).
5. Introduced into a 30% sodium hypochlorite solution for 3 min.
6. Wash the seeds three or four times with sterile distilled water to remove residues of disinfectant solutions.
7. Establishing seeds in the growing medium.
8. The seedlings generated in vitro are selected for the multiplication of the material.

3.2 Shoots Proliferation

1. Melina shoots are cut at 15 mm height and placed in a semi-solid medium with growth regulators (see Table 1) to encourage the formation of new shoots. Kept in this medium for 30 days.
2. The incubation conditions of the explants are: a temperature of 24 ± 2°C and a photoperiod of 16 h of light and 8 h of darkness with a light intensity of 2000 lux.
3. After 30 days, the generated shoots are separated from the initial shoot and are subsequently used in the next processes.

3.3 Minimum Growth

1. The shoots are required to be 15 mm long and contain at least one growth point (bud), either lateral or apical (*see* **Note 6**), and are placed in the minimum growth medium (see Table 1).
2. The shoots are incubated at a temperature of 18 ± 2°C, with a photoperiod of 16 h of light and 8 h of darkness with a light intensity of 2000 lux.
3. The shoots are kept in the same medium for a period of 12 months (Fig. 1).

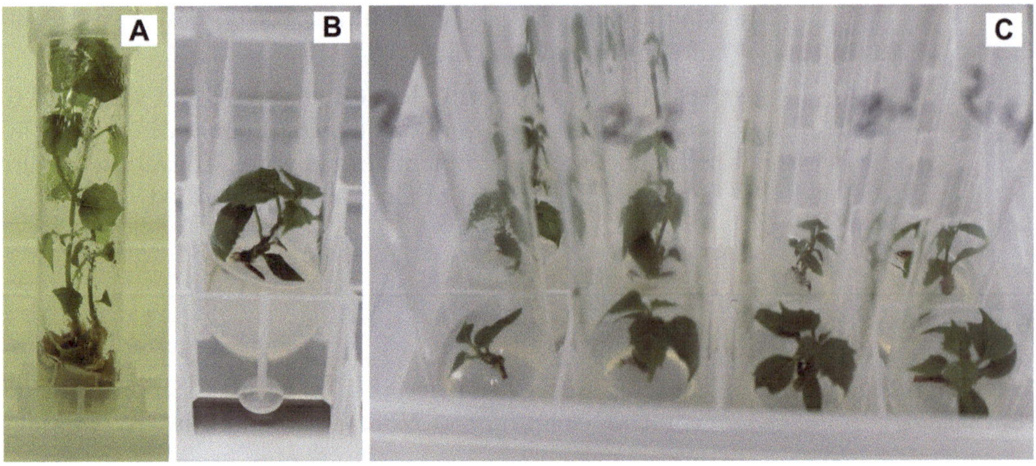

Fig. 1 (**A**) *Gmelina arborea* Roxb. seedling in optimal in vitro culture conditions for its development, (**B** and **C**) seedlings in culture medium for minimum growth

4 Notes

1. Powder detergent is a brand name used for washing dishes.
2. Disinfectant solutions are prepared with distilled water.
3. The water is sterilized in an autoclave at 1.5 kg/cm^2 of pressure and at a temperature of 121°C for 15 min.
4. Captan® is a commercial fungicide.
5. The disinfection protocol using the 70% ethanol solution is carried out inside the laminar flow hood.
6. Tissues are dissected, 15 mm in size, within the laminar flow hood and established in the culture medium (see Table 1).

Acknowledgments

This work was supported by the National Genetic Resources Center of National Institute of Forestry, Agriculture and Livestock Research.

References

1. Ramírez G, Lozano-Contreras MG, Ramírez JH (2019) Potential areas for growing *Gmelina arborea* Roxb., under rainfed conditions in Tabasco, Mexico. Agric Sci 10(9):1206–1216. https://doi.org/10.4236/as.2019.109090
2. Kumar R, Kumar A, Saikia P (2022) Deforestation and forests degradation impacts on the environment. Environmental degradation: challenges and strategies for mitigation. Springer International Publishing, Cham, pp 19–46
3. Freitas TR, Santos JA, Silva AP, Fraga H (2021) Influence of climate change on chestnut trees: a review. Plants 10(7):1463

4. George JP, Yrjänä L, Beuker E, Rusanen M (2024) *Ex situ* conservation of forest genetic resources in Finland: Evaluation and adaptation of the current strategy. In: Natural Resources and Bioeconomy Studies 27/2024. Natural Resources Institute Finland, Helsinki. p 40
5. Kabir CR, Islam M, Hassan MB, Rahman M (2025) Impact of local climate and large-scale climate drivers on the growth of *Gmelina arborea* Roxb. in a South Asian moist tropical forest. Trees 39(2):41
6. Sahu R, Vinchurkar K, Mishra DK, Dixit P (2022) *In vitro* mutagenicity of *Gmelina arborea* Roxb (gambhari) extract by ames assay. In: Jain S, Mishra DK, Dixit P, Shukla R (eds) Pharmaceutical research and innovations. RFI, New Delhi, p 29.
7. Murashige T, Skoog F (1962) A revised medium for rapid growth and bio assays with tobacco tissue cultures. Physiol Plant 15(3):473–497. https://doi.org/10.1111/j.13993054.1962.tb08052.x

Chapter 18

In Vitro Conservation of *Agave potatorum* Zucc and *Agave marmorata* Roezl

Daniel Aguilar-Jiménez, Gina Kelly Mendez-Castillo, Benito Reyes-Trejo, Ernestina Valadez-Moctezuma and José Luis Rodríguez-De la O

Abstract

In Mexico, mezcal agaves, especially wild ones, are overexploited, and there is a lack of conservation efforts. Furthermore, plantations of various agaves are currently being established, and if they reach flowering, multiple crosses could arise, and with them, the possible loss of species that have been well identified until now. Plant tissue culture represents an environmentally friendly biotechnology alternative that promotes the rescue and conservation of valuable plant genetic resources, preserving plant biodiversity and quality. The use of MS medium reduced to 50% and added with sorbitol creates a slow-growing culture medium, allowing in vitro plants of *Agave potatorum* and *Agave marmorata* to be maintained in excellent condition for more than 24 months.

Keywords Sorbitol, Slow growth, Biotechnology, Wild species

1 Introduction

Agave spp. species in Mexico are going through an important and critical stage, especially those used to make mezcal, a very popular alcoholic beverage around the world. The above has led to some species being considered in danger of extinction [1]. In the case of *Agave potatorum*, it is considered the most valuable agave species due to the quality of mezcal obtained [2]. Despite being small in size, it takes between 8 and 12 years to reach maturity to be harvested [3], causing its value to increase because it produces less mezcal compared to *Agave angustifolia* (the most common mezcal agave in Mexico), but *Agave potatorum* mezcal is considered to be of higher quality, and therefore the demand increases. However, despite being considered a wild plant genetic resource (*Agave potatorum*), it has a transcendental place in the national mezcal industry. It is appreciated as a key piece in the local and national economy, which is historically linked to the cultures of our country [4]. However,

Marco A. Ramírez-Mosqueda et al. (eds.), *Conservation of Plant Genetic Resources*, Methods in Molecular Biology, vol. 3011,
https://doi.org/10.1007/978-1-0716-5130-8_18,

it only reproduces by seed; there are very rare cases where it is seen with offspring from rhizomes.

Similarly, another important mezcal-producing agave is *Agave marmorata*, which is also considered a wild and recently valuable phytogenetic resource for obtaining mezcal. Before the value given to mezcal, the *Agave marmorata* was only used in gastronomy and as a medicinal plant in the Mixteca Poblana (Mexico), despite being a plant that does not require fertile soils and has little demand for water, no other value was attributed to it [5]. In addition, some people consider it a dual-purpose plant to obtain alcoholic beverages: pulque or mezcal. However, it is one of the wild agaves at risk of disappearing since its main form of reproduction is by seed; per rhizome it has few, approximately 2.7 offspring. In addition, to obtain seeds, their maturation period can take between 25 and 35 years in the wild, adding that the flowers are also used as food and in religious festivities, which makes their natural preservation more difficult [6, 7]. For this reason, alternatives for massive clonal multiplication and conservation have been proposed, with the in vitro culture of plant cells, tissues, or organs being a biotechnological tool characterized by maintaining the genetic characteristics of the cultivated plants, identical to the selected mother plants. In this way, in vitro cultivation becomes an alternative that can conserve valuable plant genetic resources when the physiology of the seeds does not allow it [8].

In this context, there are several published works for the in vitro propagation of different species of the *Agave* genus with great success [9], where the objective has practically been to increase the coefficient of clonal propagation. There are even studies that evaluate the effect of different concentrations of inorganic MS salts [10] on the propagation of agaves [11] and other species [12, 13] with the aim of finding the most effective and efficient protocol for the mass micropropagation of plants. However, for the conservation of plant genetic resources, it has been decided to use slow growth or minimum growth media, where the objective is not a massive and accelerated clonal propagation as in the previous case, but on the contrary, to reduce growth ensuring that the genetic and morphological characteristics of the plants are conserved [8]. For this case, there are some studies on the conservation of agaves in vitro [14]. In these studies, growth regulators and osmoregulators are mainly evaluated as inhibitors of plant growth, with abscisic acid (ABA), paclobutrazol (PAC), ancymidol (ANC), mannitol, and sorbitol being some of the most studied in different plant species [15–17]. However, no study was found where plant growth inhibitory substances were assessed in combination with different concentrations of MS medium. Therefore, the objective proposed in this study was to develop an efficient slow-growth protocol for the in vitro conservation of *Agave potatorum* Zucc and *Agave marmorata* Roezl.

2 Materials

2.1 Plants and Culture

1. *Agave potatorum* and *Agave marmorata* plants are collected between 8 and 12 months old. Correctly identify the species by experts before selecting the agave mother plants (Fig. 1a, b) (*see* **Note 1**).
2. Solution of 1 g/L of fungicide and 1 g/L of bactericide systemics action (Fig. 1c, d) (*see* **Note 2**).
3. Solution of 1 g/L of insecticide.
4. Ethanol 70% (v/v), sodium hypochlorite 0.6% (v/v), commercial detergent, and osmosis water.
5. Murashige and Skoog (MS) medium (1962) was used as the basic medium [10] (*see* **Note 3**).
6. Constituents of the nutrient medium for the different stages of culture: Basic medium with inorganic salts MS adding different constituents according to the stage of culture, including the establishment of explants (EE), shoot induction by direct organogenesis (IB), shoot multiplication (MB), and slow growth medium (MLC) (see Table 1).
7. Solutions to adjust pH: 1 N NaOH and 1 N HCl.

Fig. 1 Identification and selection of parent plants. (**a** and **b**) Identification of donor plants, *Agave potatorum* and *Agave marmorata*, respectively, and (**c** and **d**) *Agave potatorum* and *Agave marmorata* plants, respectively, in a nursery with phytosanitary treatment

Table 1
Culture medium additives used for the different stages of in vitro agave conservation

Constituents of the culture medium	Cultivation stage			
	Establishment of explants (EE)	Shoot induction (SI)	Multiplication of shoots (MSs)	Slow-growing medium (SGM)
MS concentration	100%	100%	100%	50%
Sucrose (g/L)	20	20	30	30
Agar-agar (g/L)	7	7	7	7
Indole-3-acetic acid (IAA) (mg/L)	–	1	0.3	–
6-Benzylaminopurine (BAP) (mg/L)	–	1	3	–
Sorbitol (g/L)	–	–	–	10

2.2 Equipment

1. Culture area: laminar air flow hood and stereoscopic microscope.
2. Dissection tools: stainless steel forceps, surgical scalpel, sterile surgical knives, and sterile Petri dishes.
3. Area for preparing culture medium: analytical balance, digital granular balance, magnetic stirrer, hot plate with stirring, pH meter, micropipettes with sterile tips, refrigerator, and autoclave.
4. Culture vessels: reagent bottles, Erlenmeyer flasks (500 mL and 1000 mL), beakers, graduated glass cylinders, and culture flasks (5.3 × 5 × 5.5 cm) with transparent polypropylene lids.
5. Wash glassware with a liquid detergent solution and rinse thoroughly with tap water.
6. Rinse the glassware with distilled water and dry it at room temperature for 24 h or for 30 min at 70°C in drying ovens.

3 Methods

3.1 Selection of Mother Plants

1. Collected plants are placed in the nursery for 4 weeks with preventive phytosanitary treatment (applying once a week a solution fungicide and bactericide systemic action, and solution insecticide) (*see* **Note 2**).

2. After the 4-week quarantine, 1-year-old plants are used. The plants are stripped of their developed leaves, and the basal and apical ends are cut off, leaving only the stem with the plant apex (Fig. 2a) (*see* **Notes 4** and **5**).

3.2 Establishment of In Vitro Explants

1. After isolating the agave tips for each species separately, wash them with commercial detergent and tap water for 3 min. Then, perform three rinses with osmosis water and immerse in 70% (v/v) ethanol for 3 min (*see* **Note 5**).
2. Then, immerse the explants in a 0.6% (v/v) sodium hypochlorite solution for 15 min (Fig. 2b). Before the 15 min are up, disinfect the surface of the laminar flow hood with 70% (v/v) alcohol and introduce the containers with culture medium for establishment, and sterile dissection material.
3. After 15 min of exposure to a 0.6% (v/v) sodium hypochlorite solution, rinse three times with sterile osmosis water (*see* **Note 6**).

Fig. 2 Obtaining explants and disinfestation. (**a**) explant preparation, (**b**) disinfestation with sodium hypochlorite, (**c**) in vitro establishment, and (**d**) 1 week of explant with response

4. Subsequently, the agave apices (explants) are placed in flasks separated by species (Fig. 2c); these flasks contain MS medium without growth regulators.
5. After 1 week, the explants of each *Agave* species without the presence of necrosis, mycelium, and/or bacterial colonies are transferred to the MS medium for shoot induction for 8 weeks (Fig. 2d), remaining incubated at 25 ± 2°C, under irradiation of 40 ± 5 μmol m^{-2} s^{-1} and a photoperiod of 16 h (*see* **Note 7**).

3.3 In Vitro Multiplication of Agaves

1. For multiplication, the shoots obtained in the MS medium for induction (Fig. 3a) were transferred to flasks with culture medium containing 25 mL of MS medium supplemented with 30 g/L of sucrose, 3 mg/L of benzylaminopurine (BAP), and 0.3 mg/L of IAA (indole-3-acetic acid) (see Table 1) (*see* **Notes 7 and 8**) [5].
2. The pH of the culture medium is adjusted to 5.7 ± 0.01 before adding 7 g/L of micropropagation agar as a gelling agent; the medium is then autoclaved at 121°C and 1.5 kg/cm^2 of pressure for 20 min.

Fig. 3 In vitro multiplication and the effect of slow-growth media on *Agave potatorum* and *Agave marmorata*. (**a**) Shoot induction, (**b**) Multiplication by organogenesis, (**c**) Individual separation of shoots of uniform size, and (**d**) From left to right: *Agave potatorum* and *Agave marmorata* in slow-growth media

3. The shoots of *Agave potatorum* and *Agave marmorata* are incubated at 25 ± 2°C, under irradiation of 40 ± 5 μmol m^{-2} s^{-1} and photoperiod of 16 h, also for 8 weeks each subculture (Fig. 3b). After three subcultures, the conservation phase begins in a slow growth medium.

3.4 Establishment of Shoots in a Slow-Growth Medium

1. The shoots of each species obtained in the multiplication phase are separated individually and in homogeneous sizes of approximately 1.5 cm (Fig. 3c). Subsequently, the root is removed and one is placed in each flask with 25 mL of 50% reduced MS medium, supplemented with 30 g/L of sucrose and 10 g/L of sorbitol (see Table 1) (see **Notes 9 and 10**) [15].
2. The pH of the culture medium is adjusted to 6.0 ± 0.01, 7 g/L of agar-agar for micropropagation is added as a gelling agent, and the medium is sterilized in an autoclave at 121°C and 1.5 kg/cm^2 of pressure for 20 min.
3. Shoots of *Agave potatorum* and *Agave marmorata* are incubated at 25 ± 2°C, under irradiation of 40 ± 5 μmol m^{-2} s^{-1} and photoperiod of 16 h for 24 months (Fig. 3d) (*see* **Note 11**).

4 Notes

1. To avoid uncertainty with agave species, experts must identify each taxonomic species and label.
2. Choose young plants that are visibly free of pests and diseases to avoid biotic stress.
3. Prepare MS stock solutions A-F separately using reverse osmosis water and analytical grade reagents. Pour each stock, as needed, into a 1 L flask with 900 mL of osmosis water; keep stirring constantly with a magnetic stirrer, add the sucrose and dissolve. Adjust the pH to 5.7 or 6 with a pH meter by adding 0.1, 0.5, or 1 N NaOH or HCl with constant stirring. Add 7 g of micropropagation agar and make up to 1 L with osmosis water using a graduated cylinder. Heat the culture medium until it boils, pour 25 mL per culture container, and cover the flasks. Sterilize in an autoclave for 20 min at 121°C and 1.5 kg/cm^2 pressure, remove the culture medium from the autoclave, and let it cool and gel at room temperature. Strictly follow the rules for disposing of all waste materials.
4. Stems approximately 3–5 cm long are obtained for in vitro establishment.
5. Wash the stems with tap water for 5 min to remove dirt and contaminants.

6. Inside the laminar flow hood, remove leaves with the help of dissection material and a stereoscopic microscope, until locating the apical micro-apes of approximately 0.5 cm long for in vitro establishment.
7. Ensure that the stainless-steel forceps used to transfer the tips are at room temperature to avoid burns to the explants.
8. Separate the shoots individually by agave species, remove the root, and reduce apical dominance.
9. For the preparation of slow growth MS medium, add the constituents as appropriate (see Table 1). Sterilize for 20 min at 121°C and 1.5 kg/cm^2 pressure.
10. Separate the shoots individually by agave species, remove the roots, and arrange them in a homogeneous size without reducing the apical dominance, and place one per culture container.
11. It is suggested to transfer the in vitro plants to a fresh, slow-growth medium, selecting the most vigorous shoot, which is usually the main shoot, especially in *Agave marmorata*, which forms new shoots in this conservation strategy (see Fig. 3d).

References

1. Alducin-Martínez C, Ruiz MKY, Jiménez Barrón O, Aguirre-Planter E, Gasca-Pineda J, Eguiarte LE et al (2023) Uses, knowledge and extinction risk faced by agave species in Mexico. Plan Theory 12:1–124. https://doi.org/10.3390/plants12010124
2. Barrientos-Rivera G, Esparza-Ibarra EL, Segura-Pacheco HR, Talavera-Mendoza O, Sampedro-Rosas ML, Hernández-Castro E (2019) Caracterización morfológica de *Agave angustifolia* y su conservación en Guerrero, México. Rev Mex Cienc Agríc 10(3):655–668. https://doi.org/10.29312/remexca.v10i3.1554
3. Martínez-Ramírez S, Trinidad-Santos A, Bautista-Sánchez G, Pedro Santos EC (2013) Crecimiento de plántulas de dos especies de mezcal en función del tipo de suelo y nivel de fertilización. Rev Fitotec Mex 36(4):387–393. https://doi.org/10.35196/rfm.2013.4.387
4. Villavicencio-Gutiérrez EE, Torres-Tapia MaA, Méndez-González J, Curiel-López C, Sánchez-Pérez F (2024) Germinación de semillas en *Agave potatorum* Zucc. Rev Mex Cienc For 15(86):31–51. https://doi.org/10.29298/rmcf.v15i86.1403
5. Aguilar-Jiménez D, Rodríguez-De-la-O JL (2018) Micropropagación y aclimatación de Maguey Pitzometl (*Agave marmorata* Roezl) en la Mixteca poblana. Rev Colomb Biotecnol 20(2):124–131. https://doi.org/10.15446/rev.colomb.biote.v20n2.77084
6. Nogales L (2017) Tepeztate, el maguey del monte. https://mezcologia.mx/tepeztate/. Fecha de consulta: 26 de marzo de 2025
7. Alvarez-Aragón C, Arzate-Fernández AM, Martínez-Martínez SY, Martínez-Velasco I (2020) Regeneración de plantas de *Agave marmorata* Roezl, vía embriogénesis somática. Trop Subtrop Agroecosyst 23(36):1-13. https://doi.org/10.56369/tsaes.3117
8. Martínez-Estevéz M (2021) Conservación *in vitro* de germoplasma, un método para salvar el acervo genético en plantas. Editores: Germán Carnevali Fernández-Concha y José Luis Tapia Muñoz. Desde el Herbario CICY 13:83-86. o de Investigación Científica de Yucatán, A.C. http://www.cicy.mx/sitios/desde_herbario/
9. Domínguez-Rosales MS, González-Jiménez MaL, Rosales-Gómez C, Quiñones- Valles C, Delgadillo-Díaz-de-León S, Mireles-Ordaz SJ, Pérez-Molphe-Balch E (2008) El cultivo *in vitro* como herramienta para el aprovechamiento, mejoramiento y conservación de especies del género *Agave*. Investigación Y Ciencia 41:53–62

10. Murashige T, Skoog F (1962) A revised medium for rapid growth and bio assays with tobacco tissue cultures. Physiol Plant 15:473–497
11. Miguel-Luna ME, Enríquez-del-Valle JR, Velasco-Velasco VA, Villegas-Aparicio Y, Carrillo-Rodríguez JC, Rodríguez-Ortíz G (2013) Composición del medio de cultivo y la incubación para enraizar brotes de *Agave*. Rev Mex Cienc Agríc 6:1151–1159
12. Martínez-Villegas YM, Andrade-Rodríguez M, Colinas-León MT, Villegas-Torres OG, Castillo-Gutiérrez A, Alia-Tejacal I (2015) Efecto de las sales inorgánicas del medio de cultivo en el crecimiento de pascuita (*Euphorbia leucocephala* Lotsy). Rev Fitotec Mex 38(4):369–374
13. Rodríguez M, Chacón M, Carrillo R (2014) Efecto de la concentración y de los componentes del medio de cultivo MS sobre la germinación *in vitro* de *Ugni molinae*. BOSQUE 35(1):119–122. https://doi.org/10.4067/S0717-92002014000100012
14. Pérez-Molphe-Balch E, Esparza-Araiza MJ, Pérez-Reyes ME (2012) Conservación *in vitro* de germoplasma de *Agave* spp. bajo condiciones de crecimiento retardado. Rev Fitotec Mex 35(4):279–287
15. Mancilla-Álvarez E, Ramírez-Mosqueda MA, Arano-Avalos S, Núnez-Pastrana R, Bello-Bello JJ (2019) *In vitro* techniques to the conservation and plant regeneration of malanga (*Colocasia esculenta* L. Schott). HortScience 54(3):514–518. https://doi.org/10.21273/HORTSCI13835-18
16. Baltazar-Bernal O, Spinoso-Castillo JL, Mancilla-Álvarez E, Muñoz-Márquez TRA, Bello-Bello JJ (2023) *In vitro* conservation and regeneration of potato (*Solanum tuberosum* L.): role of paclobutrazol and silver nanoparticles. Horticulturae 9:676. https://doi.org/10.3390/horticulturae9060676
17. García-Merino GF, Ramírez-Mosqueda MA, Mata-Alejandro H, López-Larios AV, López-Aguilar R (2024) *In vitro* conservation of *Notylia barkeri* Lindl. Revista Bio Ciencias 11:e1633. https://doi.org/10.15741/revbio.11.e1633

Check for updates

Chapter 19

Asymbiotic Germination, Micropropagation, and Slow Growth In Vitro Culture for Conservation of *Epidendrum stamfordianum* Bateman

Guadalupe López-Puc, Ángeles Sánchez-Contreras, Keila Anel Pool-Chab, Diego Alfredo Olvera-Casanova and Gilbert José Herrera-Cool

Abstract

Epidendrum stamfordianum Bateman is an epiphytic orchid native to Mexico and Central America. It faces a real threat of extinction due to rampant habitat destruction and widespread looting. In vitro orchid culture is a highly effective strategy for efficient propagation that is essential for conserving these remarkable plants. Asymbiotic in vitro propagation has been examined in various studies to establish effective strategies for the reproduction and conservation of endangered orchids. In vitro, slow-growth storage has been crucial in preserving valuable materials. Thus, asymbiotic germination, a slow growth method, and micropropagation are effective strategies to ensure the survival of *Epidendrum Stamfordianum* Bateman.

Keywords Micropropagation, Sorbitol, Orchids, Bank germplasm, Ex situ conservation

1 Introduction

Orchids are renowned for their extensive species diversity and variety of habitats. Their floral structures are so distinctive among angiosperms that many orchids hold significant economic and cultural importance in human society [1].

Globally, orchids are declining because of habitat loss and climate change, which are expected to worsen in the coming decades [2]. Preserving orchids is crucial to prevent the loss of genetic diversity and ensure ecosystems' overall resilience [3]. *Epidendrum stamfordianum* Bateman is an epiphytic orchid native to Mexico and Central America. This species grows up to 80 cm in length and produces large, fragrant flowers measuring between 20 cm to 25 cm in width (Fig. 1). It is characterized by prominent pseudobulbs and several long, thick roots [4, 5]. In Soconusco, Chiapas, Mexico, this species usually thrives in the shaded environments of coffee plantations.

Marco A. Ramírez-Mosqueda et al. (eds.), *Conservation of Plant Genetic Resources*, Methods in Molecular Biology, vol. 3011, https://doi.org/10.1007/978-1-0716-5130-8_19,

Fig. 1 *Epidendrum stafordianum* Bateman flowers

However, it is at risk of extinction due to habitat destruction and widespread looting [6].

Orchids can be propagated through two main methods: asexual and sexual reproduction. Asexual reproduction is a lengthy process that generates new pseudobulbs [7]. Sexual reproduction is slower and less common because orchid seeds lack endosperm, necessitating a symbiotic relationship with specific mycorrhizal fungi for successful germination [3, 8].

In vitro orchid culture may be a strategy for achieving efficient propagation and thus contributing to their conservation [9].

Asymbiotic germination is carried out on nutrient-rich media containing minerals; it does not require the association of mycorrhizal fungi [10]. These media are commonly referred to as basal media. Asymbiotic germination of orchids has been observed using various basal media, including Murashige and Skoog (MS) [11–14]; Kudnson C (KC), and Vacin Went [15]. Other formulations specifically for orchids include Phytamax™ [16] and Hyponex medium (N:P:K ratio of 6.5:6:19) [11, 17]. In some cases, asymbiotic germination requires the addition of other compounds, such as plant growth regulators (PGRs), carbon activated, vitamins, coconut water, banana puree, and casein hydrolysate [14, 16, 17]. Asymbiotic germination is the most effective method for conserving and commercially cultivating orchids [18]. On the other hand, asymbiotic germination of orchid seeds has proven successful for many species and could play a crucial role in generating seedlings to help restore threatened populations [9, 10].

In situ conservation is costly and requires space with inherent risks of pests or disease and natural disasters, such as drought, storms, and fires. In vitro conservation techniques are alternative tools for germplasm collection to complement conservation and rescue genetic diversity; they make it possible to preserve long-term genetic collections of wild and cultivated plant species [19, 20].

There are in vitro collections of potatoes, clones of genotypes of deciduous woody plants (All-Russian Research Institute of Forest Genetics, Breeding and Biotechnology, Russia), in vitro collection of the Royal Botanical Gardens (Royal Botanic Gardens, Kew, UK) [21]. Maintaining plants in vitro can be quite expensive to reduce costs and improve the preservation of genotypes shoots in vitro cultures are utilized for medium-term conservation of plant germplasm.

Established shoot cultures can slow down their growth and development by adjusting the culture conditions that influence normal metabolic processes. This technique, known as slow growth or minimal growth storage, uses various physical, chemical, and nutritional factors to restrict the growth of the plantlets [22]. In vitro slow growth protocols are widely used as backups for field collections, ensuring the safe conservation of valuable species [23].

Slow growth enables plant material to be stored without frequent and labor-intensive sub-cultivation on fresh media, allowing for preservation from two months to several years. Various chemical and physical factors are crucial in effectively reducing the metabolic activity of in vitro plants. Among these, the mineral composition of the culture medium and the strategic use of osmotic agents and growth retardants significantly impact plant development. Notably, the formation of new shoots is intricately influenced by the type and concentration of the osmotic substance utilized [24]. Among these agents, mannitol and sorbitol have become extensively used to restrict the growth rate of diverse plant species, including orchids [20]. Sorbitol reduces mineral uptake due to the change in osmotic pressure, resulting in the inhibition of growth and the formation of new shoots [25].

Individualized protocols are essential for the successful propagation of each orchid species [9]. Shoot regrowth must be evaluated after in vitro conservation, and it has been conducted in all the studies assessing slow growth protocols [23].

The strategy for in vitro conservation of *Epidendrum Stamfordianum* Bateman involves several key steps. First, plants are obtained through asymbiotic seed germination. Next, these plants are cultured in vitro under slow growth conditions to preserve the germplasm for a year. Finally, a protocol is established for multiplying the conserved material in vitro, ensuring an adequate number of plants are available for use in reintroduction programs (Fig. 2).

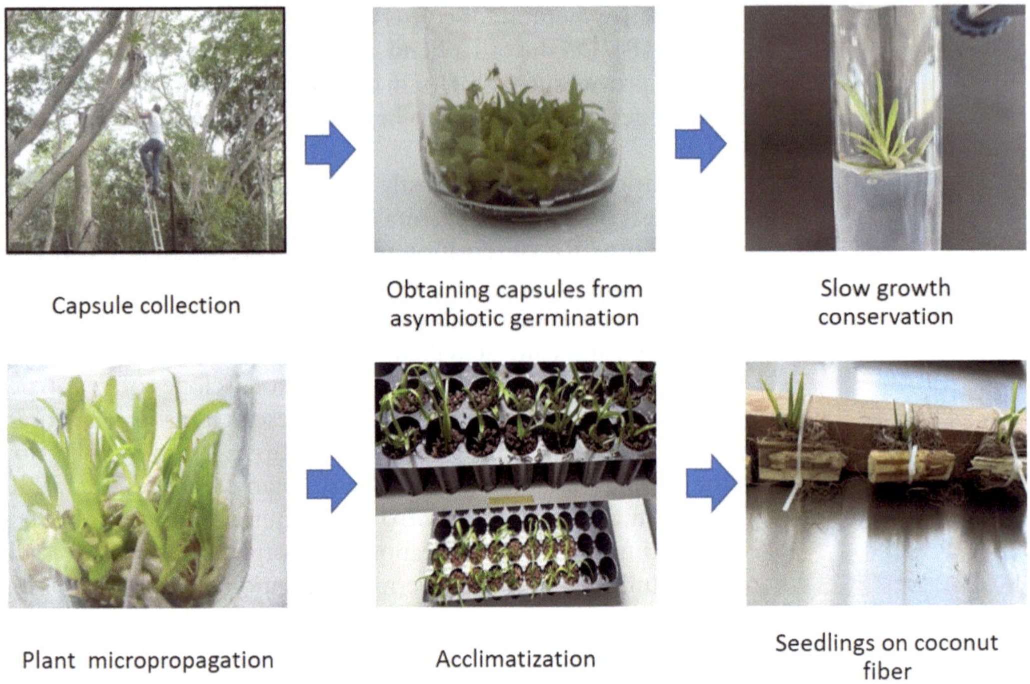

Fig. 2 Strategy for in vitro conservation of *Epidendrum Stamfordianum* Bateman

2 Materials

2.1 Plant Material and Surface Sterilization

1. Capsules mature of *Epidendrum stamfordianum* are used for the establishment of aseptic plants.
2. Disinfectant solution: 1% commercial sodium hypochlorite solution with one drop of detergent.

2.2 Culture Medium

1. Basal Murashige and Skoog (MS) [26] culture medium is used (*see* **Notes 1–4**).
2. Medium additives used for various culture stages: MS basal medium supplemented with various additives according to the culture stage, including asymbiotic germination medium (AGM), slow growth medium (SGM), and micropropagation medium (MM).
3. pH adjustment solutions: 1 N NaOH or 1 N HCl.
4. Gelrite.

2.3 Equipment

1. Culture room: Laminar airflow hood.
2. Surgical tools: stainless steel forceps, surgical scalpels, removable sterile surgical blades, and sterile Petri dishes.

3. Medium preparation: weighing balances, magnetic stirrer, hot plate, pH meter, microwave, micropipettes, magnets, refrigerator, and autoclave.
4. Culture vessels: universal bottle, and magenta vessel.

2.4 Acclimation of Plantlets to Ex Vitro Conditions

1. Plastic trays.
2. 5% solution of Murashige and Skoog without vitamins (*see* **Note 5**).
3. Barks and coconut fiber.

3 Methods

3.1 Culture Medium

3.1.1 Asymbiotic Germination Medium

1. Pour 500 mL of distilled water into a universal bottle while stirring continuously (*see* **Note 1**).
2. Add 4.4 g/L of basal medium Murashige and Skoog without vitamins (*see* **Note 2**).
3. Add 30 g/L of sucrose.
4. Add 0.22% of Gelrite.
5. Fill to a total volume of 1 L.
6. Adjust the pH to 5.8 while stirring continuously by adding 1 N NaOH or 1 N HCl dropwise.
7. Dispense 25 mL asymbiotic germination medium to each magenta vessel.
8. Sterilize the asymbiotic germination medium using an autoclave for 15 min at 121°C.

3.1.2 Slow Growth Medium

1. Pour 500 mL of distilled water into a universal bottle while stirring continuously (*see* **Note 1**).
2. Add 2.2 g/L of basal medium Murashige and Skoog with vitamins (*see* **Note 3**).
3. Add 30 g/L of sorbitol.
4. Add 0.26% (w/v) of Gelrite.
5. Fill to a total volume of 1000 mL.
6. Adjust the pH to 5.8 while stirring continuously by adding 1 N NaOH or 1 N HCl dropwise.
7. Dispense 20 mL of slow growth medium into glass culture tubes (150 × 25 mm).
8. Sterilize slow growth medium using an autoclave for 15 min at 121°C.

3.1.3 Micropropagation Medium

1. Pour 500 mL of distilled water into a universal bottle while stirring continuously (*see* **Note 1**).
2. Add 4.4 g/L of basal medium Murashige and Skoog with vitamins (*see* **Note 4**).
3. Add 30 g/L of sucrose.
4. Add 8.88 μM BAP.
5. Add 150 ml/L of coconut milk.
6. Add 0.22% (w/v) of Gelrite.
7. Fill to a total volume of 1 L.
8. Adjust the pH to 5.8 while stirring continuously by adding 1 N NaOH or 1 N HCl dropwise.
9. Dispense the 20 mL multiplication media into glass culture tubes (150 × 25 × 25 mm).
10. Sterilize the micropropagation medium using an autoclave for 15 min at 121°C.

3.2 Asymbiotic Germination

1. Collect green capsules (*see* **Note 6**).
2. To allow maturation, store them in kraft paper envelopes and keep them in a desiccator at 25°C for 5–10 days (*see* **Note 7**).
3. Wash the capsules with detergent, then rinse them with sterile distilled water.
4. Disinfect the capsules by immersing them in a 1% commercial sodium hypochlorite solution with one drop of detergent for 10 min. Then, rinse thrice with sterile distilled water, each for 5 min.
5. Immerse the capsules in 96% alcohol and quickly flame to sterilize.
6. Place the disinfected capsules in a sterile Petri dish. Using a sterile blade, make a longitudinal slit on each capsule and carefully scoop out the healthy aseptic seeds. Dust the seeds over the culture medium. Repeat this process until all the missing seeds have been used. Then, add a few drops of water to each jar to help separate and distribute the seeds evenly over the medium for germination.
7. The seeds are inoculated aseptically in the asymbiotic germination medium.
8. Incubate seeds in darkness in a culture room at 25 ± 1°C for 4 weeks at 65% humidity.
9. Transfer seed cultures under a 16/8 photoperiod (light/dark; photosynthetic photon flux density = 60 $\mu mol/m^2/s^1$). Light source was provided by LED lamps.
10. Protocorm-like bodies must be observed 2 weeks later.
11. Seedlings must be observed 3 weeks later.

3.3 Slow Growth

1. Select seedlings with two leaves and two to three roots.
2. Place 2 cm long seedlings with two leaves and two to three roots into glass culture tubes (150 × 25 × 25 mm) containing 20 mL of sterile minimal growth medium.
3. Incubate under a 16/8 photoperiod (light/dark; photosynthetic photon flux density = 60 μmol/m²/s¹) at 25°C, 65% humidity.
4. Seedlings can remain in good condition for up to twelve months without replanting (*see* **Note 8**). After this period, the medium culture must be replenished.

3.4 Micropropagation from Basal Apices

1. Seedlings from slow growth conditions are the source material for the seedlings multiplication (*see* **Note 9**).
2. Cut the basal apices of the seedlings.
3. Place the basal apices inside magenta boxes containing micropropagation media for development, growth, and rooting (*see* **Note 10**).
4. Incubate under a photoperiod 16/8 (light/dark; photosynthetic photon flux density = 60 μmol/m²/s¹) at 25°C, 65% humidity.
5. Shoots must be observed 6 weeks later.

3.5 Acclimatization

1. In vitro-developed seedlings were thoroughly washed with distilled water to remove agar residues (*see* **Note 11**).
2. Next, the seedlings were placed in trays filled with sterilized coconut fiber and bricks for 30 days. During this time, the plants were maintained in a grow room with adequate humid ity, proper ventilation, and supplemental lighting.
3. To support their growth, they were regularly misted with water containing trace amounts of macronutrient salts (MS).
4. Each plant was positioned on the coconut fiber and secured with a small piece of bark. Finally, the seedlings were placed on trees.

4 Notes

1. The quantities needed to prepare the germination, slow growth, and micropropagation media are specified for one liter of culture medium.
2. Murashige and Skoog without vitamins, from SIGMA-ALDRICH, catalog number 5524, is the basal medium for asymbiotic germination.

3. Murashige and Skoog, with vitamins from SIGMA-ALDRICH, catalog number 5519, is the basal medium for slow-growth culture. The basal medium used for slow growth is half its ionic strength.
4. Murashige and Skoog, with vitamins from SIGMA-ALDRICH, catalog number 5519, is the basal medium for micropropagation. The basal medium used for micropropagation, MS5519, is of full ionic strength.
5. Murashige and Skoog, without vitamins from SIGMA-ALDRICH, catalog number 5524, is the basal medium for misting the seedlings during acclimatization. This formula is prepared without agar.
6. Green capsules must be completely closed and collected with the peduncle, ensuring no damage or signs of contamination.
7. The capsules should be placed in the desiccator for 5–7 days, or until they begin to turn yellow. If the capsules open slightly, they will no longer suit the disinfection methodology described here.
8. The seedlings can be kept in a slow growth medium for up to twelve months. However, it is advisable to check them monthly for changes in the growing medium's color or signs of contamination.
9. Ensure the seedlings show no abnormal growth or contamination.
10. The same cultivation medium formulation was successful in helping the sprouts develop, grow, and root.
11. To ensure the healthiest transition of regenerated plants to compost, it is essential to trim the roots to 2–3 cm. Additionally, be sure to remove any dead or broken lateral roots. This practice promotes robust growth and enhances the overall vitality of the plants.

Acknowledgments

The authors thank Antonio Sánchez-Martínez for his invaluable assistance collecting the plant material. The authors sincerely appreciate CIATEJ for generously providing the facilities and equipment that significantly contributed to this work.

References

1. Chen YY, Sun Y, Li CI et al (2025) OrchidBase 6.0: increasing the number of Cymbidium (Orchidaceae) genomes and new bioinformatic tools for orchid genome analysis. BMC Plant Biol 25(1):1–13. https://doi.org/10.1186/s12870-024-06024-1

2. Kindlmann P, Kull T, McCormick M (2023) The distribution and diversity of orchids. Diversity 15(810):1–3. https://doi.org/10.3390/d15070810
3. Tinoammini N, Aazhivaendhan G, Senthil Kumar T (2024) In vitro germination and optimization of basal media for protocorm-like bodies proliferation in *Dienia ophrydis* (J. Koenig) Seidenf. Rhizosphere 29:1–7. https://doi.org/10.1016/j.rhisph.2024.100854
4. Sánchez-Martínez A, Sarmiento, M, Andrews J (2002) *Epidendrum Stamfordianum* Bateman. In: Orquídeas de Campeche, 1st edn. Instituto Nacional de Investigaciones Forestales, Agrícolas y Pecuarias, Campeche
5. Bertolini V, Cruz-Blasi J, Damon A et al (2014) Seasonality and mycorrhizal colonization in three species of epiphytic orchids in southeast Mexico. Acta Bot Brasilica 28(4):512–518. https://doi.org/10.1590/0102-33062014abb3436
6. Damon A (2019) ¡Orquídeas en peligro! Guía de manejo sustentable en el Soconusco. In: Damon A (ed) El colegio de la Frontera sur, 1st edn
7. Vera-Aguilar A, Ramírez-Mosqueda MA, Lee-Espinosa HE et al (2022) Efficient protocol for *in vitro* propagation of *Laelia anceps* ssp. anceps white variant from asymbiotic seed germination. S Afr J Bot 149:376–380. https://doi.org/10.1016/j.sajb.2022.06.013
8. Abraham S, Augustine J, Thomas TD (2012) Asymbiotic seed germination and *in vitro* conservation of *Coelogyne nervosa* A. Rich. an endemic orchid to Western Ghats. Physiol Mol Biol Plants 18(3):245–251. https://doi.org/10.1007/s12298-012-0118-6
9. Valencia-Glushchenko N, Oña-Arias CG, Orellana M et al (2024) In vitro asymbiotic seed germination and seedling development of four endangered Ecuadorian orchids: *Epidendrum Jamiesonis*, *Pleurothallis pulchella*, *Oncidium pentadactylon*, and *Elleanthus capitatus*. Plant Cell Tissue Organ Cult 158(60):1–13. https://doi.org/10.1007/s11240-024-02841-2
10. van der Walt K, Lehnebach CA, Alderton-Moss J (2024) Asymbiotic germination, seedling establishment and fungal uptake of *Pterostylis montana* and *P. paludosa*, two orchid species endemic to New Zealand. N Z J Bot 63(4):859-866. https://doi.org/10.1080/0028825X.2023.2294840
11. Kang HJ, Kang KW, Kim DH et al (2020) In vitro propagation of *Gastrochilus matsuran* (Makino) schltr., an endangered epiphytic orchid. Plants 9(524):1–10. https://doi.org/10.3390/plants9040524
12. Solís-Zanotelli FY, Baltazar-Bernal O, Cruz-Huerta N et al (2022) *In vitro* germination and development of "Canelita" (*Lycaste aromatica* (Graham) Lindl.) in gravity immersion bioreactors. In Vitro Cell Develop Biol Plant 58:1117–1125. https://doi.org/10.1007/s11627-022-10314-4
13. Hussien M, Kryuchkova V, Raeva-Bogoslovskaya E et al (2023) Clonal micropropagation of *Cymbidium erythrostylum* Rolfe. Int J Plant Biol 14:28–38. https://doi.org/10.3390/ijpb14010003
14. Borbolla-Pérez V, Iglesias-Andreu LG, Luna-Rodríguez M (2024) Effect of zeatin and casein hydrolysate on in vitro asymbiotic germination of immature seeds of *Vanilla planifolia* Jacks ex Andrews (Orchidaceae). S Afr J Bot 171:802–811. https://doi.org/10.1016/j.sajb.2024.07.005
15. Gayatri PEP, Astarini IA, Pharmawati M (2023) In vitro culture of wild Vanda tricolor Lindley seeds from Kintamani, Bali, as initial step for conservation. IOP Conf Ser: Earth Environ Sci 1255:1–6. https://doi.org/10.1088/1755-1315/1255/1/012026
16. Gutiérrez-Zavala JT, Ávila-Díaz I, Magaña-Lemus RE (2021) *In vitro* development of the Mexican endemic twig epiphyte *Erycina hyalinobulbon* (Orchidaceae) to promote its conservation. Acta Bot Mex 128:1–10. https://doi.org/10.21829/abm128.2021.1808
17. Hwang JE, Park H, Jeon DY et al (2024) Effect of different basal media and organic supplements on in vitro seedling development of the endangered orchid species *Dendrobium moniliforme* (L.) Swartz. Plants 13(2721):1–13. https://doi.org/10.3390/plants13192721
18. Li Y, Chen H, Kong X et al (2025) Excessive accumulation of auxin inhibits protocorm development during germination of *Paphiopedilum spicerianum*. Plant Cell Rep 44:23. https://doi.org/10.1007/s00299-024-03419-0
19. Ivanova NN, Tsiupka VA, Mitrofanova OV et al (2022) Rare and endemic plant species of Mountain Crimea flora: a morphogenic response to conservation in the gene bank *in vitro*. Turczaninowia 25(1):24–30. https://doi.org/10.14258/turczaninowia.25.1.3
20. Linjikao J, Inthima P, Limmongkon A et al (2024) Mannitol and sorbitol concentration optimization for effective *Epipactis flava* Seidenf. *in vitro* slow growth storage. In Vitro Cell Develop Biol Plant 60:496–507. https://doi.org/10.1007/s11627-024-10437-w
21. Vidyagina EO, Shestibratov KA (2018) Efficient technique for long-term storage of transgenic Aspen genotypes. Am J Plant Sci 9(13):2593–2600. https://doi.org/10.4236/ajps.2018.913188
22. Benelli C, Tarraf W, Izgu T et al (2022) In vitro conservation through slow growth storage technique of fruit species: an Overview of the

last 10 years. Plants 11(3188):1–18. https://doi.org/10.3390/plants11233188
23. Lu X, Sun P, Liu R et al (2024) In vitro slow-growth conservation, acclimatization, and genetic stability of virus-free apple plants. Hortic Adv 2(30):1–12. https://doi.org/10.1007/s44281-024-00049-0
24. Armijos-González R, Ramón P, Cueva-Agila A (2024) *Cinchona officinalis* L. ex situ conservation by in vitro slow growth and cryopreservation techniques. Plant Cell Tissue Organ Cult 158(6):1–17. https://doi.org/10.1007/s11240-024-02784-8
25. Ambrožič-Dolinšek J, Oblonšek S, Ciringer T (2022) Slow growth storage of *Berula erecta* in vitro-effect of sucrose, sorbitol and temperature. Agric Scientia 19(1):33–41. https://doi.org/10.18690/agricultura.19.1.33-41.2022
26. Murashige T, Skoog F (1962) A revised medium for rapid growth and bio assays with tobacco tissue cultures. Physiol Plant 15:473–497

Chapter 20

Medium-Term Preservation by the In Vitro Minimum Growth Method of *Swietenia humilis* Zucc

Esmeralda J. Cruz-Gutiérrez, Claudia Berenice Espitia-Flores, Liliana Muñoz-Gutiérrez, Luis Alberto Gómez-Reyes, Gabriela Sandoval-Cancino and Juan M. Pichardo-González

Abstract

Swietenia humilis Zucc. is a precious tropical hardwood species belonging to the Meliaceae family. It's planted as a living fence along the edges of land used as pastures or at property boundaries. Its wood is of excellent quality, and it's also attributed with medicinal properties. However, its natural habitat is experiencing fragmentation, which is why this species is included in the International Union for Conservation of Nature's Red List of Threatened Species. The loss of this species' populations justifies the need for its conservation in germplasm banks, ensuring its medium and long-term ex situ conservation and providing material for various purposes, such as reforestation. A method for medium-term conservation is the minimal growth technique in vitro conditions, which consists of reducing the growth rate of the tissue or seedling grown under controlled conditions, in vitro culture.

Keywords Médium-term conservation, Genetic resources, Ex situ conservation

1 Introduction

Agriculture and livestock farming have contributed to the loss of natural areas, in addition to the overexploitation of wood from species classified as precious woods, such as *Swietenia humillis* Zucc. This species is distributed on the Pacific Coast from Central America to Mexico [1], and exhibits allopatric speciation, concentrating in humid and dry forests [2].

Species conservation requires ensuring the greatest possible variability, ensuring continued use, and saving time, money, and effort. In addition, useful information must be available on the species' distribution, risk of extinction, and genetic variability [3].

A method for medium-term conservation is the minimal growth technique in vitro conditions, which consists of reducing the growth

Marco A. Ramírez-Mosqueda et al. (eds.), *Conservation of Plant Genetic Resources*, Methods in Molecular Biology, vol. 3011, https://doi.org/10.1007/978-1-0716-5130-8_20,

rate of the tissue or seedling grown under controlled conditions, in vitro culture [4].

2 Materials

2.1 Plant Material

For the in vitro establishment of *Swietenia humilis* Zucc., seeds are collected from young, healthy, vigorous, and pest-free plants.

2.2 Sanitizing Agents

- 1% Powder Detergent Util® Solution (*see* **Note 1**).
- 70% ethanol solution (*see* **Note 2**).
- 30% sodium hypochlorite solution (*see* **Note 2**).
- Sterile distilled water (*see* **Note 3**).
- Captan® Solution (3 $g{\cdot}L^{-1}$).

2.3 Culture Medium

1. Woody Plant Medium [5], Schenk and Hildebrandt [6].
2. Different additives can be used in the culture medium to promote the different stages of in vitro culture (Table 1), Seed Germination (GS), Multiplication (M), and Minimum Growth (CM).

2.4 Equipment

1. Culture room: Laminar airflow hood.
2. Surgical tools: stainless steel forceps, surgical scalpels, removable sterile surgical blades, and sterile Petri dishes.
3. Preparation of the culture medium: granular and analytical scales for weighing, spatulas, magnetic stirrers, hot plate with stirring, potentiometer, microwave oven, graduated beaker, and autoclave.

Table 1
Culture medium additives used for the different culture stages of *Swietenia humilis* Zuc

Culture medium	Components
Seed establishment medium	Woody Plant Medium [5], or Schenk and Hildebrandt [1], added with 20 g/L of sucrose, 1 g/L of activated carbon, 1 mL/L of PVP, 8 g/L of agar, and a pH of 5.7
Propagation medium	Murashige and Skoog medium [7], added with 30 g/L of sucrose, 1.5 $mg{\cdot}L^{-1}$ of benzylaminopurine (BA), 0.1 mg/L of naphthaleneacetic acid (NAA), 8 g/L of agar, and a pH of 5.7
Minimum growth medium	Murashige and Skoog medium [7] at 50% of its salt concentration, with 15 g/L of sucrose, 15 g/L of Mannitol®, 8 g/L of agar, and a pH of 5.7

4. Culture vessels: Beakers (capacity 1000 mL), reagent cups, culture tubes, and Petri dishes.
5. Miscellaneous materials: To clean glassware, it is suggested to scrub the glassware with a commercial liquid detergent solution and rinse it with plenty of tap water. To seal the culture tubes, use plastic wrap or film for packaging food.

3 Methods

3.1 Disinfection Plant Material

1. The seeds are washed with a 1% powder detergent Util® solution (*see* **Note 1**) for 10 min, followed by two or three rinses.
2. Left to soak for 16 h in double-distilled water.
3. Immersed in a Captan® 3 g/L solution for 30 min (*see* **Note 4**).
4. Placed in a 70% ethanol solution for 1 min (*see* **Note 5**).
5. Introduced into a 30% sodium hypochlorite solution for 9 min.
6. Wash the seeds three or four times with sterile distilled water to remove residues of disinfectant solutions.
7. Establishing seeds in the growing medium.
8. The material obtained from the germination of seeds in vitro free of fungus and bacteria is selected for the multiplication of the material.

3.2 Proliferation of Shoots

1. The explants are dissected to 15 mm in length (*see* **Note 6**) and placed in a semi-solid medium supplemented with growth regulators (*see* Table 1). Kept in this medium for 30 days.
2. The incubation conditions of the explants are: a temperature of 24 ± 2°C and a photoperiod of 16 h of light and 8 h of darkness with a light intensity of 2000 lux.
3. Once 30 days have passed, the generated shoots are separated from the initial explant and are subsequently used in the next processes.

3.3 Minimum Growth

1. The explants must be 20 mm long and contain at least one growth point (bud) and are placed in the minimum growth medium (Table 1).
2. The sprouts are incubated at a temperature of 18 ± 2°C, with a photoperiod of 16 h of light and 8 h of darkness with a light intensity of 2000 lux.
3. The shoots can be kept in the same medium for up to a year, as observed in Fig. 1.

Fig. 1 (**A**) *Swietenia humilis* Zucc. seedling with minimal growth. (**B**) Treatments for minimal growth. (**C**) Conservation of *Swietenia humilis* Zucc

4 Notes

1. Powdered detergent is a commercial product for washing dishes.
2. Disinfectant solutions are prepared with distilled water
3. The water is sterilized in an autoclave at 1.5 kg/cm^2 of pressure and at a temperature of 121°C for 15 min.
4. Captan® is a commercial fungicide.
5. The disinfection protocol using the 70% ethanol solution is carried out inside the laminar flow hood.
6. The tissues are dissected, with a size of 15 mm and 20 mm depending on the process requirement, within the laminar flow hood and established in the culture medium (Table 1).

Acknowledgments

This work was supported by the National Genetic Resources Center of National Institute of Forestry, Agriculture and Livestock Research.

References

1. Rosas FV, Fuchs EJ, Quesada M, Lobo JA (2011) Effects of habitat fragmentation on pollen flow and genetic diversity of the endangered tropical tree Swietenia humilis (Meliaceae). Biol Conserv 144(12):2919–2926
2. Flores L. ZY, A. P. Campos D, E. H. Quiroga S, Y. Gutiérrez G, E. Salas O, G. García S, A. Cuellar C (2019) Características fitoquímicas y toxicológicas de la semilla *Swietenia humilis* Zuccarini y su efecto hipoglucemiante. Rev Cubana Farm 52(1):e129. http://www.revfarmacia.sld.cu/index.php/far/article/view/129. Accessed 5 May 2025
3. Centro Internacional de Agricultura Tropical (CIAT) (2007) Multi-institutional distance learning course on the ex situ conservation of plant genetic resources. (Online). CIAT publication No 360, Palmira, Colombia. p 235. https://cgspace.cgiar.org/items/47656996-a579-4620-a2e6-4059ae85cac8. Accessed 4 Dec 2025
4. Montiel-Castelán P, Castillo-Martínez CR*, Gómez-Reyes LA, Valle-Arizaga M, Jasso-Mata J (2016) Conservación *in vitro* por crecimiento mínimo de Swietenia macrophylla King, y Tectona grandis L. Agroproductividad 9(2):20–25
5. Lloyd G, McCown B (1981) Woody plant medium: a mineral nutrient formulation for microculture of woody plant species. HortScience 16:453
6. Schenk RU, Hildebrandt AC (1972) Medium and techniques for induction and growth of monocotyledonous and dicotyledonous plant cell cultures. Can J Bot 50:199–204. https://doi.org/10.1139/b72-026
7. Murashige T, Skoog F (1962) A revised medium for rapid growth and bio assays with tobacco tissue cultures. Physiol Plantarum 15(3):473–497. https://doi.org/10.1111/j.13993054.1962.tb08052.x

Chapter 21

In Vitro Storage Techniques for Orchids Conservation: A Review

Evelia Guadalupe Mora-González, Obdulia Baltazar-Bernal and Marco A. Ramírez-Mosqueda

Abstract

In vitro culture supports the ex situ maintenance of the Orchidaceae family, which is valuable for its ecological and economic roles. Due to over-extraction and habitat destruction, it is essential to develop preservation strategies to maintain a germplasm reservoir. Two main approaches are currently applied: slow-growth techniques that alter culture conditions for short-term storage, and low-temperature and cryogenic methods aimed at long-term storage. This review presents the application of in vitro storage techniques for orchid conservation. It discusses 53 studies concerning explants, culture media, slow growth, protocorms, synthetic seeds, cryopreservation, plant regeneration, and genetic stability.

Keywords Orchid fruit, Orchid seeds, Slow growth, Protocorm, PLBs, Synthetic seed, Cryopreservation, Plant regeneration, Gene stability

1 Introduction

The Orchidaceae is one of the biggest families in the world with 30,000 native species and more than 100,000 orchid hybrids [1]. Orchids play a crucial ecological role with pollinators [2]. Seventy-three percent of orchid species are epiphytes [3]; thus, they need mycorrhiza to germinate and a phorophyte to develop [4, 5]. A study made in the Colombian Andes showed that deforestation at a local scale caused a loss of over 70% of orchid species in the region and a transformed habitat displayed 23 times fewer individuals versus a natural habitat [6]. Consequently the entire family is listed in Appendix II of the Convention on International Trade in Endangered Species of Wild Fauna and Flora [7]. Currently, India, Brazil, and Mexico lead in in vitro conservation research in orchids, which is a vital strategy to keep a germplasm bank (Fig. 1). Each country has performed research involving 19, 11, and 8 orchid species or hybrids, respectively (Table 1).

Marco A. Ramírez-Mosqueda et al. (eds.), *Conservation of Plant Genetic Resources*, Methods in Molecular Biology, vol. 3011,
https://doi.org/10.1007/978-1-0716-5130-8_21,

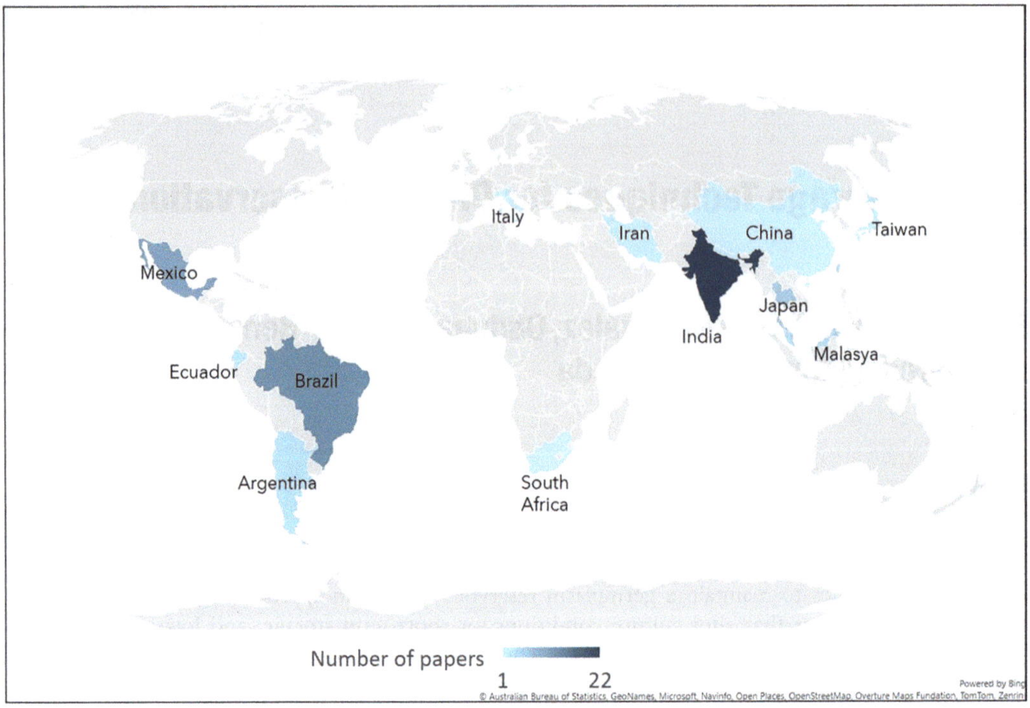

Fig. 1 Works in in vitro conservation per country

The most outstanding orchids flowering has been used to generate commercial hybrids as ornamental plants for commercialization [68]. Besides, some important species such as *Bletilla striata* and *Dendrobium nobile* [34] are important for medical purposes, and some *Vanilla* species are used for flavor industry (Fig. 2) [64], these genera have been studied to be managed in slow growth. Long-term storage at below zero temperatures stops biological activity which means that in theory, preservation for unlimited periods of time is possible [69]. Subzero temperatures ranging from −20°C to −80°C fall into the conventional and ultra freeze storage category, respectively, the latter using Ultralow-temperature (ULT) refrigeration to achieve such temperatures [70]. Meanwhile, cryopreservation is defined as the storage of biological material at ultra-low or cryogenic temperatures that can be met by immersion in liquid nitrogen (LN) that can reach −196°C or liquid nitrogen vapor (LNV) with temperatures between −165°C and −195°C [71].

Despite the capacity of cryopreservation to prevent the aging of the stored germplasm, viability may be compromised, which is why many strategies have been developed with this objective. These techniques include encapsulation-dehydration, vitrification, droplet-vitrification, and cryo-plate methods, all of them aiming to improve survival and regrowth capacity [72].

Table 1
Orchid species in vitro storage techniques by osmotic substances, PGR, temperature, and cryopreservation

Species or hybrids	Explant	Main results	Recovery, regrowth/aclimtization	References
Ansellia africana Lindl.	PLB encapsulation	Synseeds were effectively stored for 75 days at 8°C with a conversion frequency of 86.21%.	Acclimatization was 89% at 3 months.	[8]
Anacamptis	Seeds	Seeds of the four *Anacamptis* species proved to be desiccation tolerant and have orthodox storage behavior, stored *A. laxiflora*, *A. morio*, *A. palustris*, and *A. papilionacea* at −24°C for 3 years	Germination was 80%, 93%, 11%, and 68% for each species, respectively.	[9]
Bletilla formosana (Hayata) Schltr.	Seeds	Seed germination decreased to 47%–57% after storage at 3°C for 6 months		[10]
Bletilla formosa Rchb. F.	Embryos	Embryos were exposed to a glycerol (2 M) and sucrose (0.3M) solution for 15 min at 25°C and dehydrated with PSV2 for LN immersion. Plantlets were obtained from these embryos at a 60% rate.	Embryos were supplemented with a liquid ND medium for a successful recovery and subsequent development into plantlets.	[11]
Bletilla striata		Immature seeds were subjected to direct vitrification and preculture vitrification for LN immersion. Precultured-vitrified seeds showed greater survival rates of 92% and 81% after 30 min of LN exposure.	Germination of treated cryopreserved seeds was possible after rapid warming at 38°C and sowed in semisolid ND medium supplemented with 88 mM.	[12]
Bletilla striata	Seeds	Seeds were desiccated by silica gel and by air-drying and were exposed to LN for 24 h. Both treatments showed viability and germination rates above 84% and 68%, respectively.	Cryopreserved seeds were germinated in a half-strength MS medium with 2% sucrose and 0.1% activated charcoal.	[13]

(continued)

Table 1
continued

Species or hybrids	Explant	Main results	Recovery, regrowth/aclimtization	References
Bletilla formosana	Seeds	Germination rate of cryopreserved seeds by vitrification using a pretreatment of glycerol and sucrose and vitrification with PVS2 was 91%. Germination failed in seeds that skipped vitrification by PVS2.	Cryopreserved seeds were sowed in half-strength, supplemented MS medium. Then, an 8-week-old protocorm. After three sub-cultures in SD medium, plantlets were transferred to a greenhouse in peat moss and fern chip substrate.	[14]
Brassavola nodosa (L.) Lindl.	Protocorm encapsulated	ABA pretreatments were beneficial to cryo-conserved encapsulated protocorms.	Protocorms were regrowth on solid half-strength MS incubated at 25°C.	[15]
Caladenia	Protocorm	Conditions that showed greater survival percentage are as follows: 4 mm large, no glycerol treatment, 30 min PSV2 treatment, and 15 min PSV4 treatment.	Recovery media did not change between treatments.	[16]
Catasetum integerrimum	Seedlings	Six months in half MS supplemented with 2% sorbitol was the best treatment.	–	[17]
Catasetum macrocarpum Rich. ex Kunth.	Seedlings	Survival was above 93% at 450 days at 25°C no matter osmotic regulators. The minimal growth was at 18°C.	Aclimatization reached 100%.	[18]
Cattleya tigrina	Plantlets	The plantlets were conserved under slow growth for 730 days using 25% MS salts, and temperatures of 18°C or 25°C.	Aclimatization	[19]

Species or hybrids	Explant	Main results	Recovery, regrowth/aclimtization	References
Cattleya leopoldii	PLB encapsulation	Germination of 100 and 92% was observed after 15 and 60 days, respectively, at 4°C.	Survival 40%–100% after 8 weeks	[20]
Cattleya harrisoniana × *Cattleya walkeriana* hybrid	Seeds	For seed survival after LN immersion, vitrification with PVS2 is required.	12-month-old plantlets were transferred to a greenhouse in trays with sphagnum and shade net. After 6 months survival rate was reported at 100%.	[21]
Cattleya tigrina	Seeds	No vitrification treatment with PSV2 is required for seed survival after LN immersion. This result is attributed to the low water content of the seed.	After acclimation, cryopreserved plants did not differ with the control plants.	[21]
Cephalanthera rubra (L.) Rich	Encapsulation shoot tips	The survival percentage at 4°C or cryopreservation for 6 months was 82% in MS+ Na-alginate + air-desiccation.	–	[22]
Coelogyne breviscapa Lindl	PLBs Synthetic seeds	Synthetic Seeds stored at 4°C for 60 days showed no reduction in viability.	–	[23]
Cohniella cepula	Pollinia/seed	Pollinia stored at +4°C or −20°C their longevity was reduced (60–120 days), but at −70°C and −196°C showed high fertilizing ability (94%–100%) in 1 year. Finally, when seeds were stored at +27°C and +4°C, germination decreased to 0% within 1–6 months of storage.	–	[24]
Cymbidium aloifolium (L.) Sw	Artificial seed protocorms	Syneeds were successfully stored at 4°C till 28 days.	Acclimatization 85%	[25]

(continued)

Table 1
continued

Species or hybrids	Explant	Main results	Recovery, regrowth/aclimtization	References
Cymbidium bicolor Lindl.	Encapsulated protocorms	Encapsulated protocorms stored at 25°C were green and germinated (52%) after 360 days. The encapsulated protocorms stored at 4°C remained viable up to 30 days, the germination decreased drastically.	Acclimatization 90%	[26]
Cymbidium eburneum Lindl		MS with 0.7 M sucrose was used as osmoprotected and seeds were encapsulated in alginate beads, then desiccated for 6 h. Submersion in LN was performed for 1 h.	Regeneration of 72% was achieved in MS medium.	[27]
Cymbidium finlaysonianum Lind	Artificial seeds PLB	The four temperature of storage (0 ± 2°C, 4 ± 2°C, 8 ± 2°C, and 25 ± 2°C), artificial seeds were successfully stored at 8 ± 2°C, 105 days with percentage of conversion frequency at 44.0% on a VW medium with 2% (w/v) sucrose with 0.2% (w/v) AC.	Acclimatization was 90% survival; 12 months after reestablishment in the forest was 75.8%.	[28]
Cymbidium hybrid	PLB	After 1 year of cryopreservation, survival rate was 37% when removed from alginate beads and 21% when not.	–	[29]
Dendrobium draconis	Protocorm	High viability during storage at 25°C for 6 months	–	[30]
Dendrobium friedericksianum Rchb.f.	Seedlings	Plants cultured with 10% sucrose were preserved for 6 months without subculture, and plant height was decreased.	–	[31]

Species or hybrids	Explant	Main results	Recovery, regrowth/aclimtization	References
Dendrobium lituiflorum Lindl.	Encapsulated PLBs	Encapsulated PLBs stored at 8°C for 90 days had 80% regeneration.	77% plants survived and bloomed after 2.5 years.	[32]
Dendro-bium nobi-le Lindl.	PLBs	Concentrations of 7.5% and 12.5% of sucrose and mannitol were used for successful short-term storage.	PLB regeneration was performed in half-strength MS with BAP and NAA, with shoot and root formation after 9 weeks.	[33]
Dendrobium nobile Lindl.		Encapsulation-vitrification showed a survival and regrowth rate of 78.1% and 75.9%, respectively, after direct immersion in LN for 1 h.	–	[34]
Dendrobium Shavin White	Encapsulated PLBs	Synthetic seed survived longer at 25 ± 2°C compared to 4°C, 10°C, and 30 ± 2°C, storage up to 75 days had 80%–92% survival. Further storage up to 135 days had 52% survival.	Acclimatization 100%	[35]
Dendrobium 'Sonia'	PLB	Germination of 100% and 88% was observed after 60 and 75 days, respectively, at 4°C.	Survival 60%–100% after 8 weeks	[20]
Dendro-bium officinale	Seedlings	Seedlings were conserved in vitro on 1/2 MS and 1/4 MS media for 12 months at 4°C in dark.	Survival 90%	[36]
Dendrobium wardianum Warner	Synseeds PLB	The beads stored for 180 days at 4°C, the percentage regeneration declined to 10%.	Survival 65%	[37]

(continued)

Table 1
continued

Species or hybrids	Explant	Main results	Recovery, regrowth/aclimtization	References
Epidendrum anderssonii	Seeds	Gradual and rapid freezing was performed, the former at 0°C, −22°C, and −60°C for an hour and a final −196°C exposure and the later at sudden −196°C. No cryopreservatives were needed for a 93.5% viability.	–	[38]
Epidendrum chlorocorymbos Schltr	Seedlings	After 6 months, 50% MS supplemented with 1% sorbitol had the lowest amount of growth.	–	[39]
Epidendrum quitensium	Seeds	Gradual freezing without cryoprotectants showed 83.2% viability while rapid freezing showed 93.5%.	–	[38]
Epipactis flava Seidenf	Seedlings	MS medium with 3.0% mannitol is optimal for conserving and recovering growth plantlets for up to 24 weeks.	–	[40]
Geodorum densiflorum (Lam) Schltr.	Artificial seeds PLB	Artificial seeds stored at 4°C for 120 days showed no reduction in viability.	28% viability after transferring to nonsterile soil condition	[41]
Gomesa bifolia	Seeds	Seeds stored at +27°C and +4°C, germination rapidly decreased to 0% in 1–6 months, but seeds maintained at −196°C germination percentages were similar to fresh seeds, even after 3 years.	–	[42]

Species or hybrids	Explant	Main results	Recovery, regrowth/aclimtization	References
Grandiphyllum divaricatum, *Gomesa praetexta*, *Gomesa forbesii*, *and Gomesa recurva*	Seeds	Up to 12 months storage was performed at −20°C and −80°C for all four species. Treatment before ULT storage consisted of drying using silica gel. Best results were at −80°C for 6 and 12 months for *G. divaricatu*; −80°C for 6 months for *G. praetexta*; −20°C during 1 and 6 months for *G. forbesii*; −80°C for 12 months for *G. recurva*.	Seeds germinate into protocorm with shoot apex in WPM culture medium.	[43]
Ipsea malabarica	Plantlets	50% MS and sugar free during storage at 25°C for 27 months had high viability.	Aclimatization and reintroduction	[44]
Laelia anceps ssp. dawsonii	Somatic embryoids encapsulated	4°C, 20°C, and 25°C on the germination percentage was 9%, 80%, and 95% after 30 days storage.	Acclimatization and reintroduction	[45]
Laelia anceps Lindl	Plantlets	The highest survival (90%) was in 2 mg/L PBZ or 0.5 mg/L ABA, and 75% MS for 168 days.	BAP 2 mg/L was used to regenerated, 7.8 shoots per explant	[46]
Luisia macrantha	Pollen	After 668 days storage in LN, pollinia treated with vitrification solution (PVS2) for 10 min showed 56% germination rate. These same conditions allowed generation of viable hybrid seeds.	–	[47]
Mokara sayan × *Ascocenda wangsa* gold	Encapsulated PLB	Synthetic seeds stored at 25°C for 180 days exhibited 71.2% germination at 14 days. Random amplified polymorphic DNA marker-based analysis post-storage showed genetic fidelity.	–	[48]

(continued)

Table 1
continued

Species or hybrids	Explant	Main results	Recovery, regrowth/aclimtization	References
Notylia barkeri Lindl.	Plantlets	ABA 1 mg/L, the shoot length was reduced without affecting survival for 180 days, but ancymidol 1 and 2 mg/L affect survival.	Regeneration in MS with 2 mg/L 6-benzylaminopurine. Aclimatization	[49]
Oeceoclades maculate Lindl.	Seedlings	Survival was above 90% at 450 days at 18°C or 25°C no matter osmotic regulators.	Aclimatization 80%–100%	[18]
Oncidium bifolium Sims	Seeds and protocorm	Beads containing ½ MS medium plus 3% sucrose and 3% calcium alginate were pretreated in medium plus sucrose at different concentrations and periods. Desiccation was carried out with silica gel for a 19.2% moisture content. Beads were then immersed in LN for 1 h. 4.8% cryopreserved seeds produced complete plants. The same procedure was performed for in vitro culture protocorms with 11.3% regrowth.	–	[50]
Oncidium flexuosum Sims	Seeds	1% phloroglucinol addition after vitrification for 120 min using PSV2 showed the best survival rate with 78.44% after 30 min of LN immersion.	–	[51]
Oncidium golden anniversary		Dehydration with silica gel for 4 h resulted in plasmolysis.	–	[52]
Oncidium "Gower ramsay"	PLB encapsulation	Germination of 100% and 92% was observed after 45 and 60 days, respectively, at 4°C.	Survival 78%–100% after 8 weeks	[20]

Species or hybrids	Explant	Main results	Recovery, regrowth/aclimtization	References
Paphiopedilum niveum (Rchb. f.) Stein	Somatic embryos	Cryopreserved somatic embryos, plunged into LN for 1 h, had 20% survival after precondition with Vacin and Went medium and a two-step preculture. AA improved survival percentage to 39%.	–	[53]
Paphiopedilum niveum (Rchb. F.) Stein	PLBs	22% survival of cryopreserved PLBs with no ploidy level change after a preculture with VWM with daily sucrose concentration increase during 5 days and PVS2 exposure for 90 min.	–	[54]
Paphiopedilum villosum Lind	Seeds	Seed storage in LN (−196°C) was effective	–	[55]
Phalaenopsis	Encapsulated PLBs	The best storage was at 25°C for 180 days under LED lights.	–	[56]
Polystachya estrellensis Rchb. f.	Seedlings	Survival was above 88% at 450 days at 18°C or 25°C no matter osmotic regulators.	–	[18]
Rhyncholaelia digbyana (Lindley)	Plantlets	Slow growth with MS 2.2 g/L with 1% mannitol	–	[57]
Spathoglottis plicata Blume	Synthetic seeds/somatic embryo	The SS stored for 60 days at −4°C with 97.4% germination, and shoot proliferation using MS with 2.0 mg/L BAP and 0.25 mg/L indole-3-acetic acid.	Acclimatization with a 90% survival	[58]

(continued)

Table 1
continued

Species or hybrids	Explant	Main results	Recovery, regrowth/aclimtization	References
Stanhopea tigrina	Plantlets	In vitro conservation for 180 days, with 2 mg/L ABA generated, the lowest shoot height (1.6 cm), the lowest number of shoots (4.3), and the lowest number of roots (3.7). Regeneration of 3.2 shoots per explant was achieved with 2 mg/L BAP.	Regeneration of 3.2 shoots per explant was achieved with 2 mg/L BAP. Acclimatization 80%	[59]
Sobralia rosea	Seeds	Out of three species tested, *S. rosea* had the lowest viability percentage. Yet, non-cryoprotectant showed the best result: 55.93% in gradual freezing and 65.6% in rapid freezing.	–	[38]
Vanda coerulea Grifft. ex. Lindl.	Encapsulated PLBs	Synseeds stored at 4°C maintain their viability up to 100 days.	–	[60]
Vanda tessellata (Roxb.) Hook. ex G. Don	Encapsulated somatic embryos	Encapsulated was successfully conserved at –4°C for 12 months and 91% regenerated with 0.5 (mg L^{-1}) BAP, Kinetin, and IAA.	90% survival in the greenhouse	[58]
Vanilla sp.	Shoot-tips and root tips	Droplet-vitrification, V cryoplate, and D cryoplate were tested for 1 h LN storage. Shoot-tips performed better than root-tips, while this only survived cryopreservation with a PVS3 dehydration on a D-V procedure.	Regrowth happened on MS semisolid medium with 1 mg L^{-1} BAP and 0.5 mg^{-1} IBA in the dark for shoots and 1 mg L^{-1} KIN for root-tips.	[61]

Species or hybrids	Explant	Main results	Recovery, regrowth/aclimtization	References
Vanilla sp.	Shoot buds and protocorms synseeds	The synthetic seeds stored at 5°C and 15°C lost their viability within 30 days, but at 22°C, they were stored successfully for 10 months in sterile water. The beads germinated normally on MS with BAP (1 mg/L) and IBA (0.5 mg/L), maximum germination (80%) was observed in 2 weeks, and in 60 days developed normal plantlets. MS or 50% MS, mannitol and sucrose in equal proportions at 10 or 15 g/L, could maintain the cultures for 1 year in vitro for more than 7 years with yearly subculture.	–	[62]
Jacks. (Salisb.) Ames	Nodal segments	MS + 0.25 mg/L putrescine, the shoot cultures were conserved for 290 days, and MS + 1 mg/L spermidine for 320 days without subculture.	–	[63]
Jacks. ex Andrews	Nodal segments	MS + ABA 3 mg/L the shoots showed lower growth values. This prolongs the period between subcultures every 180 days, without affecting the viability and phenotype of the plants.	90% survival.	[64]
Jacks. ex Andrews	Mature somatic embryos	Encapsulated to form synseed and stored at –4°C for 1 year with a proliferation and germination rate of 17.2 shoots per SE and 95.3%, respectively.	Regenerated plantlets were hardened in coco peat substrate during 4 weeks under greenhouse conditions with a survival rate of 97% after 3 months.	[65]

(continued)

Table 1
continued

Species or hybrids	Explant	Main results	Recovery, regrowth/aclimtization	References
Vanda tessellata (Roxb.) Hook. ex G. Don	SE	Artificial seeds were stored at −4°C during 12 months with a germination rate of 91%.	Encapsulated somatic embryos stored under dark at −4°C for 12 months.	[58]
Vanda spathulata	Pollinia	Dehydration in laminar airflow for 3 h for moisture content below 20% results in successful pollen germination in vivo after overnight LN exposure.	–	[66]
Vanilla walkeriae Wight	Shoot	Shoots were stored for 13 months on low-nutrient medium with a thin overlay of liquid paraffin.	Chromosome number $2n = 24$ was the same.	[67]

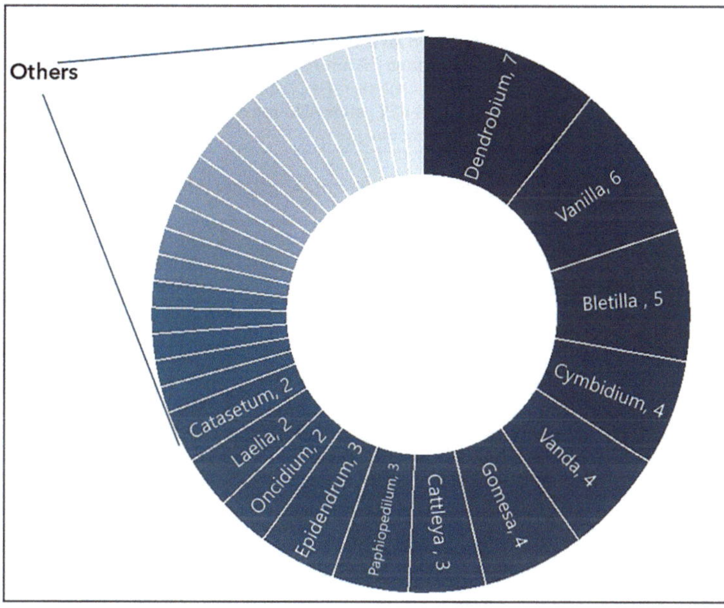

Fig. 2 Most studied orchid genera for in vitro conservation. Orchid hybrids were not included

Unlike short- and medium-term storage, long-term storage eliminates the risk of genetic and phenotypic variability due to continuous growth and propagation caused by repeated subcultures [73] and oxidative stress that results from the in vitro aging process [74].

2 Method

The queue shown in Fig. 3 was entered into the Scopus search engine. Eighty-three articles were found and manually filtered, excluding reviews and works that did not discuss in vitro storage methods. Additional articles found by Google Academic were included and a database of 56 papers was formed and classified in (a) slow-growth method and (b) long-term storage method (Fig. 4) that involved preservation at 0°C or lower. Table 1 summarizes the content of the articles included in the database.

To establish the number of protocols developed by each country, a chart (not shown) and a map (Fig. 1) were created from the database considering multiple entries if more than one species per paper was studied.

(TITLE-ABS-KEY (**cryopreservation** AND **conservation**) OR TITLE-ABS-KEY (**seed** AND **storage** OR **pollen** AND **storage** OR **seed** AND **storage** OR **embryos** AND **storage** OR **apical** AND **storage** OR **axillary** AND **buds** AND **storage** OR **microshoots** AND **storage** OR **nodal** AND **segments** AND **storage** OR **microplantlets** AND **storage** OR **calli** AND **storage** OR **plb** AND **storage** OR **germplasm** AND **storage**) OR TITLE-ABS-KEY (**slow** AND **growth**) AND TITLE-ABS-KEY (**in** AND **vitro**) AND TITLE-ABS-KEY (**orchid** OR **orchidaceae**))

Fig. 3 Input for Scopus data search. Eighty-three articles were found

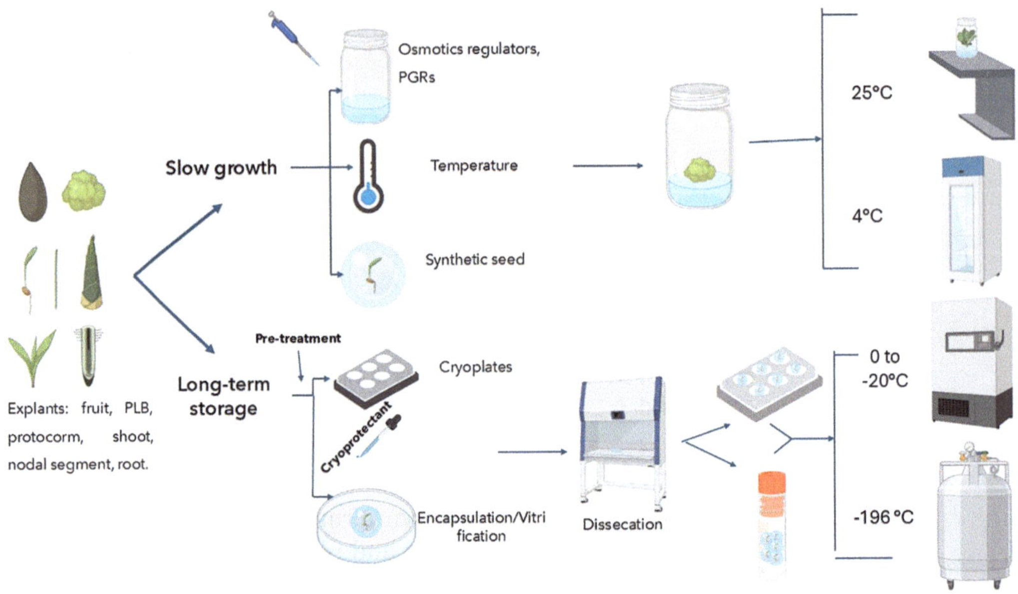

Fig. 4 General schematic representation of slow-growth and cryopreservation protocols for orchids presented in the reviewed articles

3 In Vitro Slow-Growth Storage

In vitro slow growth occurs when an explant is not growing at the average rate for their age; it has the purpose to reduce explant metabolism and to conserve it for short-term storage. It could be managed by osmotic substances, PGR, and temperatures (4–27°C) [75, 76].

3.1 In Vitro Storage: Osmotic Regulators

Osmotic substances such as mannitol, sorbitol, polyethylene glycol, sucrose, and salt culture media have been used for reducing growth and increasing the storage life of in vitro grown explants of different plant species [77] including orchids [39, 62] (Fig. 5).

3.1.1 Shoots

Vanilla walkeriae shoot tips were stored in vitro for 13 months on low-nutrient medium with thin overlay of liquid paraffin; cytological studies were made in root tips of in vitro plantlets before and after storage. They showed an identical chromosome number ($2n = 24$) [67].

A B C D E F G

Fig. 5 Effect of ancymidol (Anc) and paclobutrazol (Pbz) on the in vitro conservation of *Laelia anceps* var. alba. (**A**) Control treatment, (**B**) MS + 0.25 mg/L of Anc, (**C**) MS + 0.50 mg/L of Anc, (**D**) MS + 0.75 mg/L of Anc, (**E**) MS + 1 mg/L of PBZ, (**F**) MS + 2 mg/L of PBZ, (**G**) MS + 3 mg/L of PBZ

In another study, shoots 2 cm long were cultured on MS basal medium in full and half strength, with sucrose (30, 20, 15, 10, 0 g L^{-1}), the subculture intervals were yearly with the addition of mannitol at 10 and 15 g L^{-1} in the culture media. These cultures were maintained as in vitro explants for more than 7 years [62].

3.1.2 Plantlets

Rhyncholaelia digbyana seedlings were used for conservation in vitro, to slow down the metabolism, they were in MS basal media at 2.2 g L^{-1} with 1% mannitol [57]. *Dendrobium friedericksianum* Rchb.f. plantlets were put in Vacin and Went (1949) (VW) medium and sucrose (0%, 2%, 4%, 6%, 8%, and 10%) to reduce growth in vitro preservation; plant height decreased in 8% and 10% sucrose and planets were preserved with 10% sucrose for 6 months [31]. *Epidendrum chlorocorymbos* Schltr seedlings were placed in MS with mannitol (0%, 1%, 2%, and 3%), sucrose (0%, 1%, 2%, and 3%) or sorbitol (0%, 1%, 2%, and 3%); 50% MS with 1% sorbitol give the lowest amount of growth after 6 months [39].

The use of different concentration of osmotic regulators in the media slows the growth; *Catasetum integerrimum* seedlings were conserved in half MS supplemented with 2% sorbitol during 6 months [17]; similarly, sorbitol was the best option for *Epidendrum chlorocorymbos* and *Catasetum integerrimum* conservation [17, 39]. *C. tigrina* plantlets were established in vitro conservation using different MS concentrations (25%, 50%, 75%, and 100%) and place at 18°C (A) and 25°C (B), and the osmotic regulators (sucrose 20 g L^{-1}, 10 g L^{-1} and mannitol or sorbitol 5 g L^{-1}, the best to conserve the seedlings for 730 days was 25% of the MS [19].

C. macrocarpum, O. maculata, and *P. estrellensis* plantlets were used for in vitro conservation for 450 days, the use of 25% of MS or 20 g L^{-1} sucrose at 25°C is recommended. *C. macrocarpum* and *O. maculata* seedlings were acclimatized reaching 100% survival [18].

3.2 In Vitro Culture: Plant Growth Regulators (PGRs)

ABA is a PGR that participates in the suppression of development and growth, as well as in the adaptive response to water stress survival [78]. For these reasons it is used to preserve various species of plants including orchids [79]. Other substances that are used are polyamides, which interferes with gibberellin and cellulose cells like ancymidol [80]; and PBZ which is a triazole that blocks the synthesis of gibberellins in plant cells, reducing shoot elongation during in vitro conservation [81].

3.2.1 Nodal Segments

The in vitro conservation of some species of agricultural importance like vanilla, involves the reduction of growth by polyamines on in vitro conservation of has been reported by Thyagi et al. [82] to conserved for until 320 days in spermidine. Besides, Bello-Bello [64] conserved *V. planifolia* using ABA 3 mg/L for 180 days, the shoots had lower growth values and 90% survival; this period did not affect the viability and phenotype of the plants.

3.2.2 Plantlets

It is important to rescue and conserve valuable genetic ornamental orchids resource, such as *Laelia anceps* seedlings and cultured in vitro with 2 mg L^{-1} of PBZ to reduce the development and growth, without affecting the survival of the plants after 168 days [46].

In vitro conservation of *Stanhopea tigrina* for 180 days, with 2 mg/L ABA, generated the lowest growth and regeneration of 3.2 shoots per explant was achieved with 2 mg/L BAP [59]; similar behavior as in *Notylia barkeri* Lindl. with ABA 1 mg/L the shoot length was reduced without affecting survival for 180 days [49].

3.3 In Vitro Culture Storage at Temperature: 4–27°C

Short-term storage of explants has been successfully achieved by reducing the temperature at which the cultures are grown [63]. Orchid in vitro storage at low temperature for short periods has been done with pollen, fruits, seed, protocorm, callus, protocorm-like bodies (PLBs), synthetic seed, buds, root tips, meristem, and plantlets storage.

3.3.1 Pollinia

Cohniella cepula pollinia were collected and transferred to cryovials to be stored at +4°C; in vitro seed germination showed their longevity only last 60 days [24].

3.3.2 Fruits

Orchids fruits of *Catasetum integerrimum* Hook., *Encyclia parviflora* (Regel) Withner, *Laelia anceps* Lindl., *Lycaste aromatica* (Graham) Lindl., *Myrmecophyla grandiflora* (Lindl.) Carnevali, Tapia-Muñoz y I. Ramírez, *Notylia Barkeri* Lindl., and *Oncidium sphacelatum* Lindl have been stored at +4°C for 6 weeks before germination in the laboratory of the authors.

3.3.3 Seeds

Cohniella cepula, *C. jonesiana*, and *Gomesa bifolia* seeds were collected and immediately transferred to cryovials and stored at 27°C and 4°C for 36 months; however, germination in vitro decreased to 0% within 1 and 6 months of storage, respectively, in all species [42].

3.3.4 Protocorms

Protocorms of *Dendrobium draconis* are reported to have high viability during storage at 25°C for 6 months [30].

3.3.5 Synthetic Seed Storage 3–27°C

Artificial seed was first coined by Murashige, it is now known as manufactured seed, synthetic seed, or synseed [83]. Synseeds are produced by encapsulation technology, a process that can be used for threatened species' conservation, followed by mass propagation and *ex vitro* establishment [25, 84]. Synseeds are capable of germinate in MS medium supplemented with BAP and IBA [62]; germination has been reported to reach up to 100 % [37] while development into normal plantlets ranged from 30 to 90 % [35], depending on storage conditions. Temperatures used for synseeds storage varies from 3°C [10] to 27°C [24] and storage period from 28 days [25] to 1 year [26, 36]. Synseed technology was ideal for germplasm conservation in orchids species such as *Cymbidium bicolor* [26], *C. finlaysonianum* [28], *Dendrobium lituiflorum* [32], *D. Shavin*White [35], *Dendrobium* Sonia [20], *D. officinale* [36], *D. wardianum* [37], *Geodorum densiflorum* [41], *Gomesa bifolia* [42], *Laelia anceps* ssp. Dawsonii [45], *Mokara sayan* × *Ascocenda wangsa* gold [48], *Oncidium* "Gower ramsay" [20], *Phalaenopsis* [56], and *Vanda coerulea* Grifft. ex. Lindl. [60]. Their acclimatization was reported from 60% [20]to 100% [35].

3.3.6 Microplantlets

Plantlets are explants that could be stored in low nutrients in the culture media to reduce growth [41]. For instance, *Ipsea malabarica* was maintained in 1/2 MS and sugar-free during storage at 25°C for 27 months, it had high viability [44]. *Dendrobium officinale* was conserved in vitro on 1/2 MS and 1/4 MS media for 12 months at 4°C in dark [36].

4 Conventional, Ultra-Freezed and Cryopreserved

Similar to short-term storage techniques, medium- and long-term storage is typically done at 0°C to −20°C, −70°C, and −180°C and can be performed using different explants (Fig. 6). Desirable effects at below-zero temperatures such as enzyme inactivation, photosynthesis inhibition and overall shift to a dormant state [85] help extend the period of storage, nonetheless damage to the cell membrane must be avoided to have acceptable regrowth from the explant used, which is why long-term storage techniques relay on the use of cryoprotectants to ensure the survival of the stored tissue or seed [86].

When non-adapted plants are exposed to temperatures below the freezing point of water, a series of stress responses can lead to freezing injuries that end up affecting germination, growth, development, and reproduction of the plant [87].

Cryoprotectant molecules synthesized by resistant plants such as AFP (antifreeze proteins) or IRI proteins (Ice Recrystallization Inhibition) help to mitigate cell dehydration and ice crystallization maintaining the liquid state of the intracellular region granting cold tolerance in these plants [88].

For storage purposes, synthetic cryoprotectants must be added to tropical plant species. Cryoprotectants must secure an adequate control of the physiological changes that take place during freezing temperatures. On the other hand, a correct desiccation avoids the formation of ice crystallization and while this strategy is successful for orthodox seeds, a pretreatment is required for other explants with high water content in which this sudden dehydration might lead to membrane rupture [71]; specifically in orchids, PLBs, protocorms, or even seeds of specific species are tissues or explants that require protective agents preventing over drying and cell membrane damage [86]. Most common methods for subzero temperature storage are encapsulation-dehydration, vitrification, droplet-vitrification, V-cryo plate and D-cryo plate method, all of them further explained below [71].

4.1 0°C to −20°C (Conventional Freezer)

Conventional freeze storage is a cheaper alternative to ULT refrigeration and does not require special facilities or a constant supply of LN, making it more available for underfunded laboratories. However, it has been observed that explants stored at this temperature lost viability as early as six months post-storage [89].

4.2 Encapsulation-Dehydration

The encapsulation-dehydration method can be used for both conventional and ultra-low temperature storage. For this protocol, seeds, protocorms, somatic embryos, PLBs, and pollen have been employed [29, 89–91]. It consists of the formation of beads by putting the desired, sterile explant or tissue in a sodium alginate solution varying from 2% to 5% w/v and the respective culture medium. Then a

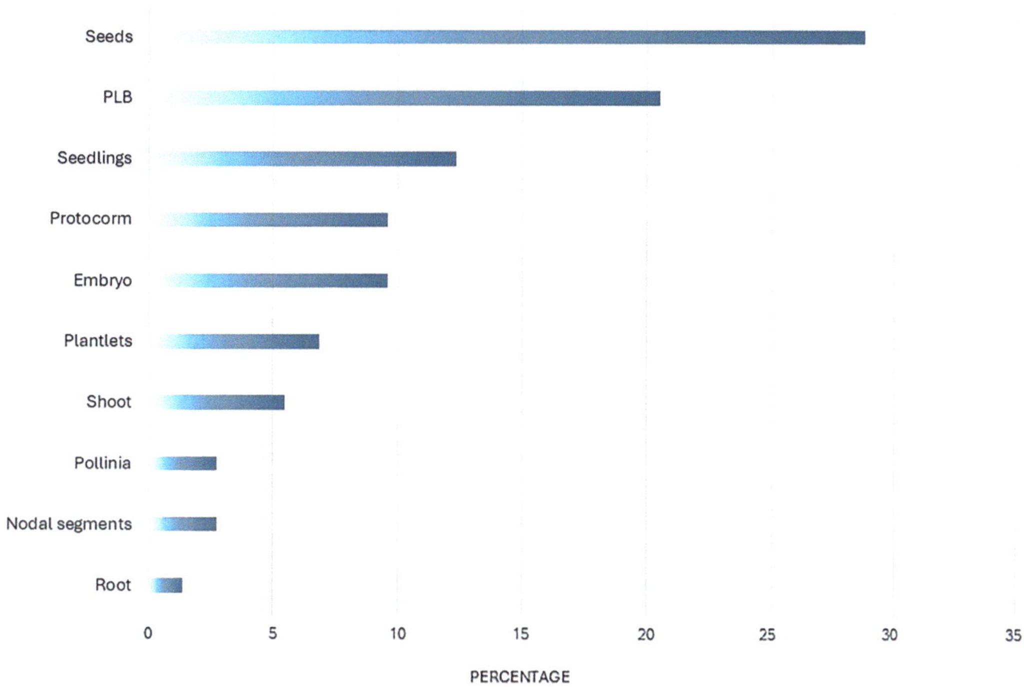

Fig. 6 Type of explant used for in vitro conservation protocols

drop-size aliquot is dropped into a calcium chloride solution [58]. During polymerization, the solution can be stirred at intervals to prevent adherence between beads [29]. Alginate, thus, works as a protectant from freezing stress. A preculture step can be performed by adding the beads to a sucrose, culture medium, DMSO, or glycerol [29, 92]. For the dehydration process alginate beads are surface-dried on filter paper and placed in petri dishes and exposed to laminar airflow for up to 6 h [91]. The treated beads are then placed into cryotubes and then put in a conventional or ULT refrigeration or in LN.

4.2.1 Seeds

Soomerville et al. [89] performed an adapted encapsulation protocol and tested it for both −18°C and −196°C storage and kept it there for 6 months. Interestingly, seeds from *Peterostylis saxicola* and *Diuris arenaria* were cold-stored alongside fungal isolates obtained from seed baiting. Apart from being less limited by lack of infrastructure or equipment, conventional freezer preservation showed equal success when compared to −196°C storage in this experiment, since the germination of beads stored at −18°C exhibited no significant difference from either the control or storage at −196°C after 6 months at their respective temperatures.

4.2.2 Embryos

Higher temperature storage using an encapsulation technique is also viable as demonstrated by Manokari et al. [58]. In their work, somatic embryos of *Spathoglottis plicata* were stored in the dark for 60 days at −4°C using the encapsulation method minus the desiccation step. Different sodium alginate concentrations were evaluated with differences in the consistency of the beads: higher concentration resulted in thicker, harder beads whereas 2% sodium alginate was more fragile. With this protocol, somatic embryos achieved a 97.4% germination frequency and subsequently were able to proliferate shoots in MS medium supplemented with 2 mg L^{-1} of BAP and 0.5 mg L^{-1} of IAA. This outcome proved the efficacy of −4°C for short-term storage, although larger periods of time ought to be explored.

4.2.3 Dehydration

Conversely, preservation at −18°C of *Dactylorhiza fuxhsii* and *Disa uniflora* pollen by Custodio et al. [90] was possible without an encapsulation method. Instead, moisture content was controlled modifying relative humidity (RH) to assess the optimal pollen humidity for cold preservation keeping an adequate viability. They tested four different relative humidity percentages and fresh pollinia. Their results showed that *D. uniflora* pollinia stored at −20°C and with a RH of 5% and 33% had similar germination index (GI) when compared to storage at −196°C. Similar results were obtained for *D. fuchsii*, although this species showed a less desiccant tolerance at a 5% RH.

4.3 196°C by Immersion in LN

Ultra-low temperature achieved by LN decreases the molecular motion of biological beings allowing to prolong the storage of these. The strategies for cryopreservation at ULT aim to reduce cryopreservation-induce damage [93].

4.3.1 Droplet-Vitrification

This method consists of treatment in a solution often composed of glycerol and culture medium called "loading solution" [92]. Then, treated explants are exposed to a Plant Vitrification Solution or PVS [94]. PSV variants [2–4] exist although PSV2 is most used. The PVS2 solution consists of 30% (w/v) glycerol, 15% (w/v) ethylene glycol, and 15% (w/v) dimethyl sulfoxide (DMSO) [34]. Using a syringe, droplets of PSV with the explant embedded are transferred into aluminum foils [95] and immediately placed in LN.

An improvement to both the droplet-vitrification and dehydration method is the use of cryoplates developed [96]. Cryoplates are devices made of A1050 aluminum grade allowing an optimal temperature transference. Cryoplates fit inside cryotubes making easier the manipulation of the explant during the cryopreservation procedure. D cryo-plates are used for dehydration procedures [72] while V cryoplates are used for vitrification techniques [97].

4.3.2 Pollen Dehydration

A general protocol for pollen dehydration and LN exposure was described by Decruse et al. [66] for the following orchid species: *Acampe praemorsa, Calanthe masuca, Cymbidium bicolor, C. ensifolium,*

Dendrobium ovatum, Eulophia cullenii, Rhynchostylis retusa, Vanda wightii, and *V. spathulate*. Dehydration takes place in a laminar airflow chamber. Pollinia is placed in open petri dishes and times vary depending on the species, but it ranges from 1 to 3 h for a 6.5%–19.37% moisture content. Pollen germination was not reduced.

5 Genetics Stability

Since the main objective of storage techniques of species which may be endangered is their germplasm conservation, several papers had assessed the genetic stability of slowed grown, stored or cryopreserved recovered plants.

Hirano [12] evaluated the relative DNA content by flow cytometry (FCM) of immature *Bletilla striata* that were subjected to a vitrification process and cryopreserved. Compared with the control group, there was no difference in the ploidy level of seeds. Additionally, when evaluating the morphology of the resulting seedlings from cryopreserved seed after a 6-month culture there were no noticeable abnormalities, further enforcing the assumption that the molecular stability after cryopreservation was preserved. These results were similar to the ones shown by Galdiano in *Oncidium flexuosum* Sims [51] and a *Dendrobium* hybrid [98].

The genetic stability after cryopreservation was also evaluated in seeds of *Cyrtopodium hatschbachii* [99] but at a chromosomic level. The cryopreservation was made by the encapsulation-dehydration technique and stored in liquid nitrogen for 12 h and the chromosomal evaluation was made in 195-day-old plantlets derived from these seeds. DNA from control plants, plants derived from cryopreserved seed and donor plant was tested showing an identical number of chromosomes in all three conditions ($2n = 46$). Yet, there was a difference when the average centromeric index was established, indicating that the total length of the plant chromosomes from cryopreserved seeds was greater than that from the control group. This decrease in the ability to condense chromatin, as pointed out by the author, may suggest an abnormal methylation of DNA.

Although DNA methylation levels are not evaluated in this study nor in more recent ones, it is an important parameter worth to be study since DNA methylation has several molecular functions such as gene regulation, transposon silencing, chromosome interaction, and it also has various roles in plant development, for example seed development, vegetative growth and pattern formation as well as the response to external stimuli [100]. Johnson et al. [101] reported the induction of DNA demethylation in non-cryotolerant *Ribes* shoots recovered from cryopreserved meristems although this process fails to be reflected phenotypically.

In orchids DNA methylation is linked to floral pigmentation in *Oncidium* and stress response to cold and drought in *Dendrobium officinale* [102]; therefore, it is important to consider the effects of cryopreservation in DNA methylation in orchids.

5.1 Molecular Markers to Assess Gene Fidelity

Alternatively, the molecular fidelity of stored or cryopreserved tissues and the plantlets derived from these can be measured using molecular markers such as ISSR (inter simple sequence repeat) [103–105], DAMD (direct amplification of minisatellite DNA) [105], or RAPD (randomly amplified polymorphic DNA) [48, 106–109].

The post-storage genetic stability of plantlets, which were previously stored as synthetic seeds, from the hybrid *Aranda* Wan Chark Kuan "Blue" × *Vanda coerulea* was tested by Gantait [103]. Encapsulated shoot tips were stored at 4°C for a period of time ranging from 40 to 200 days and DNA from the subsequent plantlets was extracted. When observed, the PCR products, given by 9 ISSR primers, showed 51 scorable bands. These bands were compared among the stored and non-stored derived plantlets indicating no polymorphic amplification product, and therefore, suggesting the lack of detectable mutations as results of the vitrification process.

Likewise, ISSR analysis has been performed in cryopreserved tissues of *Pleione bulbocodioides*, , and *Ludisia discolor* [61, 104, 105]. Gonzalez-Arnao [24] showed a high genetic stability in cryopreserved PLBs using the droplet-vitrification and the V-cryo-plate (V Cp) method, the latter being better in retaining molecular fidelity (8.5% and 3% polymorphic percentage respectively). Since the DNA extraction was performed in every step for each method, it was possible to observe that the PSV2 treatment was the step that caused greater genetic variability and not the liquid nitrogen exposure itself. Contrary to this, Poothaby et al. [107] analyzed amplification bands for RAPD primers and found that DNA extracted from encapsulated, dehydrated, and cryopreserved PLBs of *Dendrobium Sonia 28* were mostly unable to produce PCR products suggesting that genomic changes took place at the liquid nitrogen submersion stage, since non-cryopreserved encapsulated-dehydrated PLBs got high similarity indices at 8 out 12 primers. Their results suggest that the vitrification-cryopreservation method preserves genetic fidelity better than the encapsulation-dehydration-cryopreserved method since the similarity indices from the former were 1 for 9 out of 12 primers.

When choosing a molecular marker to assess the genetic stability of stored tissues, it is important to consider that a good DNA or RNA marker must be able to display a high number of polymorphisms. Comparing diverse types of markers may be useful to determine which one is suitable for each species. Kirutika et al. [105] tested three different types of markers, ISSR, DAMD, and SCoT, and established that, for cryopreserved axillary buds of *Ludisia discolor*, DAMD and SCoT worked better for a genetic fidelity screening. Their results reflected that DAMD was more sensitive in detecting polymorphic bands when compared control versus cryopreserved axillary buds (12.5% polymorphism) and SCoT had higher sensitivity in detecting polymorphic bands when compared DNA amplification bands of control versus untreated axillary buds (10% polymorphism), ISSRs markers being less efficient in exhibiting polymorphisms among cryopreserved and non-cryopreserved axillary buds versus the control axillary buds (4.17% and 5.77%, respectively). As mentioned, they did report some genetic variations due to the cryopreservation process in all three markers, although the precise stage at which these changes took place was not assessed. Similarly, Gantait et al. [48] tried SCoT and RAPD markers but were unable to identify polymorphisms between control and vitrified-cryopreserved PLBs of *Aranda* Broga Blue, therefore determining the genetic stability of regenerated PLBs.

ISSR, RAPD, DAMD, and SCoT markers are convenient, PCR-based techniques that allow the verification of DNA conservation during and after short- and long-term storage, to ensure that the procedures won't be damaging enough to alter the genetic fidelity of the treated explant. It is worth noting that amplified regions by these markers ideally do not code for proteins and will not reflect phenotypically.

An important advantage to the use of these markers is that no prior knowledge of the complete genome is necessary for these assays [110–112], so they can be used among different orchid species or cultivars though reproductivity relies on a strict PCR protocol.

6 Conclusion

In vitro storage techniques are valuable methods for germplasm preservation and have proven to be effective in conserving genetic fidelity. Each protocol is suitable for different needs and the physiological differences of orchid species, and the type of explant must be considered before applying them. Few orchid species have been submitted to long-term storage, with 3 years being the longest time for which an explant has been stored in the studies reviewed.

References

1. The International Orchid Register/RHS Gardening. https://apps.rhs.org.uk/horticulturaldatabase/orchidregister/orchidregister.asp. Accessed 6 Nov 2024
2. Ackerman JD, Phillips RD, Tremblay RL et al (2023) Beyond the various contrivances by which orchids are pollinated: global patterns in orchid pollination biology. Bot J Linn Soc

202:295–324. https://doi.org/10.1093/botlinnean/boac082
3. Atwood JT (1986) The size of the orchidaceae and the systematic distribution of epiphytic orchids. Selbyana 9:171–186
4. Rasmussen HN, Dixon KW, Jersáková J, Těšitelová T (2015) Germination and seedling establishment in orchids: a complex of requirements. Ann Bot 116:391–402. https://doi.org/10.1093/aob/mcv087
5. Rasmussen HN, Rasmussen FN (2018) The epiphytic habitat on a living host: reflections on the orchid-tree relationship. Bot J Linn Soc 186:456–472. https://doi.org/10.1093/botlinnean/box085
6. Parra-Sanchez E, Freckleton RP, Hethcoat MG et al (2024) Transformation of natural habitat disrupts biogeographical patterns of orchid diversity. Biol Conserv 292:110538. https://doi.org/10.1016/j.biocon.2024.110538
7. Fay MF (2018) Orchid conservation: how can we meet the challenges in the twenty-first century? Bot Stud 59:16. https://doi.org/10.1186/s40529-018-0232-z
8. Bhattacharyya P, Kumar V, Van Staden J (2018) In vitro encapsulation based short term storage and assessment of genetic homogeneity in regenerated *Ansellia africana* (Leopard orchid) using gene targeted molecular markers. Plant Cell Tissue Organ Cult 133:299–310. https://doi.org/10.1007/s11240-018-1382-0
9. Magrini S, De Vitis M, Torelli D et al (2019) Seed banking of terrestrial orchids: evaluation of seed quality in Anacamptis following 4-year dry storage. Plant Biol 21:544–550. https://doi.org/10.1111/plb.12936
10. Chang C, Sung PG, Chang C et al (2006) Seed development and storage of *Bletilla formosana* (Hayata) Schltr. Seed Nurs 8:29–38
11. Ishikawa K, Harata K, Mii M et al (1997) Cryopreservation of zygotic embryos of a Japanese terrestrial orchid (*Bletilla striata*) by vitrification Plant Cell Reports 16(11) 754-757. https://doi.org/10.1007/s002990050314
12. Hirano T, Godo T, Mii M, Ishikawa K (2005) Cryopreservation of immature seeds of *Bletilla striata* by vitrification Plant Cell Reports 23(8):534-539. https://doi.org/10.1007/s00299-004-0893-9
13. Wu R-Y, Chang S-Y, Hsieh T-F, Chang Y-S (2013) Cryopreservation of *Bletilla formosana* seeds (Orchidaceae) by desiccation Scientia Horticulturae 157:108-112. https://doi.org/10.1016/j.scienta.2013.03.033
14. Hu WH, Yang YH, Liaw SI, Chang C (2013) Cryopreservation the seeds of a Taiwanese terrestrial orchid *Bletilla formosana* (Hayata) Schltr. by vitrification. Bot Stud 54:33. https://doi.org/10.1186/1999-3110-54-33
15. Mata-Rosas M, Lastre-Puertos E (2015) Long-term conservation of protocorms of *Brassavola nodosa* (L) Lind. (orchidaceae): effect of ABA and a range of cryoconservation techniques. Cryo Lett 36:289–298
16. Watanawikkit P, Tantiwiwat S, Bunn E et al (2012) Cryopreservation of in vitro-propagated protocorms of Caladenia for terrestrial orchid conservation in Western Australia: cryopreservation for terrestrial orchid conservation. Bot J Linn Soc 170:277–282. https://doi.org/10.1111/j.1095-8339.2012.01284.x
17. López Puc G, Herrera Cool GJ (2022) Germinación asimbiótica, conservación ex situ e in vitro regeneración de plantas de *Catasetum integerrimum* Hook. Polibotánica 53:135-149. https://doi.org/10.18387/polibotanica.53.9
18. Menezes TSA, Arrigoni- Blank MDF, Costa ASD et al (2019) In vitro conservation and acclimatization of Epidendroideae (Orchidaceae) from Sergipe, Brazil. Biosci J 356–366. https://doi.org/10.14393/BJ-v35n2a20198-38776
19. Menezes-Sá TSA, Costa ASD, Arrigoni-Blank MDF et al (2022) *In vitro* propagation and conservation of *Cattleya tigrina* A. Rich. Ciênc Rural 52:e20200517. https://doi.org/10.1590/0103-8478cr20200517
20. Saiprasad GVS, Polisetty R (2003) Propagation of three orchid genera using encapsulated protocorm-like bodies. In Vitro Cell Dev Biol Plant 39:42–48. https://doi.org/10.1079/IVP2002360
21. Vettorazzi RG, Carvalho VS, Teixeira MC et al (2019) Cryopreservation of immature and mature seeds of Brazilian orchids of the genus Cattleya. Sci Hortic 256:108603. https://doi.org/10.1016/j.scienta.2019.108603
22. Zargar Azad M, Kaviani B, Sedaghathoor S (2024) Cold storage and cryopreservation by encapsulation-dehydration of *Cephalanthera rubra* (L.) Rich., a threatened orchid species. Acta Agrobot 77:1–13. https://doi.org/10.5586/aa/189312
23. Mohanraj R, Ananthan R, Bai VN (2009) Production and storage of synthetic seeds in *Coelogyne breviscapa* Lindl. Asian J Biotechnol 1:124–128. https://doi.org/10.3923/ajbkr.2009.124.128
24. Dolce NR, González-Arnao MT (2020) Cryopreservation as a tool for long-term storage of *Cohniella cepula* (Orchidaceae) seeds. In: Acta Hortic. International Society for Horticultural Science, pp 113–120 https://doi.org/10.17660/ActaHortic.2020.1297.16

25. Pradhan S, Tiruwa B, Subedee BR, Pant B (2014) In vitro germination and propagation of a threatened medicinal orchid, *Cymbidium aloifolium* (L.) Sw. through artificial seed. Asian Pac J Trop Biomed 4:971–976. https://doi.org/10.12980/APJTB.4.2014APJTB-2014-0369
26. Mahendran G (2014) Encapsulation of Protocorm of *Cymbidium bicolor* Lindl. for Short-Term Storage and Germplasm Exchange. Journal of Ornamental Plants 4(4):17–27
27. Gogoi K, Kumaria S, Tandon P (2013) Cryopreservation of *Cymbidium eburneum* Lindl. and *C. hookerianum* Rchb. f., two threatened and vulnerable orchids via encapsulation–dehydration. In Vitro Cell Dev Biol Plant 49:248–254. https://doi.org/10.1007/s11627-013-9505-0
28. Klaocheed S, Rittirat S, Thammasiri K, Prasertsongskun S (2017) Alginate-encapsulation, short-term storage and plantlet regeneration from encapsulated protocorm-like bodies (PLBs) of *Cymbidium finlaysonianum* Lindl.: an endangered orchid of Thailand. Walailak J Sci Technol WJST 15:725–737. https://doi.org/10.48048/wjst.2018.3325
29. Teixeira Da Silva JA (2013) Cryopreservation of hybrid *Cymbidium* protocorm-like bodies by encapsulation–dehydration and vitrification. In Vitro Cell Dev Biol Plant 49:690–698. https://doi.org/10.1007/s11627-013-9550-8
30. Rangsayatorn N, Jina K, Pampasit S (2009) Slow growth in vitro conservation of dendrobium draconis rchb. F.: effect of black kwao krua (mucuna collettii lace). Acta Hortic (829):355–358. https://doi.org/10.17660/ActaHortic.2009.829.56
31. P., Sukhumpinij (2019) Effect of sucrose on in vitro germplasm preservation in *Dendrobium friedericksianum* Rchb.f. Acta Horticulturae (1262) 63-66. https://doi.org/10.17660/ActaHortic.2019.1262.9
32. Das MC, Kumaria S, Tandon P (2008) *In vitro* propagation and conservation of *Dendrobium lituiflorum* Lindl through protocorm-like bodies. J Plant Biochem Biotechnol 17:177–181. https://doi.org/10.1007/BF03263281
33. Mohanty P, Nongkling P, Das MC et al (2013) Short-term storage of alginate-encapsulated protocorm-like bodies of *Dendrobium nobile* Lindl.: an endangered medicinal orchid from North-east India. 3 Biotech 3:235–239. https://doi.org/10.1007/s13205-012-0090-4
34. Mohanty P, Das MC, Kumaria S, Tandon P (2012) High-efficiency cryopreservation of the medicinal orchid *Dendrobium nobile* Lindl. Plant Cell Tissue Organ Cult 109:297–305. https://doi.org/10.1007/s11240-011-0095-4
35. Bustam S, Sinniah UR, Kadir MA et al (2013) Selection of optimal stage for protocorm-like bodies and production of artificial seeds for direct regeneration on different media and short term storage of *Dendrobium* Shavin White. Plant Growth Regul 69:215–224. https://doi.org/10.1007/s10725-012-9763-6
36. Yongzhong S, Rui-chi P, Xiao-jing W et al (2000) *In vitro* conservation of *Dendrobium officinale* at low temperature. Chin J Appl Environ Biol 6:326–330
37. Sharma A, Tandon P, Kumar A (1992) Regeneration of Dendrobium wardianum Warne (orchidaceae) from synthetic seeds. Indian J Exp Biol 30:747–748
38. Cerna M, Valdivieso P, Cella R et al (2018) Cryopreservation of orchid seeds through rapid and step freezing methods. F1000Res 7:209. https://doi.org/10.12688/f1000research.13622.1
39. Lopez-Puc G (2013) An effective *in vitro* slow growth protocol for conservation of the orchid *Epidendrum chlorocorymbos* SCHLTR. Trop Subtrop Agroecosyst 16:61–68
40. Linjikao J, Inthima P, Limmongkon A, Kongbangkerd A (2024) Mannitol and sorbitol concentration optimization for effective Epipactis flava Seidenf. in vitro slow growth storage. In Vitro Cell Dev Biol Plant 60:496–507. https://doi.org/10.1007/s11627-024-10437-w
41. Datta KB, Kanjilal B, Sarker DD (1999) Artificial seed technology: development of a protocol in Geodorum densiflorum (Lam) Schltr. – an endangered orchid. Curr Sci 76:1142–1145
42. Dolce NR, Medina RD, Terada G et al (2020) In vitro propagation and germplasm conservation of wild orchids from South America. In: Khasim SM, Hegde SN, González-Arnao MT, Thammasiri K (eds) Orchid biology: recent trends & challenges. Springer Singapore, Singapore, pp 37–94. https://doi.org/10.1007/978-981-32-9456-1_4
43. Franceschi CRB, Smidt EC, Vieira LN, Ribas LLF (2019) Storage and in vitro germination of orchids (Orchidaceae) seeds from Atlantic Forest – Brazil. An Acad Bras Ciênc 91:e20180439. https://doi.org/10.1590/0001-3765201920180439
44. Martin KP (2003) Simple strategy for the *in vitro* conservation of Ipsea malabarica an endemic and endangered orchid of the Western Ghats of Kerala, India. Plant Cell

Tissue Organ Cult 74:197–200. https://doi.org/10.1023/A:1023971625994
45. Lee-Espinosa HE, Murguía-González J et al (2009) Somatic embryoids encapsulation of *Laelia anceps* ssp. dawsonii to produce synthetic seed. Rev Chapingo Ser Hortic XV:33–40. https://doi.org/10.5154/r.rchsh.2009.15.047
46. Ramírez-Mosqueda MA, Cruz-Cruz CA, Atlahua-Temoxtle J, Bello-Bello JJ (2019) *In vitro* conservation and regeneration of *Laelia anceps* Lindl. South Afr J Bot 121:219–223. https://doi.org/10.1016/j.sajb.2018.11.010
47. Ajeeshkumar S, Decruse SW (2013) Fertilizing ability of cryopreserved pollinia of Luisia macrantha, an endemic orchid of Western Ghats. Cryo Lett 34:20–29
48. Gantait S, Subrahmanyeswari T, Sinniah UR (2022) Leaf-based induction of protocorm-like bodies, their encapsulation, storage and post-storage germination with genetic fidelity in Mokara Sayan × Ascocenda Wangsa gold. South Afr J Bot 150:893–902. https://doi.org/10.1016/j.sajb.2022.08.012
49. García-Merino GF, Ramírez Mosqueda MA, Mata-Alejandro MA et al (2024) Conservación *in vitro* de *Notylia barkeri* Lindl.: Rescate de *Notylia barkeri*. Rev Bio Cienc. 11:e1633. https://doi.org/10.15741/revbio.11.e1633
50. Flachsland E, Terada G, Scocchi A et al (2006) Cryopreservation of seeds and in vitro-cultured protocorms of Oncidium bifolium Sims. (Orchidaceae) by encapsulation-dehydration. Cryo Lett 27:235–242
51. Galdiano RF, De Macedo Lemos EG, Vendrame WA (2013) Cryopreservation, early seedling development, and genetic stability of *Oncidium flexuosum* Sims. Plant Cell Tissue Organ Cult 114:139–148. https://doi.org/10.1007/s11240-013-0304-4
52. Zakaria S, Subramaniam S, Mubbarakh SA, James Antony JJ (2020) Effect of encapsulation-dehydration cryopreservation on histological analysis of oncidium golden anniversary orchid PLBs. IOP Conf Ser Earth Env Sci 596:012082. https://doi.org/10.1088/1755-1315/596/1/012082
53. Soonthornkalump S (2019) In vitro propagation and cryopreservation of lady's slipper orchid (Paphiopedilum niveum (Rchb.f.) Stein). Thesis
54. Chaireok S, Thammasiri K, Meesawat U (2016) Vitrification-based cryopreservation of protocorm-like bodies of an endangered lady's slipper orchid: Paphiopedilum niveum (Rchb.f.) Stein. Cryo Lett 37:154–162
55. Diengdoh RV, Kumaria S, Das MC (2019) Antioxidants and improved regrowth procedure facilitated cryoconservation of Paphiopedilum insigne Wall. Ex. Lindl. – an endangered slipper orchid. Cryobiology 87:60–67. https://doi.org/10.1016/j.cryobiol.2019.02.003
56. Mahdavi Z, Daylami SD, Fadavi A, Vahdati K (2023) Artificial seed production of Phalaenopsis orchid: effect of encapsulation materials, temperature, light spectra, and storage period. Plant Cell Tissue Organ Cult 155:797–808. https://doi.org/10.1007/s11240-023-02600-9
57. Chi-Ramírez Maura R, Herrera-Cool GJ, Sánchez-Contreras A, Lopez-Puc G (2024) Clonal micropropagation and *ex-situ* conservation of *Rhycncholaelia digbyana* (Lindley) Schltr. Agro Product. : 2024: EARLY ACCESS (Vol. VI). https://doi.org/10.32854/agrop.v17i6.2832
58. Manokari M, Latha R, Priyadharshini S et al (2021) Short-term cold storage of encapsulated somatic embryos and retrieval of plantlets in grey orchid (*Vanda tessellata* (Roxb.) Hook. ex G.Don). Plant Cell Tissue Organ Cult 144:171–183. https://doi.org/10.1007/s11240-020-01899-y
59. Cruz-Cruz CA, González-Arnao MT, Bautista-Aguilar JR, Ramírez-Mosqueda MA (2022) *In vitro* short-term storage of *Stanhopea tigrina* Bateman ex Lind. South Afr J Bot 151:334–338. https://doi.org/10.1016/j.sajb.2022.10.014
60. Sarmah D, Borthakur MP, Borua PK (2010) Artificial seed production from encapsulated PLBs regenerated from leaf base of Vanda coerulea Grifft. ex. Lindl. – an endangered orchid. Curr Sci 98:686–690
61. González-Arnao MT, Hernández-Ramírez F, Dolce NR et al (2020) Cryobiotechnological studies in Vanilla: the orchid of multi-industrial uses. In: Khasim SM, Hegde SN, González-Arnao MT, Thammasiri K (eds) Orchid biology: recent trends & challenges. Springer Singapore, Singapore, pp 21–35
62. Divakaran M, Babu KN, Peter KV (2006) Conservation of Vanilla species, *in vitro*. Sci Hortic 110:175–180. https://doi.org/10.1016/j.scienta.2006.07.003
63. Tandon, P. and Kumaria, S., Prospects of Plant Conservation Biotechnology in India with special reference to Northeastern Region, In: Tandon, P., Sharma, M. and Swarup, R., (eds) Biodiversity: status and prospects, Narosa Publishing House, New Delhi, 2005, 79-92
64. Bello-Bello JJ, García-García GG, Iglesias-Andreu L (2015) Conservación de vainilla (*Vanilla planifolia* Jacks.) bajo condiciones de lento crecimiento *in vitro*. Rev Fitotec

Mex 38:165. https://doi.org/10.35196/rfm.2015.2.165
65. Manokari M, Priyadharshini S, Cokulraj M et al (2022) Exogenous implications of silver nitrate on direct and indirect somatic embryogenesis and germination of cold stored synseeds of *Vanilla planifolia* Jacks. ex Andrews. South Afr J Bot 150:129–138. https://doi.org/10.1016/j.sajb.2022.07.019
66. Decruse SW, Ajeeshkumar S, Vargheese S (2023) Pollinia Cryopreservation of selected orchid species of Western Ghats, India. In: Rajasekharan PE, Rohini MR (eds) Pollen cryopreservation protocols. Springer US, New York, NY, pp 401–414
67. Agrawal DC, Morwal GC, Mascarenhas AF (1992) In vitro propagation and slow growth storage of shoot cultures of Vanilla walkeriae Wight: an endangered orchid. Lindleyana 7:95–99
68. Hinsley A, De Boer HJ, Fay MF et al (2018) A review of the trade in orchids and its implications for conservation. Bot J Linn Soc 186:435–455. https://doi.org/10.1093/botlinnean/box083
69. Panis B (2019) Sixty years of plant cryopreservation: from freezing hardy mulberry twigs to establishing reference crop collections for future generations. Acta Hortic 1234:1–8. https://doi.org/10.17660/ActaHortic.2019.1234.1
70. Mota-Babiloni A, Mastani Joybari M, Navarro-Esbrí J et al (2020) Ultralow-temperature refrigeration systems: configurations and refrigerants to reduce the environmental impact. Int J Refrig 111:147–158. https://doi.org/10.1016/j.ijrefrig.2019.11.016
71. Malhotra EV, Bansal S, Gupta S (2024) Plant cryopreservation: a molecular perspective. Plant Cell Tissue Organ Cult 158:19. https://doi.org/10.1007/s11240-024-02803-8
72. Wang M-R, Lambardi M, Engelmann F et al (2021) Advances in cryopreservation of in vitro-derived propagules: technologies and explant sources. Plant Cell Tissue Organ Cult 144:7–20. https://doi.org/10.1007/s11240-020-01770-0
73. Bunn E, Turner S, Panaia M, Dixon KW (2007) The contribution of *in vitro* technology and cryogenic storage to conservation of indigenous plants. Aust J Bot 55:345. https://doi.org/10.1071/BT06065
74. Benson EE, Harding K (2012) Cryopreservation of shoot tips and meristems: an overview of contemporary methodologies. In: Loyola-Vargas VM, Ochoa-Alejo N (eds) Plant cell culture protocols. Humana Press, Totowa, NJ, pp 191–226
75. Engelmann F (2011) Use of biotechnologies for the conservation of plant biodiversity. In Vitro Cell Dev Biol Plant 47:5–16. https://doi.org/10.1007/s11627-010-9327-2
76. Nasiruddin M, Islam AR (2018) *In vitro* slow-growth conservation for two genotypes of *Solanum tuberosum* L. Bangladesh J Bot 47:369–380. https://doi.org/10.3329/bjb.v47i3.38654
77. Shibli RA, Smith MAL, Spomer LA (1992) Osmotic adjustment and growth responses of three Chrysanthemum morifolium Ramat. cultivars to osmotic stress induced in vitro. J Plant Nutr 15:1373–1381. https://doi.org/10.1080/01904169209364404
78. Zandvoort EA, Hulshof MJH, Staritsky G (1994) *In vitro* storage of *Xanthosoma* spp. under minimal growth conditions. Plant Cell Tissue Organ Cult 36:309–316. https://doi.org/10.1007/BF00046088
79. Cruz-Cruz C, González-Arnao M, Engelmann F (2013) Biotechnology and conservation of plant biodiversity. Resources 2:73–95. https://doi.org/10.3390/resources2020073
80. Chen J-T, Chang W-C (2003) Effects of GA3, ancymidol, cycocel and paclobutrazol on direct somatic embryogenesis of *Oncidium in vitro*. Plant Cell Tissue Organ Cult 72:105–108. https://doi.org/10.1023/A:1021235700751
81. Seesangboon A, Gruneck L, Pokawattana T et al (2018) Transcriptome analysis of Jatropha curcas L. flower buds responded to the paclobutrazol treatment. Plant Physiol Biochem 127:276–286. https://doi.org/10.1016/j.plaphy.2018.03.035
82. Tyagi RK, Yusuf A, Jeyaprakash P, Dua P (2001) Effects of polyamines on in vitro conservation of *Vanilla planifolia* (Salisb.) Ames. Indian J Plant Genet Resour 14:300–302
83. Murashige T (1977) Plant cell and organ cultures as horticultural practices. Acta Hortic (78):17–30. https://doi.org/10.17660/ActaHortic.1977.78.1
84. Lal A, Pant M, Palni LMS, Kumar A (2020) Development of rapid micropropagation protocol for germplasm conservation of two orchid species - *Aerides multiflora* Roxb. and *Rhynchostylis retusa* (L.) Blume. Asian. J Conserv Biol 9:341–347
85. Źróbek-Sokolnik A (2012) Temperature stress and responses of plants. In: Ahmad P, Prasad MNV (eds) Environmental adaptations and stress tolerance of plants in the era of climate change. Springer New York, New York, NY, pp 113–134
86. Huehne PS, Bhinija K (2012) Application of cryoprotectants to improve low temperature storage survival of orchid seeds. Sci Hortic 135:186–193. https://doi.org/10.1016/j.scienta.2011.11.026

87. Aslam M, Fakher B, Ashraf MA et al (2022) Plant low-temperature stress: signaling and response. Agronomy 12:702. https://doi.org/10.3390/agronomy12030702
88. Jahed KR, Saini AK, Sherif SM (2023) Coping with the cold: unveiling cryoprotectants, molecular signaling pathways, and strategies for cold stress resilience. Front Plant Sci 14:1246093. https://doi.org/10.3389/fpls.2023.1246093
89. Sommerville KD, Siemon JP, Wood CB, Offord CA (2008) Simultaneous encapsulation of seed and mycorrhizal fungi for long-term storage and propagation of terrestrial orchids. Aust J Bot 56:609. https://doi.org/10.1071/BT08008
90. Custodio CC, Machado-Neto NB, Singer RB et al (2020) Storage of orchid pollinia with varying lipid thermal fingerprints. Protoplasma 257:1401–1413. https://doi.org/10.1007/s00709-020-01514-z
91. Maneeratta P, Bunnag S, Monthatong M (2007) In vitro conservation of *Cleisostoma areitinum* (Rchb.f.) Garay, Rare Thai orchid species by an encapsulation-dehydration method. Asian J Plant Sci 6:1235–1240. https://doi.org/10.3923/ajps.2007.1235.1240
92. Galdiano RF, Lemos EGM (2018) Encapsulation-vitrification and encapsulation-dehydration cryopreservation of *Cattleya labiata* Lindley, a threatened Brazilian orchid. Acta Hortic (1224):135–138. https://doi.org/10.17660/ActaHortic.2018.1224.18
93. Murray KA, Gibson MI (2022) Chemical approaches to cryopreservation. Nat Rev Chem 6:579–593. https://doi.org/10.1038/s41570-022-00407-4
94. Bustam BM, Dixon K, Bunn E (2016) A cryopreservation protocol for *ex situ* conservation of terrestrial orchids using asymbiotic primary and secondary (adventitious) protocorms. In Vitro Cell Dev Biol Plant 52:185–195. https://doi.org/10.1007/s11627-015-9732-7
95. Popova E, Kim H-H (2020) Cryobiotechnology of Korean orchid biodiversity: a case study using *Cymbidium kanran*. In: Khasim SM, Hegde SN, González-Arnao MT, Thammasiri K (eds) Orchid biology: recent trends & challenges. Springer Singapore, Singapore, pp 119–135
96. Niino T, Yamamoto S, Matsumoto T, Engelmann F, Valle Arizaga M, Tanaka D (2019) Development of V and D cryo-plate methods as effective protocols for cryobanking. Acta Hortic (1234):249–262. https://doi.org/10.17660/ActaHortic.2019.1234.33
97. Dolce NR, Hernández-Ramírez F, González-Arnao MT (2019) Cryopreservation of vanilla (*Vanilla planifolia*) root-tips: a new alternative for in vitro long-term storage of its germplasm. Acta Hortic (1234):203–210. https://doi.org/10.17660/ActaHortic.2019.1234.27
98. Galdiano RF, De Macedo Lemos EG, De Faria RT, Vendrame WA (2014) Seedling development and evaluation of genetic stability of cryopreserved dendrobium hybrid mature seeds. Appl Biochem Biotechnol 172:2521–2529. https://doi.org/10.1007/s12010-013-0699-8
99. Surenciski M, Dematteis M, Flachsland E (2007) Chromosome stability in cryopreserved germplasm of Cyrtopodium hatschbachii (Orchidaceae). Fenn Ann Bot Fenn 44:287–292
100. Zhang H, Lang Z, Zhu J-K (2018) Dynamics and function of DNA methylation in plants. Nat Rev Mol Cell Biol 19:489–506. https://doi.org/10.1038/s41580-018-0016-z
101. Johnston JW, Benson EE, Harding K (2009) Cryopreservation induces temporal DNA methylation epigenetic changes and differential transcriptional activity in Ribes germplasm. Plant Physiol Biochem 47:123–131. https://doi.org/10.1016/j.plaphy.2008.10.008
102. Lucibelli F, Valoroso MC, Aceto S (2022) Plant DNA methylation: an epigenetic mark in development, environmental interactions, and evolution. Int J Mol Sci 23:8299. https://doi.org/10.3390/ijms23158299
103. Gantait S, Sinniah UR (2013) Storability, post-storage conversion and genetic stability assessment of alginate-encapsulated shoot tips of monopodial orchid hybrid Aranda Wan Chark Kuan 'Blue' × Vanda coerulea Grifft. ex. Lindl. Plant Biotechnol Rep 7:257–266. https://doi.org/10.1007/s11816-012-0257-9
104. Cheng W, Li H, Zhou F et al (2020) Cryopreservation of *Pleione bulbocodioides* (Franch.) Rolfe protocorm-like bodies by vitrification. Acta Physiol Plant 42:82. https://doi.org/10.1007/s11738-020-03074-4
105. Subramaniam S, Selva Rajan K, Burkhan H et al (2022) Comparative analysis of DAMD, ISSR and SCOT molecular markers on cryopreserved ludisia discolor axillary buds. Malays J Sci 41:1–15. https://doi.org/10.22452/mjs.vol41no2.1
106. Antony JJJ, Poobathy R, Danial M et al (2012) Polymorphism analysis of cryopreserved Dendrobium Bobby Messina protocorm-like bodies (PLBs) using RAPD markers. Plant Omics 5:427–431
107. Poobathy R, Xavier R, Sinniah UR, Subramaniam S (2013) Molecular stability of protocorm-like bodies of Dendrobium sonia-28 after encapsulation-dehydration and vitrification. Aust J Crop Sci 7:189–195

108. Duangrath Jariyajirawattana (2020) The effects of vitrification method on the seed survival and genetic stability of Rhynchostylis gig antea (Lindl.) Ridl. Asia-Pac J Sci Technol 25:19. https://doi.org/10.14456/APST.2020.25
109. Khor SP, Yeow LC, Poobathy R et al (2020) Droplet-vitrification of Aranda Broga Blue orchid: role of ascorbic acid on the antioxidant system and genetic fidelity assessments via RAPD and SCoT markers. Biotechnol Rep 26:e00448. https://doi.org/10.1016/j.btre.2020.e00448
110. Bidyananda N, Jamir I, Nowakowska K et al (2024) Plant genetic diversity studies: insights from DNA marker analyses. Int J Plant Biol 15:607–640. https://doi.org/10.3390/ijpb15030046
111. Collard BCY, Mackill DJ (2009) Start codon targeted (SCoT) polymorphism: a simple, novel DNA marker technique for generating gene-targeted markers in plants. Plant Mol Biol Rep 27:86–93. https://doi.org/10.1007/s11105-008-0060-5
112. Amiteye S (2021) Basic concepts and methodologies of DNA marker systems in plant molecular breeding. Heliyon 7:e08093. https://doi.org/10.1016/j.heliyon.2021.e08093

Check for updates

Chapter 22

Factors Influencing In Vitro Conservation of *Vanilla* Species in Minimal Growth

Paulo César Parada-Molina, Zelzin Erendira Fernández-Villa, Ariadna Linares-Gabriel, María Esther Diaz-Martínez and José Roberto Bautista-Aguilar

Abstract

Vanilla planifolia is of great importance in the food and cosmetic industry. *V. planifolia* is currently reported to be at risk. This is due to overharvesting, habitat destruction, and loss of wild specimens. Therefore, it is important to undertake actions for the rescue, characterization, and conservation of this important phytogenetic resource. Plant biotechnology provides several very useful techniques in the development of conservation programs for endangered plant species. These include the conservation of germplasm in vitro using the slow growth technique, which allows the conservation of many individuals under controlled conditions for several months. However, the development of a protocol for in vitro preservation is a rigorous procedure that involves the optimization of different factors: chemical, physical and molecular. Therefore, in this chapter, we will review the advances in the slow-growth preservation of *V. planifolia*.

Keywords *V. planifolia*, Slow growth, Factors, Preservation

1 Introduction

In Mexico, *V. planifolia* is currently listed as at risk, subject to special protection (NOM-059-SEMARNAT-2010). Much of the genetic variability of vanilla has been lost due to the overharvesting of wild individuals, the loss of vanilla plantations due to poor management, the abandonment of vanilla plantations, and the continuous regime of vegetative propagation that it has had in recent years [1]. Because of the above, it is difficult to conserve in situ.

Plant biotechnology provides several very efficient techniques for the development of conservation programs for threatened plant species [2]. One of the most effective techniques for ex situ germplasm conservation is in vitro conservation by slow growth since it

Marco A. Ramírez-Mosqueda et al. (eds.), *Conservation of Plant Genetic Resources*, Methods in Molecular Biology, vol. 3011, https://doi.org/10.1007/978-1-0716-5130-8_22,

Table 1
Important factors in slow-growth conservation

Chemical factors	Physical factors	Molecular factors
Osmotic agents	Temperature	Genetic changes
Growth inhibitors	Light quality	Epigenetic changes
Concentration of culture medium		

allows the aseptic conservation of many individuals under controlled conditions for several months or even years in a reduced area [3].

The development of a protocol for slow-growth in vitro conservation of germplasm of plant species is a rigorous procedure that involves the optimization of distinct factors: chemical, physical, and molecular (Table 1). These include the use of reduced temperatures, osmotic agents, growth inhibitors, and light intensities [4, 5]. Distinct species have been conserved by the slow growth technique, such as apple [6], Aglaonema [7], banana [8], *Stanhopea tigrina* [9].

In this chapter, we describe the distinct factors that have been used in the slow-growth conservation of . Therefore, this chapter aims to show which factors have been effective in conserving *V. planifolia*.

2 Factors Used for the Conservation of *V. planifolia*

2.1 Chemical Factors

To reduce the growth rate of plants in vitro, optimal culture conditions are often altered by the addition of chemical substances, such as growth-retarding regulators, osmoregulatory compounds, and modification of the concentration of vitamins, minerals, and carbon sources in the culture medium. Some chemical factors have been used for the preservation of *V. planifolia*, such as the protocol based on the use of MS medium added with 10 g L^{-1} sucrose and 15 g^{-1} mannitol, keeping the cultures at 22 ± 2°C for slow growth preservation for up to 12 months [10]. In another study, a comparison was made on the effect of osmotic and growth inhibitory agents during the in vitro preservation process of *V. planifolia*, where they found the best results when growing shoots on MS medium supplemented with 3 mg L^{-1} of ABA [11].

2.2 Physical Factors

To establish an effective in vitro preservation protocol for minimal growth, it is necessary to reduce the metabolism of the plants. For this purpose, physical factors such as low temperatures and light intensities have been used. In *V. planifolia*, low temperatures (4°C) have been evaluated for slow-growth conservation; however, no positive effects were found for conservation [12]. However, the effect of various spectrums of illumination with red, blue, and red + blue light-emitting diodes (LEDs) has been evaluated for slow-growth preservation, obtaining the best results for preservation in blue LED light [13].

2.3 Molecular Factors

In vitro propagation and/or conservation methods can generate genetic and epigenetic alterations [14]. Thus, genetic stability is a crucially important aspect to consider in in vitro slow-growth conservation protocols [15]. In *V. planifolia*, genetic stability has been evaluated with ISSR (inter-simple sequence repeats) and SSR (simple sequence repeats) molecular markers in slow-growing preserved material, obtaining that the preservation protocol using 3 mg L^{-1} of ABA did not cause genetic alterations [12].

3 Conclusion

We can conclude that CTV tools are particularly useful for the conservation of *V. planifolia* germplasm under slow growth. The biotechnological factors studied make it possible to obtain the ideal conditions for the preservation of vanilla without altering its genetic stability. Based on the results of the different studies, it is recommended to use the protocol for the conservation of *Vanilla* germplasm in MS basal medium supplemented with 3 mg L^{-1} under the blue light spectrum (Fig. 1), since it allows maintaining slow growth without the need to perform numerous subcultures that would contribute to increasing the presence of somaclonal variants.

Fig. 1 Effect of preservation on slow growth (**A**) treatment with blue LED light and 3 mg L^{-1} of ABA (**B**) Control treatment. Bar = 1 cm

References

1. Soto-Arenas MA (1999) Filogeografía y recursos genéticos de las vainillas de México. Informe final SNIB-CONABIO proyecto No. J101. Inst Chinoın, AC, Herbario de la Asociación Mexicana de Orquideología, AC México, DF
2. Pérez-Molphe Balch E, Esparza-Araiza MJ, Pérez-Reyes ME (2012) Conservación in vitro de germoplasma de Agave spp. bajo condiciones de crecimiento retardado. Rev Fitotecnia Mexicana 35:279–287
3. Cruz-Cruz CA, Gonzalez-Arnao MT, Engelmann F (2013) Biotechnology and conservation of plant biodiversity. Resources 2:73–95. https://doi.org/10.3390/resources2020073
4. Trejgell A, Kaminska M, Tretyn A (2015) In vitro slow growth storage of Senecio macrophyllus shoots. Acta Physiol Plant 37:234. https://doi.org/10.1007/s11738-015-1983-83-470
5. Rodrigues PHV, Arruda F, Forti VA (2018) Slow-grown in vitro conservation of *Heliconia champneiana* cv. Splash under different light spectra. Sci Agr 75:163–166. https://doi.org/10.1590/1678-992x-2016-0394
6. Lu X, Sun P, Liu R, Wang C, Tong L, Tahir MM, Ma X, Bao X, Zhang D, Wang M, An N (2024) In vitro slow-growth conservation, acclimatization, and genetic stability of virus-free apple plants. Horticulture Adv 2:1–12. https://doi.org/10.1007/s44281-024-00049-0
7. Nugrahani P, Purnobasuki H, Sitawati S (2024) Composition of media for in vitro slow growth storage (sgs) of Aglaonema. Ornam Horticulture 30:e242696. https://doi.org/10.1590/2447-536X.v30.e242696
8. Rodrigues PHV, Oliveira EL, Demetrio CA, Ambrosano GB, Piedade SMS (2022) Effects of different light spectra on the slow-grown in vitro storage and quality of banana plantlets cv. Prata Catarina (AAB). Plant Cell Tissue Organ Cult 150:479–485. https://doi.org/10.1007/s11240-022-02280-x
9. Cruz-Cruz CA, González-Arnao MT, Bautista-Aguilar JR, Ramírez-Mosqueda MA (2022) In vitro short-term storage of *Stanhopea tigrina* Bateman ex Lind. South Afr J Botany 151:334–338. https://doi.org/10.1016/j.sajb.2022.10.014
10. Bello-Bello J, García-García G, Iglesias-Andreu L (2015) Conservacion de vainilla (*Vanilla planifolia* Jacks) bajo condiciones de lento Crecim vitro Rev Fitotecn Mex 38:165–171. http://www.scielo.org.mx/scielo.php?script=sci_arttext&pid=S0187-73802015000200006&lng=es&nrm=iso
11. Divakaran M, Babu KN, Peter KV (2006) Conservation ofVanilla species, in vitro. Sci Hort 110:175–180. https://doi.org/10.1016/j.scienta.2006.07.003
12. Bautista-Aguilar JR, Iglesias-Andreu LG, Martínez-Castillo J, Ramírez-Mosqueda MA, Ortiz-García MM (2021) In vitro conservation and genetic stability in *Vanilla planifolia* Jacks. HortScience 56:1494–1498. https://doi.org/10.21273/HORTSCI16118-21
13. Bautista-Aguilar JR, Iglesias-Andreu LG, Flores-Estévez N, Sánchez-Coello NG (2024) Effect of led lighting on minimal growth for in vitro conservation of vanilla genotypes. Propag Ornam Plants 24:13–19
14. Chiang NS, Jiménez VM (2009) Técnicas moleculares para la detección de variantes somaclonales. Agronomía mesoamericana 20:135–151. https://doi.org/10.15517/am.v20i1.4989
15. Bednarek PT, Orłowska R (2020) Plant tissue culture environment as a switch-key of (epi) genetic changes. Plant Cell Tissue Organ Cult 140:245–257. https://doi.org/10.1007/s11240-019-01724-1

Chapter 23

Long-Term Preservation by the Cryopreservation Method of *Cordia alliodora* (Ruiz & Pavón) Oken

Esmeralda J. Cruz-Gutiérrez, Claudia Berenice Espitia-Flores, Humberto Antonio López-Delgado, Luis Alberto Gómez-Reyes, Juan M. Pichardo-González and Alma Armenta-Medina

Abstract

Cryopreservation, for long-term storage in germplasm banks, at liquid nitrogen temperatures (−196°C), of plant material, somatic cells or embryogenic lines, allows them to conserve their genetic integrity, since when the material is frozen, they stop all their metabolism, remaining alive, but in suspension. Cryopreservation of woody species has been in use for a short time, but this technique facilitates conservation by lowering costs, time spent in storage, and avoiding contamination). On the other hand, the *Cordia alliodora* species Ruiz & Pavón is of economic and ecological importance, due to its precious wood and because it is a fast-growing tree in humid climates and has great potential for productive forest plantations, contributing to the local and national economy. The proposed method for cryopreserving plant material from wood-yielding species is vitrification. This involves transforming the liquid solution in a cell into a glassy state, without forming crystals. Therefore, the objective of this work is to present a methodology for cryopreserving plant tissue from *Cordia alliodora* species Ruiz & Pavón using vitrification.

Keywords Germplasm banks, Forestry, Ex situ conservation

1 Introduction

The *Cordia alliodora* species Ruiz & Pavón, also known as laurel or boxwood, is a species for its ecological, economic, and cultural value [1]. Its wood is highly valued by cabinetmakers for its resistance to moths and rot, as well as its ease of work. One of the characteristics of the tree is its rapid growth in humid climates; it has great potential for productive forest plantations, contributing to the local and national economy [2, 3]. It is stated that by 2030, between 50% and 75% of the world's wood production for industrial use will be obtained from forest plantations [3], which can be used as a great opportunity for the large-scale application of biotechnology, without neglecting the

Marco A. Ramírez-Mosqueda et al. (eds.), *Conservation of Plant Genetic Resources*, Methods in Molecular Biology, vol. 3011, https://doi.org/10.1007/978-1-0716-5130-8_23,

conservation of this genetic resource. The conservation of biological diversity is a strategy that has gained increasing importance in today's society. Biological studies of flora in particular are key to strategies for the conservation and management of forest genetic resources [4, 5].

One of the strategies for ex situ conservation in a germplasm bank is the cryopreservation of plant tissue [6]. This technique involves freezing plant tissue at a temperature of −196°C (the temperature of liquid nitrogen), which halts cellular metabolism and keeps the cell alive for a long time [7, 8]. This way, the characteristics of the mother plant are preserved without somaclonal variation, and maintenance costs within germplasm banks are reduced [7, 9]. For this reason, this work presents the protocol for cryopreserving *Cordia alliodora* Ruiz & Pavón, and for maintaining it for a long time with the same characteristics as the mother plant.

2 Materials

2.1 Plant Material

1. The germplasm used may be seeds or tissue from a healthy, vigorous, pest-free juvenile donor plant. For establishment, young shoots (with one or two buds) or seeds are used.

2.2 Disinfection of Surfaces and Equipment

1. Liquid detergent (*see* **Note 1**). 70% ethanol solution (*see* **Note 2**). 30% sodium hypochlorite solution (*see* **Note 2**). Sterile Distilled Water (*see* **Note 3**). Captan® solution (3 g/L^{-1}) (*see* **Note 4**). pH adjustment solutions: 1 N NaOH and 0.1 HCl.

2.3 Culture Medium

1. The culture medium used was a Murashige and Skoog [10] and a Woody Plant Medium [11] (see Table 1).
2. Medium additives (see Table 1) used for the different stages of cultivation: seed germination (GS), bud establishment (EY), and multiplication (M).

2.4 Equipment

1. Culture room: Laminar airflow hood.
2. Surgical instruments: stainless steel forceps, surgical scalpels, removable sterile surgical blades, and sterile Petri dishes.
3. Preparation of the culture medium: Analytical balances for weighing, spatulas, magnetic stirrers, hot plate with stirring, potentiometer, microwave oven, graduated beaker, and autoclave.
4. Culture vessels: Beakers (capacity 1000 mL), reagent bottles, culture tubes, and Petri dishes.
5. Glassware cleaning: scrub the glassware with a commercial liquid detergent solution and rinse it with plenty of tap water.
6. Various materials: Film or plastic film for packaging food.

Table 1
Solutions and culture medium additives used for the different culture stages of *Cedrela odorata* L

Culture medium	Components
Seed establishment medium	Murashige and Skoog medium with 30 g/L sucrose, 8 g/L agar, and a pH of 5.7
Explant establishment medium	Woody Plant Medium or Murashige and Skoog medium, added with 20 g/L sucrose, 1 g/L^{-1} activated carbon, 1 mL/L of Plant Preservative Mixture™, 1 g/L Polyvinylpyrrolidone, 8 g/L agar, and a pH of 5.7
Propagation medium	Murashige and Skoog medium, added with 30 g/L sucrose, 0.5 mg/L Benzylaminopurine (BAP), 8 g/L agar, and a pH of 5.7
Pre-culture	Murashige and Skoog medium, added with 0.3M sucrose and 8 g/L agar
Loading solution	Murashige and Skoog medium with 2M glycerol and 0.4 M sucrose
Plant vitrification solution 3	Murashige and Skoog medium 50% (v/v) glycerol and 50% (w/v) sucrose
Regrowth Solution	Murashige and Skoog medium with 1 M sucrose
Regrowth medium	Murashige and Skoog medium with 0.2 mg/L 6-bencilaminopurina (BAP)
Sodium alginate solution 2%	Murashige and Skoog medium 100 mL with 20 g of sodium alginate, 0.4 M sucrose.
Calcium chloride solution	0.1 M calcium chloride with 0.4 M sucrose

3 Methods

3.1 Disinfection Plant material

3.1.1 Seeds

1. The seeds are washed with liquid detergent (*see* **Note 1**) for a period of 10 min, repeating this process twice.
2. Left to soak for 16 h in double-distilled water.
3. Immersed in a Captan® 3 g·L^{-1} solution for 50 min.
4. Placed in a 70% ethanol solution for a period of 45 s.
5. Introduced into a 30% sodium hypochlorite solution for a period of 13 min.
6. Wash the seeds three or four times with sterile distilled water to remove residues of disinfectant solutions (*see* **Note 3**).
7. Establishing seeds in the growing medium (*see* **Note 5**).
8. The in vitro seedlings, free of microorganisms, and more vigorous, were selected for the multiplication of the material.

3.1.2 Explants

1. The plant material is collected from a plant in a field free of pests and diseases, vigorous and nourished, making it the perfect donor plant.

2. The plant material is washed with liquid detergent (*see* **Note 1**) for 10 min.
3. Introduced into a Captan® 3 g·L^{-1} solution for 45 min.
4. Immersed in a 70% ethanol solution for 30 s.
5. Immersed in a 30% sodium hypochlorite solution for 10 min.
6. Wash three or four times with sterile distilled water to remove residues of disinfectant solutions (*see* **Note 3**).
7. Establishment of plant material in the culture medium (*see* **Note 5**).
8. The in vitro plants, free of microorganisms, and more vigorous, were selected for the multiplication of the material.

3.2 Proliferation of Shoots

1. The shoots with a size of 15 mm are established in a semi-solid medium (*see* **Note 6**) (see Table 1), with an incubation period of 30 days.
2. The shoots are incubated at a temperature of 24 ± 2°C and a photoperiod of 16 h of light and 8 h of darkness with a light intensity of 2000 lux.
3. After 30 days of incubation, new shoots are obtained per explant, which are separated and used in subsequent processes.

3.3 Cryopreservation

1. Apical meristems measuring approximately 1.5 mm are used.
2. The shoot tips are placed in Petri dishes (90 × 20 mm) in a pre-culture medium (see Table 1), and incubated in a room with a temperature of 24 ± 2°C under darkness for 24 h.
3. Set the cryo-plates on the Petri dishes (*see* **note** 7) (Fig. 1).
4. Place a drop (3 μL) of the sodium alginate solution (see Table 1) on the wells of the aluminum plate using a micropipette. Place the pre-cultivated shoot tips in each well with the tip of a scalpel blade. Another drop of sodium alginate (see Table 1) is then placed on the shoot tips. The entire plate is coated with calcium chloride (Fig. 2, see Table 1). Let it incubate for 15 min.
5. Once the time has passed, the excess calcium chloride is removed and placed in the loading solution (see Table 1) for 30 min for osmoprotection.
6. After this time, they are placed in the PVS3 (see Table 1) solution for 40 min for vitrification. The cryo-plates are transferred to uncapped 2 mL cryotubes, which are held on cryo-canes, and directly in liquid nitrogen (LN), where they remain for at least 30 min (Fig. 3).
7. The cryotubes are removed from the liquid nitrogen and the cryo-plates with shoot tips are transferred from the cryotubes and immersed in the regrowth solution (see Table 1) for 15 min at room temperature. The shoot tips are transferred to

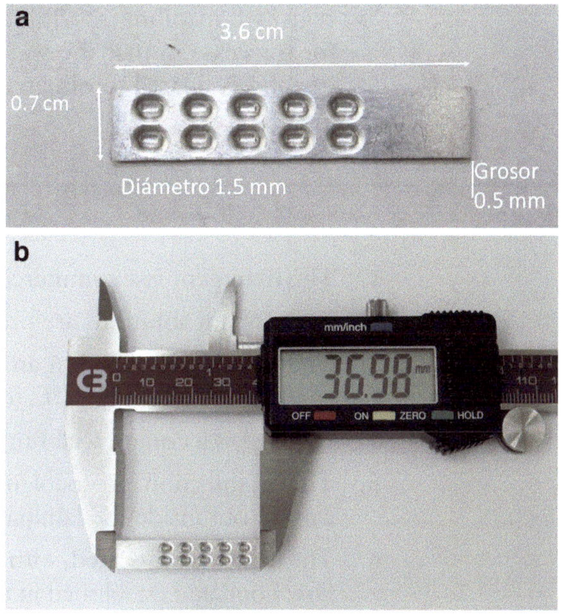

Fig. 1 (**a**) and (**b**) Shape and size of the cryo-plate

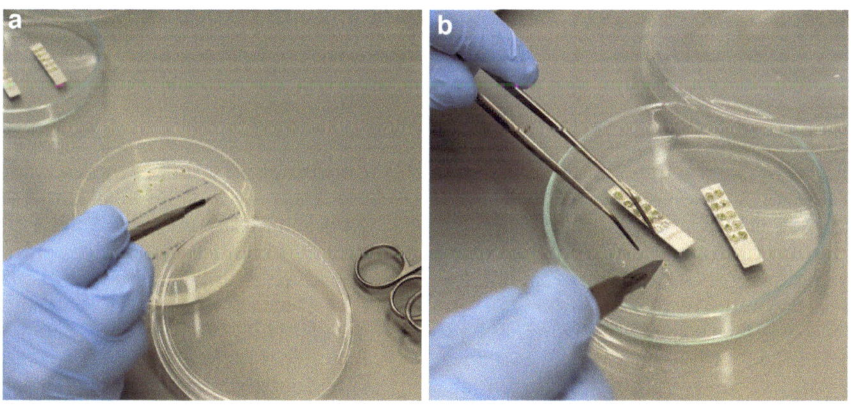

Fig. 2 (**a**) Adhesion of shoot tips on the cryo-plate and (**b**) application of a drop of sodium alginate

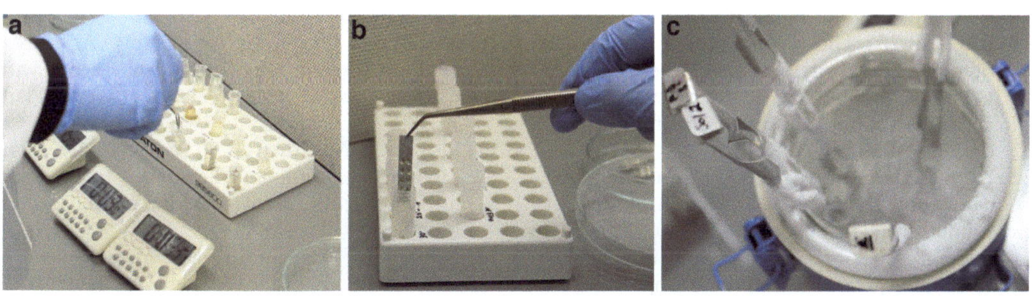

Fig. 3 (**a** and **b**) Treatment for vitrification, placement of the cryo-plates in the loading solution and the PVS3 solution. (**c**) Immersion of plant material in liquid nitrogen

regrowth medium (see Table 1) and placed in a room at 25°C for 4 weeks, so that the shoot tips that develop can be subsequently evaluated to obtain a seedling.

4 Notes

1. The detergent is a commercial brand for washing dishes.
2. Disinfectant solutions are prepared with distilled water.
3. The water is sterilized in an autoclave at 1.5 kg/cm^2 pressure at a temperature of 121°C for 15 min.
4. Captan® is a commercial fungicide.
5. The disinfection protocol using the 70% ethanol solution is carried out inside the laminar flow hood.
6. The tissues are dissected, with a size of 15 mm, within the laminar flow hood and established in the culture medium (see Table 1).
7. The aluminum cryo-plate is length of 37 mm, a width of 7 mm and thickness of 0.5 mm with 10 wells. The cryo-plate is A1050, which means that aluminium purity is over 99.5%. The plate is designed to fit in 2 mL cryotube [12].

References

1. CONABIO (Comisión Nacional para el Conocimiento y Uso de la Biodiversidad). (s.f.). Cordia alliodora. Recuperado de http://www.conabio.gob.mx/conocimiento/info_especies/arboles/doctos/16-boragl m.pdf
2. Liegel LH, Stead JW (s.f.). *Cordia alliodora* (Ruiz & Pav.) Oken Capá prieto, laurel. RNGR. Recuperado de https://www.srs.fs.usda.gov/pubs/misc/ag_654/volume_2/cordia/alliodora.htm
3. Marulanda ML, López AM, Uribe M, Ospina CM (2011) Caracterización de la variabilidad genética de progenies de *Cordia alliodora* (R. P.) Oken. Colomb Forestal 14(2):119–135
4. FAO (2025) The second report on the state of the world's forest genetic resources. FAO Commission on Genetic Resources for Food and Agriculture Assessments, Rome. https://doi.org/10.4060/cd4838en
5. Frankel OH, Brown AHD, Burdon JJ (1995) The conservation of plant biodiversity. Cambridge University Press. (Este libro es un texto clásico sobre la conservación de la biodiversidad vegetal, incluyendo la importancia de los recursos genéticos)
6. González-Arnao MT, Engelmann F (2013) Crioconservación de plantas en América Latina y el Caribe. IICA, San José, C.R., p 2013
7. Engelmann F (2011) Cryopreservation of tropical plant germplasm—current research progress and future prospects. Vitro Cell Dev Biol Plant 47(1):5–18
8. Reed BM (2008) Plant cryopreservation: a practical guide. Springer. New York. https://doi.org/10.1007/978-0-387-72276-4
9. Ballesteros D, Martínez MT, Sánchez-Romero C, Montalbán IA, Sales E, Moncaleán P, Arrillaga I, Corredoira E (2024) Current status of the cryopreservation of embryogenic material of woody species. Front Plant Sci 14:1337152. https://doi.org/10.3389/fpls.2023.1337152
10. Murashige T, Skoog F (1962) A revised medium for rapid growth and bio assays with tobacco tissue cultures. Physiologia Plant 15(3):473–497
11. Lloyd G, McCown B (1981) Woody Plant Medium: a mineral nutrient formulation for microculture of woody plant species. HortScience 16:453
12. Niino T, Yamamoto S, Matsumoto T, Engelmann F, Valle Arizaga M, Tanaka D (2018) Development of V and D cryo-plate methods as effective protocols for cryobanking. In: III International symposium on plant cryopreservation, March. pp 249–262

Check for updates

Chapter 24

Long-Term Storage of Vanilla (*Vanilla planifolia* Jacks.) by Cryopreservation Method

Marco A. Ramírez-Mosqueda, Elizabeta Hernández-Domínguez, Esmeralda J. Cruz-Gutiérrez, Flor Josefina Ramos-Cruz and José Roberto Bautista-Aguilar

Abstract

The conservation of plant genetic resources is a global issue that must be addressed through innovative and cutting-edge strategies. Cryopreservation techniques for plant tissues allow for long-term storage. Vanilla (*Vanilla planifolia*) is a plant species with wide economic interest due to the extraction of vanillin through its processed pods. Mexico is the center of origin and domestication of this crop; however, long-term conservation strategies have not been established. This chapter describes the long-term conservation process of *V. planifolia* through cryopreservation.

Keywords Cryopreservation, Cryo-plate, Vanillin, Conservation

1 Introduction

Ex situ conservation of plant genetic resources uses biotechnological tools such as plant tissue culture (short term), in vitro conservation (medium term) and cryoconservation (long term) [1]. Cryopreservation is a technique characterized by the use of liquid nitrogen to reach temperatures of −196°C [2]. This technique allows the total metabolism of the cells of the preserved tissues to stop completely; however, once the thawing is performed, the metabolism is reactivated and continues [3].

The aluminum cryo-plate technique involves the encapsulation, polymerization, dehydration, and vitrification of plant tissues. The vitrification protocol can be carried out with various solutions (V-cryoplate method) or by air dehydration (D-cryoplate method) and then, the cryo-plates are immersed directly in liquid nitrogen for cryopreservation. These two techniques guarantee very high cooling (liquid nitrogen) and heating rates, as well as recovery and regeneration of the preserved plant tissues [4]. Under this method, the

Marco A. Ramírez-Mosqueda et al. (eds.), *Conservation of Plant Genetic Resources*, Methods in Molecular Biology, vol. 3011,
https://doi.org/10.1007/978-1-0716-5130-8_24,

duration of the complete cryogenic protocol is shortened, compared to that required for encapsulation-dehydration [5].

Vanilla (Jacks.) is a crop native to Mexico, with worldwide importance due to the production of vanillin, a natural flavoring used in the food industry [6]. This crop is undergoing constant genetic erosion due to its clonal propagation method. Therefore, it requires conservation strategies for its genetic diversity that still exists in wild and/or cultivated populations [7]. In this chapter, an effective protocol for the cryoconservation of is described. This protocol can be used for the long-term conservation of this valuable plant genetic resource.

2 Materials

2.1 Plant Material and Culture Media

1. In vitro plants of , previously established. The plants used should have the least number of subcultures possible (two to three subcultures), in order to avoid genetic variations.
2. Commercial Murashige and Skoog (MS) [8] culture medium is used (supplied by any scientific supplier). Manufacturer specifications are used. The composition of the medium is shown in Table 1.
3. Medium additives used for various culture stages: MS medium supplemented with various additives according to the culture stage (see Table 2): (1) Culture (incubation), (2) Precultivation (before cryopreservation), and (3) Reheating and Recovery.
4. pH adjustment solutions: 1 N sodium hydroxide (NaOH) and 1 N hydrochloric acid (HCl).

2.2 Plant Growth Regulators (PGR) Stock Solutions and Encapsulation Solutions

1. Benzyladenine (BA, 0.1 mg/mL): dissolve 10 mg of BA in 1 mL 1 N NaOH and then add water to a volume of 100 mL.
2. 2% sodium alginate containing the mineral salts of the MS culture medium except for the inorganic component $CaCl_2 \cdot 2H_2O$. The pH should be adjusted to 5.8 in both solutions. They sterilized for 15 kPa for 5 min at a temperature of 121°C.
3. 0.1 M Calcium chloride ($CaCl_2 \cdot 2H_2O$) at added in the same way with mineral culture salts MS culture medium. They sterilized for 15 kPa for 5 min at a temperature of 121°C.

2.3 Loading Solution (Osmoprotection) and Desiccation

1. LS: 2 M glycerol, 0.4 M sucrose, and mineral salts MS.
2. The pH should be adjusted to 5.8.
3. They sterilized for 15 kPa for 5 min at a temperature of 121°C.
4. Commercial silica gel.

Table 1
Complete MS culture medium

Components	(mg/L) MS 100%
Macronutrients	
NH_4NO_3	1650
KNO_3	1900
$CaCl_2 \cdot 2H_2O$	440
$MgSO_4 \cdot 7H2O$	370
KH_2PO_4	170
Micronutrients	
H_3BO_3	6.2
$MnSO_4 \cdot 4H_2O$	22.3
$ZnSO_4 \cdot 7H_2O$	8.6
KI	0.83
$Na_2MoO_4 \cdot 2H_2O$	0.25
$CoCl_2 \cdot 6H_2O$	0.025
$CuSO_4 \cdot 5H_2O$	0.025
FeNa-EDTA	36..7
$FeSO_4 \cdot 7H_2O$	27.8
Vitamins	
Nicotinic acid	0.5
Pyridoxine	0.5
Thiamine	0.1
Myo-inositol	100
Glycine	2

Table 2
Substances that complement culture media

	Culture stage		
Media additives	**(1) Incubation**	**(2) Pre-cultivation**	**(3) Recovery**
MS	100%	100%	100%
Benzylaminopurine (BAP)	–	1 mg	1 mg
Ácido indolbutírico (AIB)	–	–	1 mg
Sucrose	30 g	100 g	400 g
Agar	8 g	8 g	8 g
pH	5.7	5.8	5.8

2.4 Reheating and Cryopreservation and Maintenance

1. 1.2 M sucrose and mineral salts MS are adjusted to a pH of 5.8.
2. Liquid nitrogen from an external distribution company is used.

2.5 Equipment

1. Culture room: airflow cabinet, bead sterilizer, and liquid nitrogen tanks.
2. Surgical tools: cryolamines (37 × 7 × 7 mm) con 10 pocillos con un diámetro de 1.5 mm. Cryocane (29 cm), stainless steel tweezers of 25 cm, surgical scalpels, removable sterile surgical blades, sterile craft paper, cotton, and adherent film.
3. Medium preparation: analytical balance, spatula, hot stirrer plate, pH meter, micropipettes, magnetic stirrer, refrigerator, and autoclave.
4. Glassware: bottles (200 mL capacity), beakers (500 mL and 1 L capacity), Reagent bottles, and measuring cylinders.
5. Micropipette of 10–100 μL.
6. Culture vessels: test tubes (4.5 cm) and Petri dishes (0.8 cm).
7. Cryopreservation containers: cryovials (4.5 cm).

3 Methods

3.1 Preparation Culture Media (Different Stages)

1. Prepare the culture medium from manufacturer specifications, add the supplementary components for each stage of the cryopreservation process (see Table 2).
2. Autoclave at 121°C with a pressure of 15 kPa for 20 min (all culture media).
3. Distribute 10 mL of culture medium in Petri dishes for all stages (*see* **Note 1**).
4. After autoclaving, the glass bottles are kept in a clean room until use.

3.2 Incubation and Pre-Cultivation

1. For incubation: shoot tips of 1 cm long (containing an apical meristem) should be extracted from in vitro plants of (Fig. 1A)
2. Must be transferred to petri dishes containing 10 mL of incubation medium (see Table 2).
3. They must be incubated for 3 weeks at a temperature of 24°C), and subsequently 2 weeks at a temperature of 29°C.
4. After 5 weeks, the shoot tips are transferred to pre-cultivation.
5. For pre-cultivation: meristems (1–1.5 mm) must be dissected using a stereoscope (Fig. 1B) (*see* **Note 2**).

Fig. 1 Crioconservación of . (**A**) Shoot tips of 1 cm long (containing an apical meristem), (**B**) meristems (1–1.5 mm) must be dissected using a stereoscope, (**C**) meristems are cultured in Petri dishes, and (**D**) Cryo-slide

6. Meristems are cultured in Petri dishes with 10 mL of pre-culture medium (see Table 2) (Fig. 1C).
7. They must be kept in pre-culture at a temperature of 24°C for 24 h, after this period, the laminar flow hood must be used to carry out the following processes.

4 Cryopreservation

4.1 Encapsulation, Osmoprotection and Desiccation

1. For encapsulation: the meristems are transferred to a cryo-slide, then one meristem must be placed per well with the help of a scalpel (Fig. 1D).
2. Encapsulation solutions must be used (*see* **Note 3**). The meristems are placed in the wells of the cryo-plate, then covered with (100 μL) of 2% sodium alginate using a 10–100 μL micropipette. Subsequently, 500 μL of 0.1 M calcium chloride is applied using a 100–1000 μL micropipette. The solution

should be allowed to polymerize for 15 min, and then the excess solution should be removed using a micropipette, and the remaining calcium chloride solution that did not polymerize should be sucked out using the micropipette.

3. Each cryo-sheet is introduced into 2 mL cryovials. These cryovials must contain 2 mL of the loading solution (LS). They must remain submerged for 30 min. This process is carried out inside the laminar flow hood.
4. For desiccation: the process for reducing water from intra and extracellular spaces is carried out by placing the cryoplates on petri dishes that contain silica gel inside (*see* **Note 4**).
5. Cryoplates are dried inside Petri dishes for 90 min. This process is carried out inside the laminar flow hood.

4.2 Freezing, Reheating, and Recovery

1. For freezing: the Cryoplates are introduced into cryovials (previously labeled), subsequently mounted on cryocane and completely immersed in liquid nitrogen (LN) at a temperature of −196°C for 1 h (*see* **Note 5**). This process is carried out inside the laminar flow hood (Fig. 2A, B).
2. To safeguard accessions within the plant collection of the National Genetic Resources Centre of the National Institute of Forestry, Agriculture and Livestock Research (CNRG-INIFAP), ten cryoplates are cryoconserved, each slide containing ten meristems. Therefore, 100 meristems are cryopreserved for each accession.
3. Subsequently, the cryocanes with the cryovials must be kept in LN tanks for an indefinite period (30–50 years) (*see* **Notes 6 and 7**).
4. For reheating and recovery: remove the cryovials from the LN and take out the cryo-slides from the cryovials.
5. Transfer to new cryovials containing 2.0 mL of reheating solution, immerse the cryoplates for 15 min.
6. Subsequently, the cryolplates are placed on sterile filter paper and with the help of a scalpel, the alginate residue (result of polymerization) is removed from the meristems.
7. The meristems are placed one by one in a Petri dish with recovery medium (Table 2).
8. The meristems are incubated at 24 ± 2°C, under 40 ± 5 μmol m^{-2} s^{-1} irradiance and a 16-h photoperiod (*see* **Note 8**) (Fig. 2A, B).

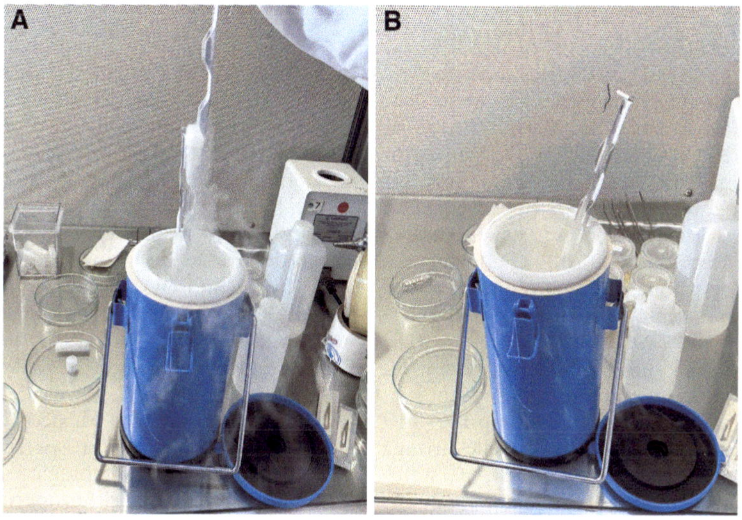

Fig. 2 Freezing in liquid nitrogen. (**A**) Cryoplates mounted on cryocane, (**B**) Cryoplates completely immersed in liquid nitrogen (LN) at a temperature of −196°C

5 Notes

1. Semi-solid medium should be used by adding a gelling agent (agar).
2. Using a stereoscope, tweezers and a scalpel, the layers of plant tissue surrounding the meristem must be cut very carefully so as not to damage it.
3. Both solutions must be kept under constant stirring and once completely dissolved, sterilization is carried out in the autoclave for 15 min at 121°C, taking care to prevent denaturation of the alginate and ensuring accuracy in the sterilization time.
4. Silica gel must be pre-treated (dried) in an oven at 60°C for 24 h.
5. Cryovials are pre-cooled in LN for 2 min.
6. It is recommended to perform reheating and recovery tests every 5–7 years to check the viability of cryopreserved plant tissues.
7. The level of the LN tanks must be checked periodically to verify that the cryovials are completely submerged in LN.
8. After 30 days of cultivation, 60% of the meristems that went through the cryopreservation process regenerate into shoots.

References

1. Ramírez M, Salcedo J (2013) Los recursos fitogenéticos y la importancia estratégica de su conservación en las Américas. In: María Teresa G-A, Florent E (eds) Crioconservación De Plantas En América Latina Y El Caribe. IICA, San José
2. Ávila-Portillo LM, Madero JI, López C, León MF, Acosta L, Gómez C, Delgado LG, Gómez C, Lozano JM, Reguero MT (2006) Fundamentos de criopreservación. Rev Colombiana de Obstetricia y Ginecología 57(4):291–300
3. Engelmann F, Dussert S (2013) Cryopreservation. In: Normah M, Chin H, Reed B (eds) Conservation of tropical plant species. Springer, New York, NY, USA, pp 107–119
4. Yamamoto S, Wunna, Rafique T, Valle-Arizaga M, Fukui K, Cruz-Gutierrez EJ, Castillo-Martínez CR, Watanabe K, Niino T (2015) The aluminum cryo-plate increases efficiency of cryopreservation protocols for potato shoot tips. Am J Potato Res 92: 250–257. https://doi.org/10.1007/s12230-014-9425-5
5. Yamamoto SI, Fukui K, Rafique T, Khan NI, Martinez CRC, Sekizawa K et al (2012) Cryopreservation of in vitro-grown shoot tips of strawberry by the vitrification method using aluminium cryo-plates. Plant Genet Resour 10(1):14–19
6. Santillán A, Salas A, Vásquez N (2018) La productividad de la vainilla (*Vanilla planifolia* Jacks. ex Andrews) en México de 2003 a 2014. Rev Mexicana de Cienc Forestales 9(47):50–69. https://doi.org/10.29298/rmcf.v9i47.165
7. Rodríguez-Deméneghi MV, Aguilar-Rivera N, Gheno-Heredia YA, Armas-Silva AA (2023) Cultivo de vainilla en México: Tipología, características, producción, prospectiva agroindustrial e innovaciones biotecnológicas como estrategia de sustentabilidad. Sci Agropecu 14(1): 93-109.
8. Murashige T, Skoog F (1962) A revised medium for rapid growth and bio assays with tobacco tissue cultures. Physiol Plant 15:473–497. https://doi.org/10.1111/j.1399-3054.1962.tb08052.x

Chapter 25

Long-Term Preservation by the Cryopreservation Method *Cedrela odorata* L.

Esmeralda J. Cruz-Gutiérrez, Claudia Berenice Espitia-Flores, Luis Alberto Gómez-Reyes, Juan M. Pichardo-González and Alma Armenta-Medina

Abstract

One of the most important timber species due to its precious wood and high commercial value is *Cedrela odorata* L. This species is overexploited for timber production, which is why it was included in NOM-059-SEMARNAT-2010 in the special protection category. For this reason, it is essential to carry out actions for the conservation of this species and its genetic diversity. In this context, a biotechnological technique for the long-term conservation of plant germplasm is cryoconservation. This technique consists of stopping cellular metabolism by freezing them at −196°C, while the cells remain alive. There are different techniques within the cryopreservation method; however, for the long-term conservation of *Cedrela odorata* L. the best method was cryo-plate vitrification. Therefore, the objective of this work is to describe the cryopreservation technique of *Cedrela odorata* L. plant tissue using the cryo-plate vitrification method.

Keywords Germplasm banks, Forestry, Ex situ conservation

1 Introduction

One of the forest species of ecological and economic importance in Mexico is *Cedrela odorata* L., commonly known as red cedar, with precious wood with high commercial value, which presents an aroma, color, and high resistance to attacks by fungi and insects [1, 2]. On the other hand, it is also a species of ecological importance since it is a pioneer and frequent species in the secondary vegetation of various forests, contributing to the recovery and maintenance of tropical and subtropical ecosystems [3]. Due to its characteristics, it is in high demand and has been overexploited, which has caused the fragmentation of natural populations, which is why it has been included in SEMARNAT NOM 059 in the special protection category, as well as in CITES Appendix III [1, 4, 5]. Due to all these characteristics, this species is key for Mexico due to the commercial

Marco A. Ramírez-Mosqueda et al. (eds.), *Conservation of Plant Genetic Resources*, Methods in Molecular Biology, vol. 3011, https://doi.org/10.1007/978-1-0716-5130-8_25,

value of its wood and its ecological importance, which is why efforts must be combined for the conservation of this genetic resource and its sustainable management [3, 6]. On the other hand, a technique for the conservation of plant tissue ex situ is cryopreservation, which consists of storage at low temperatures (−196°C, liquid nitrogen temperature), which allows stopping the metabolic and biochemical activity of cryopreserved cells, in a state of suspended life for very long periods [7, 8]. Therefore, the objective of this work is to present a protocol for the long-term conservation of *Cedrela odor*ata L.

2 Materials

2.1 Plant Material and Disinfection

1. The germplasm used may be seeds or tissue from a healthy, vigorous, pest-free juvenile donor plant. For establishment, young shoots (with one or two buds) or seeds are used.
2. Liquid detergent (*see* **Note 1**).
3. 70% ethanol solution (*see* **Note 2**).
4. 30% sodium hypochlorite solution (*see* **Note 2**).
5. Sterile Distilled Water (*see* **Note 3**).
6. Captan® solution (3 $g \cdot L^{-1}$) (*see* **Note 4**).
7. pH adjustment solutions: sodium hydroxide 1 N NaOH and hydrochloric acid 0.1 N HCl.

2.2 Culture Media

1. The culture medium used was a Murashige and Skoog [9] and a Woody Plant Medium [10] (see Table 1).
2. Seed establishment medium (see Table 1).
3. Explant establishment medium (see Table 1).
4. Propagation medium (see Table 1).
5. Pre-culture (see Table 1).
6. Loading solution (see Table 1).
7. Plant vitrification solution 3 (see Table 1).
8. Regrowth Solution (see Table 1).
9. Regrowth medium (see Table 1).
10. Sodium alginate solution 2% (see Table 1).
11. Calcium chloride solution (see Table 1).
12. Medium additives (see Table 1) used for the different stages of cultivation: seed germination (GS), bud establishment (EY), and multiplication (M).

Table 1
Solutions and culture medium additives used for the different culture stages of *Cedrela odorata* L

Culture medium	Components
Seed establishment medium	Murashige and Skoog medium [9] with 30 g/L sucrose, 8 g/L agar, and a pH of 5.7
Explant establishment medium	Woody Plant Medium [10] or Murashige and Skoog medium [9], added with 20 g/L sucrose, 1 g/L activated carbon, 1 mL/L Plant Preservative Mixture™, 1 g/L Polyvinylpyrrolidone, 8 g/L agar, and a pH of 5.7
Propagation medium	Murashige and Skoog medium [9], added with 30 g/L sucrose, 0.5 mg/L Benzylaminopurine (BA), 8 g/L agar, and a pH of 5.7
Pre-culture	Murashige and Skoog medium [9], added with 0.3 M sucrose, 8 $g{\cdot}L^{-1}$ agar [11]
Loading solution	Murashige and Skoog medium [9] with 2M glycerol, 0.4 M sucrose [11]
Plant vitrification solution 3	Murashige and Skoog medium [9] 50% (v/v) glycerol and 50% (w/v) sucrose [11]
Regrowth Solution	Murashige and Skoog medium [9] with 1 M sucrose [11]
Regrowth medium	Murashige and Skoog medium [9] with 0.2 mg/L 6-bencilaminopurina (BA) [11]
Sodium alginate solution 2%	Murashige and Skoog medium [9]100 mL with 20 g sodium alginate, 0.4 M sucrose [11]
Calcium chloride solution	0.1 M Calcium chloride with 0.1 M sucrose [11].

2.3 Equipment

1. Culture room: Laminar airflow hood.
2. Surgical instruments: stainless steel forceps, surgical scalpels, removable sterile surgical blades, and sterile Petri dishes.
3. Preparation of the culture medium: granatary and analytical scales for weighing, spatulas, magnetic stirrers, hot plate with stirring, potentiometer, microwave oven, graduated beaker, and autoclave.
4. Culture vessels: Beakers (capacity 1000 mL), reagent bottles, culture tubes, and Petri dishes.
5. Glassware cleaning: scrub the glassware with a commercial liquid detergent solution and rinse it with plenty of tap water.
6. Various materials: Film or plastic film for packaging food.

3 Methods

3.1 Disinfection Plant Material

3.1.1 Seeds

1. The seeds are washed with liquid detergent (*see* **Note 1**) for a period of 10 min, repeating this process twice.
2. Left to soak for 16 h in double-distilled water.
3. Immersed in a Captan® 3 g/L solution for 50 min.
4. Placed in a 70% ethanol solution for a period of 45 s.
5. Introduced into a 30% sodium hypochlorite solution for a period of 13 min.
6. Wash the seeds three or four times with sterile distilled water to remove residues of disinfectant solutions (*see* **Note 3**).
7. Establishing seeds in the growing medium (*see* **Note 5**).
8. The in vitro seedlings, free of microorganisms, and more vigorous, were selected for the multiplication of the material.

3.1.2 Explants

1. The plant material is collected from a plant in a field free of pests and diseases, vigorous and nourished, making it the perfect donor plant.
2. The plant material is washed with liquid detergent (*see* **Note 1**) for 10 min.
3. Introduced into a Captan® 3 g/L solution for 45 min.
4. Immersed in a 70% ethanol solution for 30 s.
5. Immersed in a 30% sodium hypochlorite solution for 10 min.
6. Wash three or four times with sterile distilled water to remove residues of disinfectant solutions (*see* **Note 3**).
7. Establishment of plant material in the culture medium (*see* **Note 5**).
8. The in vitro plants, free of microorganisms, and more vigorous, were selected for the multiplication of the material.

3.2 Shoots Proliferation

1. The shoots with a size of 15 mm (*see* **Note 6**) are established in a semi-solid medium with an incubation period of 30 days.
2. The shoots are incubated at a temperature of 24 ± 2°C and a photoperiod of 16 h of light and 8 h of darkness with a light intensity of 2000 lux.
3. After 30 days of incubation, new shoots are obtained per explant, which are separated and used in subsequent processes.

3.3 Cryopreservation

1. Apical meristems measuring approximately 1.5 mm are used.
2. The shoot tips are placed in Petri dishes (90 × 20 mm) in a pre-culture medium, and incubated in a room with a temperature of 24 ± 2°C under darkness for 24 h.

3. Set the cryo-plates on the Petri dishes (*see* **Note 7**).
4. Place a drop (3 μL) of the sodium alginate solution on the wells of the aluminum plate using a micropipette. Place the pre-cultivated shoot tips in each well with the tip of a scalpel blade. Another drop of sodium alginate is then placed on the shoot tips. The entire plate is coated with calcium chloride. Let it incubate for 15 min (Fig. 1).
5. Once the time has passed, the excess calcium chloride is removed and placed in the loading solution for 30 min for osmoprotection.
6. After this time, they are placed in the PVS3 solution for 30 min for vitrification. The cryo-plates are transferred to uncapped 2 mL cryotubes, which are help on cryo-canes, and directly in liquid nitrogen (LN), where they remained for at least 30 min (Fig. 2).

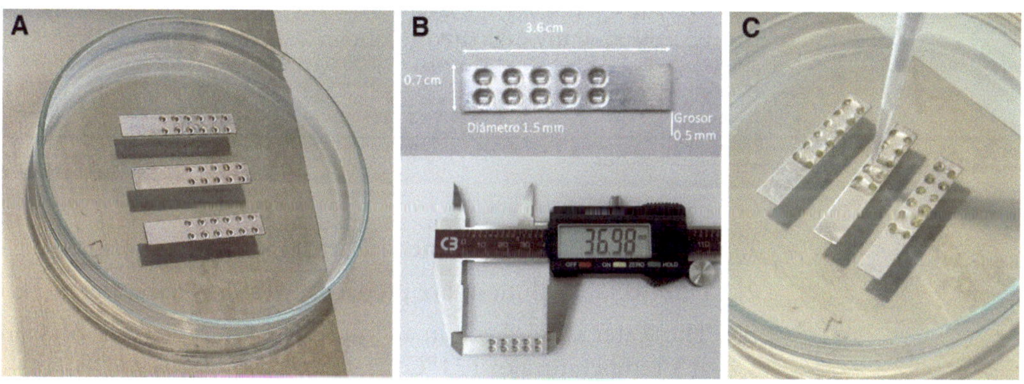

Fig. 1 (**A**) and (**B**) Shape and size of the cryo-plate. (**C**) Adhesion of shoot tips on the cryo-plate and application of a drop of sodium alginate.

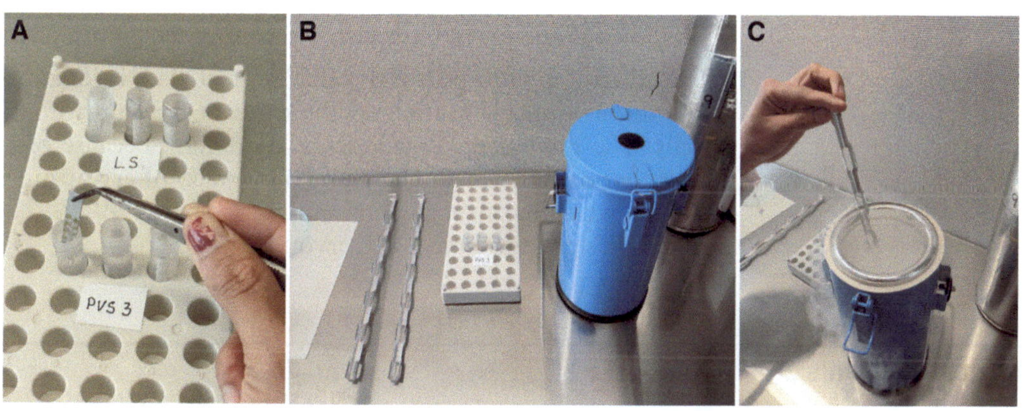

Fig. 2 (**A**) Placement of the cryo-plates in the loading solution and the PVS3 solution. (**B**) Cryocanes and liquid nitrogen tank. (**C**) Immersion of plant material in liquid nitrogen

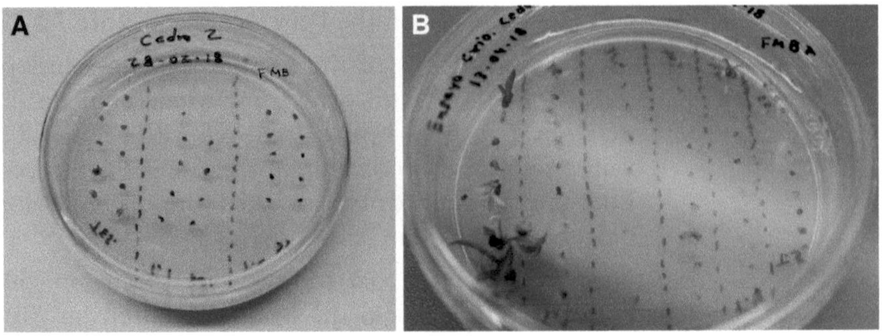

Fig. 3 (**A**) and (**B**) Shoot tips regenerating after being removed from liquid nitrogen

7. The cryotubes are removed from the liquid nitrogen and the cryo-plates with shoot tips are transferred from the cryotubes and immersed in the regrowth solution for 15 min at room temperature. The Shoot tips are transferred to regrowth medium and placed in a room at 25°C for 4 weeks so that the shoot tips that develop can be subsequently evaluated to obtain a seedling (Fig. 3).

4 Notes

1. The detergent is a commercial brand for washing dishes.
2. Disinfectant solutions are prepared with distilled water.
3. The water is sterilized in an autoclave at 1.5 kg/cm^2 pressure at a temperature of 121°C for 15 min.
4. Captan® is a commercial fungicide.
5. The disinfection protocol using the 70% ethanol solution is carried out inside the laminar flow hood.
6. The tissues are dissected, with a size of 15 mm, within the laminar flow hood and established in the culture medium.
7. The aluminum cryo-plates, the size of cryo-plate is length of 37 mm, a width of 7 mm and thickness of 0.5 mm with 10 wells. The cryo-plate is made from aluminum, because of its high thermal efficiency. The aluminum grade of cryo-plate is A1050, which means that aluminum purity is over 99.5%. The plate is designed to fit in 2 mL cryotube [11].

Acknowledgments

This work was supported by the National Genetic Resources Center of National Institute of Forestry, Agriculture and Livestock Research.

References

1. Gálvez L, Vallejo M A, Méndez C, López J (2020) *Cedrela odorata* L.: oportunidades para su conservación y mejoramiento genético. Revista Mexicana de Ciencias Forestales 11(58):4–25. https://doi.org/10.29298/rmcf.v11i58.622.
2. Jiménez JM (2023) Plantaciones forestales con cedro rojo en el trópico mexicano. Condiciones para su establecimiento comercial y cuidado. Instituto Nacional de Investigaciones Forestales, Agrícolas y Pecuarias. https://www.gob.mx/inifap/articulos/plantaciones-forestales-con-cedro-rojo-en-el-tropico-mexicano. Accessed 31 May 2025
3. Márquez-Ramírez J, Cruz H, Alba-Landa J, Mendizábal-Hernández L, Ramírez García E, Lavalle R (2019) Producción de frutos y semillas de *Cedrela odorata* L. en dos sitios de Veracruz, México. Foresta Veracruzana 1(21):23–31.
4. CITES (2007) Apéndices I, II y III. Convención sobre el Comercio Internacional de Especies Amenazadas de Fauna y Flora Silvestres. https://cites.org/esp/app/appendices.php. Accessed 27 May 2025
5. SEMARNA T (2010) NORMA Oficial Mexicana NOM-059-Protección ambiental-Especies nativas de México de flora y fauna silvestres-Categorías de riesgo y especificaciones para su inclusión, exclusión o cambio-Lista de especies en riesgo. Secretaría del Medio Ambiente y Recursos Naturales. Diario Oficial de la Federación. Cd. Mx., México. http://www.profepa.gob.mx/innovaportal/file/435/1/NOM_059_SEMARNAT_2010.pdf. Accessed 27 May 2025
6. Hernández-Ramos J, Reynoso R, Hernández A, García X, Hernández-Máximo E, Cob JV, Sumano D (2018) Distribución histórica, actual y futura de *Cedrela odorata* en México. Acta Botanica Mexicana 124.
7. CGIAR Genebanks. (s.f.). Cryopreservation | Conservation Module | Resources. Recuperado de https://genebanks.cgiar.org/the-platform/conservation-module/cryopreservation/
8. Engelmann F (2011) Cryopreservation of tropical plant germplasm—current research progress and future prospects. Vitro Cell Dev Biol Plant 47(1):5–18.
9. Murashige T, Skoog F (1962) A revised medium for rapid growth and bio assays with tobacco tissue cultures. Physiologia Plant 15(3):473–497
10. Lloyd G, McCown B (1981) Woody Plant Medium: a mineral nutrient formulation for microculture of woody plant species. HortScience 16:453
11. Niino T, Yamamoto S-I, Maki S, Tanaka D, Engelmann F (2017) Manual of cryopreservation methods using cryo-plate. Plant Tissue Culture and Cryopreservation Group. p 185

Chapter 26

Long-Term Preservation by Cryopreservation Method *Gmelina arborea* Roxb.

Esmeralda J. Cruz-Gutiérrez, Claudia Berenice Espitia-Flores, Marcelina Vélez-Torres, Luis Alberto Gómez-Reyes, Juan M. Pichardo-González and Martín Quintana-Camargo

Abstract

The conservation of plant genetic resources is carried out to protect species that are in danger of extinction, under special protection, or due to their economic and ecological importance. For the conservation of forest genetic resources, ex situ conservation in a germplasm bank is preferable to field conservation, as the latter increases conservation costs. One of the techniques for long-term conservation is cryopreservation using the vitrification method. This method allows plant tissue to be frozen at a temperature of −196°C, without the formation of ice crystals and thus avoiding damage to the cell wall. This way, plant material is preserved for a long period by halting its metabolism. Furthermore, forest species such as *Gmelina arborea* Roxb. They are genetic resources of ecological and economic importance for Mexico, so their long-term conservation is of vital importance in order to take advantage of this resource in the future.

Keywords Forest genetic resources, Ex situ conservation

1 Introduction

One of the techniques for long-term conservation is cryopreservation by vitrification. This technique consists of solidifying the water found inside the plant cell to a vitreous (amorphous) state. This prevents ice crystals from forming, thus preventing them from damaging the cell wall, dehydrating, or causing mechanical stress [1–3]. To achieve the vitreous state, high concentrations of cryoprotectants are used, which are chemical compounds such as glycerol, dimethyl sulfoxide (DMSO), ethylene glycol, sucrose, these substances are often combined to obtain specific formulas called Plant Vitrification Solution (PVS) such as PVS 2 (glycerol 30% (w/v), ethylene glycol 15% (w/v), dimethyl sulfoxide (DMSO) 15% (w/v), and sucrose 0.4 M), sucrose or PVS3 (50% glycerol and 50% sucrose) [4]. Of these two PVS, the most widely used because it is a more efficient

Marco A. Ramírez-Mosqueda et al. (eds.), *Conservation of Plant Genetic Resources*, Methods in Molecular Biology, vol. 3011,
https://doi.org/10.1007/978-1-0716-5130-8_26,

solution for cryopreservation is PVS2 [5, 6]. Cryoprotectants are used in concentrated solutions to lower the freezing point, increase cytoplasmic viscosity, and osmotically dehydrate the cell [1, 13]. The cooling rate also plays a role, as immersing them directly in liquid nitrogen (−196°C) prevents water molecules from organizing into crystals, causing them to solidify in an amorphous state [2, 7].

On the other hand, one of the forest species with high timber potential is *Gmelina arborea* Roxb. Although it is not a native species of Mexico, it has gained importance due to its rapid growth and easy adaptation to a diversity of forest areas [7–10]. This benefits the country's economy, as capital invested in wood production is quickly returned [9]. Its wood is easy to work with and has a good finish, which is why it is used to make high-quality furniture [11, 12]. For this reason, it is important to conserve the genetic resources of this species ex situ, since it would have a high maintenance cost in the field [13]. This work describes the cryopreservation technique using the vitrification method with PVS3 for *Gmelina arborea* Roxb.

2 Materials

2.1 Plant Material

1. The plant material to be used is seedlings that are in vitro conditions.

2.2 Culture Medium

1. The culture medium used was a Murashige and Skoog [14] and a Woody Plant Medium [15] (see Table 1).
2. Medium additives (see Table 1) used for the multiplication of plant material (M).

2.3 Equipment

1. Culture room: Laminar airflow hood.
2. Surgical instruments: stainless steel forceps, surgical scalpels, removable sterile surgical blades, and sterile Petri dishes.
3. Preparation of the culture medium: granatary and analytical scales for weighing, spatulas, magnetic stirrers, hot plate with stirring, potentiometer, microwave oven, graduated beaker, and autoclave.
4. Culture vessels: Beakers (capacity 1000 mL), reagent bottles, culture tubes, and Petri dishes.
5. Glassware cleaning: scrub the glassware with a commercial liquid detergent solution and rinse it with plenty of tap water.
6. Various materials: Film or plastic film for packaging food.

Table 1
Solutions and Culture medium additives used for the different culture stages of *Cedrela odorata* L

Culture medium	Components
Propagation medium	Murashige and Skoog medium [14], added with 30 g/L of sucrose, 0.5 mg/L of Benzylaminopurine (BA), 8 g/L of agar, and a pH of 5.7
Pre-culture	Murashige and Skoog medium [14], added with 0.3M of sucrose, 8 g/L of agar [16]
Loading solution	Murashige and Skoog medium [14] with 2M glycerol, 0.4 M sucrose [16]
Plant vitrification solution 3	Murashige and Skoog medium [14] with 50% glycerol and 50% sucrose [16]
Regrowth Solution	Murashige and Skoog medium [14] with 1 M of sucrose [16]
Regrowth medium	Murashige and Skoog medium [14] with 0.2 mg/L of 6-bencilaminopurina (BA) [16]
Sodium alginate solution 2%	Murashige and Skoog medium [14] 100 mL with 20 g of sodium alginate, 0.4 M sucrose [16]
Calcium chloride solution	0.1 M of Calcium chloride with 0.4 M sucrose [16]

3 Methods

3.1 Proliferation of Shoots

1. The shoots with a size of 15 mm are established in a semi-solid medium (see Table 1), with an incubation period of 30 days (*see* **Note 1**).
2. The shoots are incubated at a temperature of 24 ± 2°C and a photoperiod of 16 h of light and 8 h of darkness with a light intensity of 2000 lux.
3. After 30 days of incubation, new shoots are obtained per explant, which are separated and used in subsequent processes.

3.2 Cryopreservation

3.2.1 Material and Pre-Culture

1. Apical shoot tips measuring approximately 1.5 mm are used.
2. The shoot tips are placed in Petri dishes (90 × 20 mm) in a pre-culture medium (see Table 1), and incubated in a room with a temperature of 24 ± 2°C under darkness for 24 h.

3.2.2 Adhesion of Shoot Tips on the Cryo-plate

1. Set the cryo-plates (dimensions are shown in Fig. 1) (*see* **Note 2**) on the Petri dishes.
2. Place a drop (3 μL) of the sodium alginate solution (*see* Table 1) on the wells of the aluminium plate or cryo-plate (*see* **Note 2**) using a micropipette. Place the pre-cultivated shoot tips in each well with the tip of a scalpel blade number 11. Another

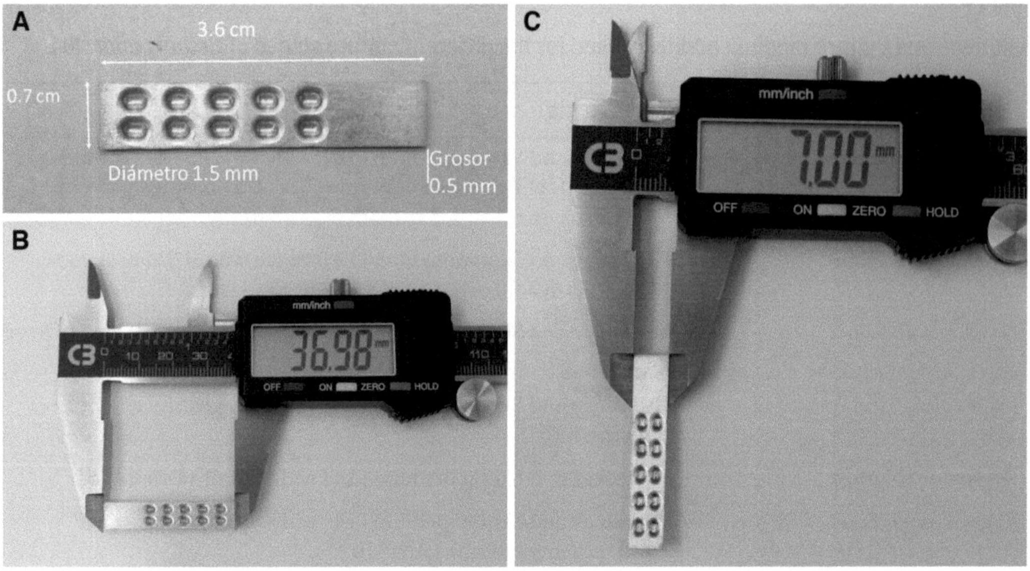

Fig 1 (**A**) and (**B**) Shape and size of the cryo-plate. (**C**) Cryo-plate width

drop of sodium alginate (see Table 1) is then placed on the shoot tips. The entire plate is covered with calcium chloride (see Table 1). Let it incubate for 15 min.

3. Once the time has passed, the excess calcium chloride is removed (it can be removed by placing the cryo-plate on sterile sanitary paper) and placed in the loading solution (see Table 1) for 30 min for osmoprotection.
4. After this time, they are placed in the Plant Vitrification solution (PVS3) (see Table 1) solution for 30 min for vitrification. The cryo-plates are transferred to uncapped 2 mL cryotubes, which are held on cryo-canes, and directly in liquid nitrogen (LN), where they remain for at least 30 min (Fig. 2).

3.2.3 Regrowth

1. The cryotubes are removed from the liquid nitrogen and the cryo-plates with shoot tips are transferred from the cryotubes and immersed in the regrowth solution (*see* Table 1) for 15 min at room temperature.
2. The shoot tips are transferred to regrowth medium (see Table 1) and placed in a room at 25°C for 4 weeks so that the shoot tips that develop can be subsequently evaluated to obtain a seedling (Fig. 3).

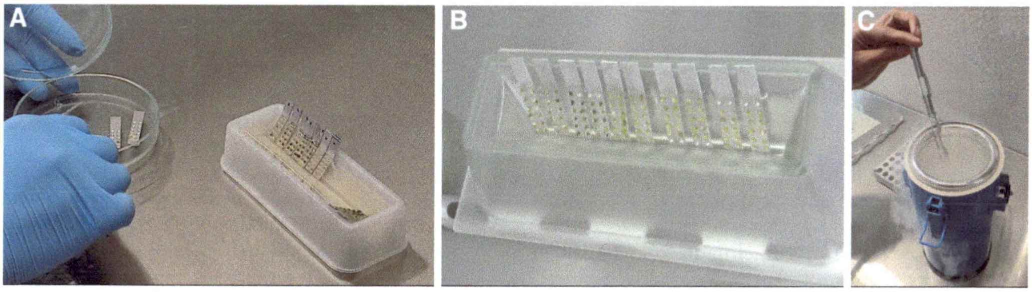

Fig 2 (**A**) Placement of shoot tips in the loading solution (LS). (**B**) Placement of shoot tips in Plan Vitrification Solution 3 (PVS3). (**C**) Immersion of plant material in liquid nitrogen.

Fig 3 (**A**) and (**B**) Regeneration of meristems after immersion in liquid nitrogen

4 Notes

1. The tissues are dissected, with a size of 15 mm, within the laminar flow hood and established in the culture medium (see Table 1).
2. The aluminum cryo-plates, the size of cryo-plate is length of 37 mm, a width of 7 mm, and thickness of 0.5 mm with ten wells. The cryo-plate is made from aluminum, because of its high thermal efficiency. The aluminum grade of cryo-plate is A1050, which means that aluminum purity is over 99.5%. The plate is designed to fit in 2 mL cryotube [16].

Acknowledgments

This work was supported by the National Genetic Resources Center of National Institute of Forestry, Agriculture and Livestock Research.

References

1. Engelmann F (2011) Cryopreservation of tropical plant germplasm—current research progress and future prospects. In Vitro Cell Dev-Pl 47(1):5–18
2. Panis B, Lambardi M (2005) Status of cryopreservation of plant germplasm *in vitro*. Acta Hortic 691:21–31
3. Villalobos VM, Engelmann F (2007) Cryopreservation of plant germplasm: a practical guide. Bioversity International.
4. Esquivel-Figueroa D, Tiznado-Hernández ME, Islas-Osuna MA, Lazo-Javalera MF, Rivera-Domínguez M (2023) Etapas del protocolo de crioconservación y su efecto sobre la viabilidad y regeneración de embriones cigóticos de vid (*Vitis vinífera* L.). Revista de Ciencias Biológicas y de la Salud XXV(3):170–175
5. Normah MN, Sulong N, Reed BM (2019) Cryopreservation of shoot tips of recalcitrant and tropical species: advances and strategies. Cryobiology 87:1–14
6. Sakai A, Kobayashi S, Oiyama I (1990) Cryopreservation of nucellar cells of navel orange (*Citrus sinensis* Osb. var. brasiliensis Tanaka) by vitrification. Plant Cell Rep 9(1):30–33
7. Nausch H, Buyel JF (2021) Cryopreservation of plant cell cultures – diverse practices and protocols. N Biotechnol 62:86–95. https://doi.org/10.1016/j.nbt.2021.02.002
8. Indira EP (2006) Provenance variations in *Gmelina arborea* with particular reference to tree form. J Trop For Sci 36–50. Recuperado de https://www.jstor.org/stable/43594644
9. Moya-Roque R, Muñoz-Acosta F, Salas-Garita C, Berrocal-Jiménez A, Leandro-Zúñiga L, Esquivel-Segura E (2010) Tecnología de madera de plantaciones forestales: Fichas técnicas. Rev For Mesoam Kurú 7:18–19. https://revistas.tec.ac.cr/index.php/kuru/article/view/383 Recuperado de
10. Rojas FR, Murillo OG (2004) Botánica y ecología. In: Rojas F, Arias D, Moya R, Meza A, Murillo O, Arguedas M (eds) Manual para productores de melina (*Gmelina arborea*) en Costa Rica. Fonafifo, Cártago, Costa Rica, p 314. Recuperado de http://www.sirefor.go.cr/Documentos/Especies_plantaciones/MELINA/Manual para los productores demelina.pdf%5Cnhttp://www.sirefor.go.cr
11. Gonzáles GT, Serrano JRM (2004) Propiedades y utilizaciones de la madera de melina (*Gmelina arborea* Roxb) procedente de árboles plantados en Costa Rica. Rev For Mesoam Kurú 6(17):61–69. http://revistas.tec.ac.cr/index.php/kuru/article/view/386 Recuperado de
12. Rejón LAS, Romero JL (2004) Plantations of *Gmelina arborea* in southern Mexico. New Forests 28(2–3):293–297. https://doi.org/10.1023/B:NEFO.0000040956.68838.97
13. González-Arnao MT, Engelmann F (ed) (2013) Crioconservación de plantas en América Latina y el Caribe. IICA, San José
14. Murashige T, Skoog F (1962) A revised medium for rapid growth and bio assays with tobacco tissue cultures. Physiol Plant 15(3):473–497
15. Lloyd G, McCown B (1981) Woody Plant Medium: a mineral nutrient formulation for microculture of woody plant species. HortScience 16:453
16. Niino T, Yamamoto S-I, Maki S, Tanaka D, Engelmann F (ed) (2017) Manual of cryopreservation methods using cryo-plate. Plant Tissue Culture and Cryopreservation Group, p 185

Chapter 27

Long-Term Preservation by the Cryopreservation Method of *Swietenia humilis* Zucc

Claudia Berenice Espitia-Flores, Esmeralda J. Cruz-Gutiérrez, Manuel de Jesús Bermúdez-Guzmán, Luis Alberto Gómez-Reyes, Juan M. Pichardo-González and José Luis Aguirre-Noyola

Abstract

The species *Swietenia humilis* Zucc., better known as mahogany, is an important forest species for Mexico, forming part of the low deciduous forest. Its wood is considered precious and is therefore used in fine cabinetmaking. Due to its importance to the country, it is crucial to conserve this species. There are different types of conservation, ex situ and in situ, with ex situ conservation being the most widely used. For long-term conservation in ex situ conditions, cryopreservation is used through the vitrification technique. This technique involves preventing the formation of ice crystals in cells that are frozen at very low temperatures, at −196°C, the temperature of liquid nitrogen. At these temperatures, cell metabolism stops, allowing plant tissue to be preserved for long periods. To carry out this method, substances called cryoprotectants must be applied, substances that increase osmolarity in plant tissue by reducing the unbound water content. The objective of this work is to describe the cryopreservation protocol using the vitrification method of plant tissue of *Swietenia humilis* Zucc.

Keywords Vitrification, Ex situ conservation, Forest plant tissue, In vitro

1 Introduction

The species *Swietenia humilis* Zucc., better known as caobilla, cóbano, gateado, or zopilote, is a highly relevant forest species in Mexico, distributed from Sinaloa to Chiapas, forming a fundamental part of the lowland deciduous and subdeciduous forests [1, 2]. Its wood is considered precious; it is durable, resistant, beautiful, and easy to work with, which is why it is primarily used in fine cabinetmaking [3]. This species is emblematic of Mexico's forests, valued for its wood, traditional uses, and ecological role. Its current importance lies in protecting it from extinction and ensuring its long-term ex situ survival for future generations [4].

Marco A. Ramírez-Mosqueda et al. (eds.), *Conservation of Plant Genetic Resources*, Methods in Molecular Biology, vol. 3011, https://doi.org/10.1007/978-1-0716-5130-8_27,

A strategy for the conservation of *Swietenia humilis* Zucc. plant material. It is by the method of cryopreservation of plant tissue, and this method consists of freezing plant cells at very low temperatures at −196°C, the temperature of liquid nitrogen, with cryoprotectants, which help to stabilize the cells and prevent them from dying when frozen, balancing the molar concentrations of the cryoprotectants, preventing the formation of crystals inside the cell. Within the cryopreservation method there are several techniques, one of which has been observed to have a good response in forest species is the vitrification technique [5, 6]. This technique consists of preventing the formation of ice crystals inside and outside the cells, solidifying the water forming a metastable amorphous solid known as glass [7]. To achieve this process, it is necessary to use substances known as cryoprotectants, which increase osmolarity in plant tissue by reducing the unbound water content [8, 9]. There are two groups of cryoprotectants: penetrating and non-penetrating. Penetrating agents are low molecular weight compounds that cross the cell membrane, acting mainly on the protoplast by replenishing intracellular water before and during cooling and lowering the freezing point of water. The most commonly used are glycerol, DMSO, and 1,2-propanediol (propylene glycol or PG) [7, 8]. Non-penetrating cryoprotectants are high molecular weight substances that form hydrogen bonds with water. As their name suggests, they cannot penetrate the cell wall, thereby reducing ionic forces on the outside of the cell and moderately decreasing the formation of ice crystals. The most common cryoprotectants are carbohydrates such as polysaccharides, sucrose and glucose being the most used, or sugar alcohols (mannitol, sorbitol), which are effective due to their osmotic power and therefore promote rapid cell dehydration [10, 11]. The objective of this work is to present a protocol for cryopreserving *Swietenia humilis* Zucc. plant tissue using the vitrification method, where the methodology is described step by step.

2 Materials

2.1 Plant Material and Culture Medium

1. The plant material to be used is seedlings that are in vitro conditions.
2. The culture medium used was Murashige and Skoog [12] (see Table 1).
3. Medium additives (see Table 1) used for the multiplication of plant material (M).

2.2 Equipment

1. Culture room: Laminar airflow hood.
2. Surgical instruments: stainless steel forceps, surgical scalpels, removable sterile surgical blades, and sterile Petri dishes.

Table 1
Solutions and culture medium additives used for the different culture stages of *Cedrela odorata* L

Culture medium	Components
Propagation medium	Murashige and Skoog medium, added with 30 g/L sucrose, 0.5 mg/L Benzylaminopurine (BAP), 8 g/L agar, and a pH of 5.7
Pre-culture	Murashige and Skoog medium, added with 0.3 M sucrose, 8 g/L of agar
Loading solution	Murashige and Skoog medium with 2 M glycerol, 0.4 M sucrose
Plant vitrification solution 3	Murashige and Skoog medium 50% (v/v) glycerol and 50% (w/v) sucrose
Regrowth Solution	Murashige and Skoog medium with 1 M sucrose
Regrowth medium	Murashige and Skoog medium with 0.2 mg/L 6-bencilaminopurina (BA)
Sodium alginate solution 2%	Murashige and Skoog medium 100 mL with 20 g sodium alginate, and 0.4 M sucrose
Calcium chloride solution	0.1 M calcium chloride with 0.4 M sucrose

3. Preparation of the culture medium: scales for weighing, spatulas, magnetic stirrers, hot plate with stirring, potentiometer, microwave oven, graduated beaker, and autoclave.
4. Culture vessels: Beakers (capacity 1000 mL), reagent bottles, culture tubes, and Petri dishes.
5. Glassware cleaning: scrub the glassware with a commercial liquid detergent solution and rinse it with plenty of tap water.
6. Various materials: Film or plastic film for packaging food.

3 Methods

3.1 Proliferation of Shoots

1. The shoots with a size of 15 mm are established in a semi-solid medium (see Table 1), with an incubation period of 30 days (*see* **Note 1**).
2. The shoots are incubated at a temperature of 24 ± 2°C and a photoperiod of 16 h of light and 8 h of darkness with a light intensity of 2000 lux.
3. After 30 days of incubation, new shoots are obtained per explant, which are separated and used in subsequent processes.

3.2 Cryopreservation

3.2.1 Material and Pre-Culture

1. Apical shoot tips measuring approximately 1.5 mm are used.
2. The shoot tips are placed in Petri dishes (90 × 20 mm) in a pre-culture medium (see Table 1), and incubated in a room with a temperature of 24 ± 2°C under darkness for 24 h.

3.2.2 Adhesion of Shoot Tips on the Cryo-plate

1. Set the cryo-plates (*see* **Note 2 and Note 3**) on the Petri dishes.
2. Place a drop (3 μL) of the sodium alginate solution (see Table 1) on the wells of the aluminium plate or cryo-plate (*see* **Note 2**) using a micropipette. Place the pre-cultivated shoot tips in each well with the tip of a scalpel blade number 11. Another drop of sodium alginate (see Table 1) is then placed on the shoot tips. The entire plate is covered with calcium chloride (see Table 1). Let it incubate for 15 min.
3. Once the time has passed, the excess calcium chloride is removed (it can be removed by placing the cryo-plate on sterile sanitary paper) and placed in the loading solution (see Table 1) for 30 min for osmoprotection (Fig. 1A).
4. After this time, they are placed in the Plant Vitrification solution (PVS3) (see Table 1) solution for 30 min for vitrification. The cryo-plates are transferred to uncapped 2 mL cryotubes (*see* **Note 4**), which are held on cryo-canes, and directly in liquid nitrogen (LN), (*see* **Note 3**) where they remain for at least 30 min (Fig. 1B).

3.2.3 Regrowth

1. The cryotubes are removed from the liquid nitrogen and the cryo-plates with shoot tips are transferred from the cryotubes and immersed in the regrowth solution (see Table 1) for 15 min at room temperature.
2. The shoot tips are transferred to regrowth medium (see Table 1) and placed in a room at 25°C for 4 weeks so that the shoot tips that develop can be subsequently evaluated to obtain a seedling.

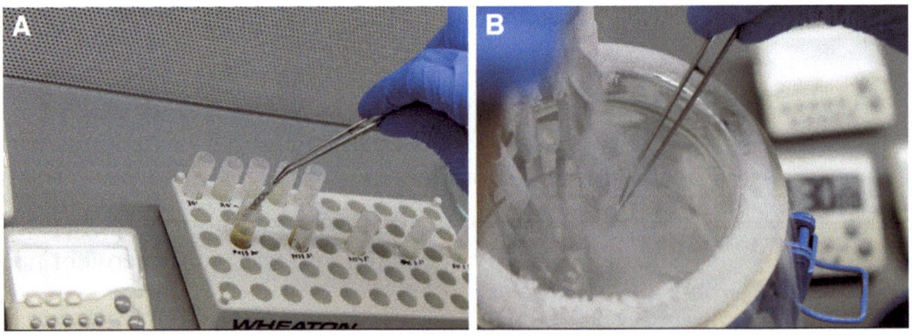

Fig. 1 (**A**) Cryo-plates in PVS3 solution and (**B**) Cryo-plates in liquid nitrogen

4 Notes

1. The tissues are dissected, with a size of 15 mm, within the laminar flow hood and established in the culture medium (see Table 1).
2. The aluminum cryo-plate is length of 37 mm, a width of 7 mm, and thickness of 0.5 mm with 10 wells.
3. The cryo-plate is made from aluminum, because of its high thermal efficiency.
4. The aluminum grade of cryo-plate is A1050, which means that aluminum purity is over 99.5%. The plate is designed to fit in 2 mL cryotube [13].

References

1. Pennington TD, Sarukhán J (2005) Árboles tropicales de México: una guía para la identificación de las especies más comunes, 3ra ed. Fondo de Cultura Económica, UNAM
2. Villaseñor CIS, Larios DHM, Hernández MJL (2022) Características morfométricas, reproductivas y germinativas del germoplasma de *Swietenia humilis* Zucc. en Guerrero. Revista Mexicana de Ciencias Forestales 13(72). Disponible en: https://cienciasforestales.inifap.gob.mx/index.php/forestales/article/download/1250/3141?inline=1
3. CONABIO (2009) Fichas de especies mexicanas incluidas en CITES: Swietenia humilis Zucc. México
4. Ledig FT (2004) Conservación y Manejo de Recursos Genéticos Forestales. In: Vargas H, Jesús J, Bermejo BV, Ledig FT (eds) Manejo de Recursos Genéticos Forestales. Servicio Forestal de los Estados Unidos de América, Organización de las Naciones Unidas para la Agricultura y la Alimentación (FAO) y la Comisión Nacional Forestal (Conafor). Zapopan, Jal., México, pp 6–20
5. Langis R, Steponkus PL (1990) Cryopreservation of rye protoplasts by vitrification. Plant Physiol 92:666–671
6. Sakai A, Kobayashi S, Oiyama I (1990) Cryopreservation of nucellar cells of navel orange (*Citrus sinensis* Osb. var. brasiliensis Tanaka) by vitrification. Plant Cell Rep 9(1):30–33. https://doi.org/10.1007/BF00232130
7. Roque-Borda CA, Kulus D, Vacaro de Souza A, Kaviani B, Vicente EF (2021) Cryopreservation of agronomic plant germplasm using vitrification-based methods: an overview of selected case studies. Int J Mol Sci 22:6157. https://doi.org/10.3390/ijms22116157
8. Elliott GD, Wang S, Fuller BJ (2017) Cryoprotectants: a review of the actions and applications of cryoprotective solutes that modulate cell recovery from ultra-low temperatures. Cryobiology 76:74–91
9. Wang S, Oldenhof H, Dai X, Haverich A, Hilfiker A, Harder M, Wolkers WF (2014) Protein stability in stored decellularized heart valve scaffolds and diffusion kinetics of protective molecules. Acta Proteins Proteom 1844:430–438
10. Bahr MM, Amer MS, Abo-El-Sooud K, Abdallah AN, El-Tookhy OS (2020) Preservation techniques of stem cells extracellular vesicles: a gate for manufacturing of clinical grade therapeutic extracellular vesicles and long-term clinical trials. Int J Vet Sci Med 8:1–8
11. Wang Q, Tanne E, Arav A, Gafny R (2000) Cryopreservation of *in vitro*-grown shoot tips of grapevine by encapsulation-dehydration. Plant Cell Tissue Organ Cult 63:41–46
12. Murashige T, Skoog F (1962) A revised medium for rapid growth and bio assays with tobacco tissue cultures. Physiol Plant 15(3):473–497
13. Niino T, Yamamoto S, Matsumoto T, Engelmann F, Valle Arizaga M, Tanaka D (2018) Development of V and D cryo-plate methods as effective protocols for cryobanking. In: III international symposium on plant cryopreservation 1234, pp 249–262

Chapter 28

Long-Term Preservation by Cryopreservation Method *Swietenia macrophylla* King

Claudia Berenice Espitia-Flores, Esmeralda J. Cruz-Gutiérrez, Luis Alberto Gómez-Reyes and Juan M. Pichardo-González

Abstract

Mahogany (*Swietenia macrophylla* King) is a highly valued Mexican forest species for its precious wood, recognized for its hardness, resistance, and beauty. It is widely distributed in various states of Mexico, and its trees are important for soil conservation and erosion control. However, intense exploitation has led to a decrease in its genetic variability, underscoring the urgent need for its ex situ conservation in plant germplasm banks.

A key strategy for ex situ conservation is in vitro culture, which allows for rapid multiplication of plant tissue, benefiting both industrial plant production and conservation. Within in vitro culture, there is a technique called cryopreservation for long-term conservation in germplasm banks. There are several methods within cryopreservation, one of which is cryopreservation by vitrification, which involves turning intracellular water vitreous (amorphous) at extremely low temperatures, such as those of liquid nitrogen. This is achieved with cryoprotectants (such as DMSO, glycerol, ethylene glycol, and sucrose), which at high concentrations reduce the freezing point and increase cytoplasmic viscosity. The objective of this work is to describe the application of this vitrification cryopreservation technique for *Swietenia macrophylla* King.

Keywords Forestry, Precious wood, Germplasm banks, Ex situ conservation

1 Introduction

King mahogany, or *Swietenia macrophylla* King, is a species distributed in the states of Campeche, Chiapas, Colima, Guanajuato, Guerrero, Hidalgo, Jalisco, the State of Mexico, Michoacán, Morelos, Nayarit, Oaxaca, Puebla, Querétaro, Quintana Roo, Sinaloa, Tabasco, Veracruz, and Yucatán [1]. This species is highly valued by cabinetmakers, as the wood is considered precious for its hardness, resistance, beauty, and quality [2]. It is also very important because the trees are used for soil conservation and erosion control [1]. However, it has been intensively exploited, resulting in a decrease in genetic variability [2], so ex situ conservation in plant germplasm banks is essential. One of the techniques for ex situ conservation is in vitro culture. This technique

Marco A. Ramírez-Mosqueda et al. (eds.), *Conservation of Plant Genetic Resources*, Methods in Molecular Biology, vol. 3011, https://doi.org/10.1007/978-1-0716-5130-8_28,

rapidly increases plant tissue, which is beneficial for both industrial plant production and conservation in germplasm banks [3]. Furthermore, the technique used for conservation in germplasm banks is cryopreservation using the vitrification method [4]. Although there are several cryopreservation techniques, this method was observed in this study to be the most feasible for the *Swietenia macrophylla* King species.

Cryopreservation by vitrification seeks to protect cells and tissues from freezing damage by preventing the formation of ice crystals. These crystals can perforate cell membranes and damage internal structures, compromising the viability of the material [5, 6]. Vitrification allows water to form glassy or amorphous forms inside and outside cells instead of forming ice crystals when it reaches very low temperatures, such as liquid nitrogen [7]. This technique involves the use of cryoprotectants, which reduce the freezing point of water and increase the viscosity of the cytoplasm, making it difficult for ice crystals to form. These solutions, such as dimethyl sulfoxide (DMSO), glycerol, and ethylene glycol, are often combined with sugars such as sucrose [8]. Subsequently, osmotic dehydration is also performed, which induces controlled dehydration of the cells before cooling. This reduces the water content that could crystallize, making vitrification more effective [5]. Ultra-rapid cooling is achieved by direct immersion in liquid nitrogen at −196°C, preserving the plant tissue for an extended period. Therefore, the objective of this work is to describe the cryopreservation technique for *Swietenia macrophylla* King.

2 Materials

2.1 Plant Materials and Disinfection

1. The germplasm used may be seeds or tissue from a healthy, vigorous, pest-free juvenile donor plant. For establishment are used young shoots (with one or two buds) or seeds.
2. Liquid detergent (*see* **Note 1**).
3. 70% ethanol solution (*see* **Note 2**).
4. 30% sodium hypochlorite solution (*see* **Note 2**).
5. Sterile distilled water (*see* **Note 3**).
6. Captan® solution ($3\,g{\cdot}L^{-1}$) (*see* **Note 4**).
7. pH adjustment solutions: 1 N sodium hydroxide (NaOH) and 0.1 N hydrochloric acid (HCl).

2.2 Culture Medium

1. The culture medium used was a Murashige and Skoog [9] and a Woody Plant Medium [10] (see Table 1).
2. Seed establishment medium (see Table 1).
3. Explant establishment medium (see Table 1).
4. Propagation medium (see Table 1).

Table 1
Solutions and culture medium additives used for the different culture stages of *Cedrela odorata* L.

Culture medium	Components
Seed establishment medium	Murashige and Skoog medium [9] with 30 g/L sucrose, 8 g/L agar, and a pH of 5.7.
Explant establishment medium	Woody Plant Medium [10] or Murashige and Skoog medium [9], added with 20 g/L sucrose, 1 g/L activated carbon, 1 mL/L Plant Preservative Mixture™, 1 g/L Polyvinylpyrrolidone, 8 g/L agar, and a pH of 5.7.
Propagation medium	Murashige and Skoog medium [9], added with 30 g/L sucrose, 0.5 mg/L Benzylaminopurine (BA), 8 g/L agar, and a pH of 5.7.
Pre-culture	Murashige and Skoog medium [9], added with 0.3 M sucrose, 8 $g{\cdot}L^{-1}$ agar [11].
Loading solution	Murashige and Skoog medium [9] with 2 M glycerol, 0.4 M sucrose [11].
Plant vitrification solution 3	Murashige and Skoog medium [9] 50% (v/v) glycerol and 50% (w/v) sucrose [11].
Regrowth Solution	Murashige and Skoog medium [9] with 1 M sucrose [11].
Regrowth medium	Murashige and Skoog medium [9] with 0.2 mg/L 6-bencilaminopurina (BA) [11].
Sodium alginate solution 2%	Murashige and Skoog medium [9] 100 mL with 20 g sodium alginate, 0.4 M sucrose [11].
Calcium chloride solution	0.1 M Calcium chloride with 0.4 M sucrose [11].

5. Pre culture (see Table 1).
6. Loading solution (see Table 1).
7. Plant vitrification solution 3 (see Table 1).
8. Regrowth Solution (see Table 1).
9. Regrowth medium (see Table 1).
10. Sodium alginate solution 2% (see Table 1).
11. Calcium chloride solution (see Table 1).
12. Medium additives (see Table 1) used for the different stages of cultivation: seed germination (GS), bud establishment (EY), and multiplication (M).

2.3 Equipment

1. The aluminum cryo-plate is length of 37 mm, a width of 7 mm, and thickness of 0.5 mm with ten wells. The cryo-plate is made from aluminum, because of its high thermal efficiency. The aluminum grade of cryo-plate is A1050, which means that aluminum purity is over 99.5%. The plate is designed to fit in 2 mL cryotube [11].

2. Culture room: Laminar airflow hood.
3. Surgical instruments: stainless steel forceps, surgical scalpels, removable sterile surgical blades, and sterile Petri dishes.
4. Preparation of the culture medium: granary and analytical scales for weighing, spatulas, magnetic stirrers, hot plate with stirring, potentiometer, microwave oven, graduated beaker, and autoclave.
5. Culture vessels: Beakers (capacity 1000 mL), reagent bottles, culture tubes, Petri dishes.
6. Glassware cleaning: scrub the glassware with a commercial liquid detergent solution and rinse it with plenty of tap water.
7. Various materials: Film or plastic film for packaging food.

3 Methods

3.1 Disinfection Plant Material

3.1.1 Seeds

1. The seeds are washed with liquid detergent (*see* **Note 1**) for a period of 10 min, repeating this process twice.
2. Left to soak for 16 h in double-distilled water.
3. Immersed in a Captan® 3 $g{\cdot}L^{-1}$ solution for 50 min.
4. Placed in a 70% ethanol solution for a period of 45 s.
5. Introduced into a 30% sodium hypochlorite solution for a period of 13 min.
6. Wash the seeds three or four times with sterile distilled water to remove residues of disinfectant solutions (*see* **Note 3**).
7. Establishing seeds in the growing medium (*see* **Note 5**).
8. The in vitro seedlings, free of microorganisms and more vigorous, were selected for the multiplication of the material.

3.1.2 Explants

1. The plant material is collected from a plant in a field free of pests and diseases, vigorous, and nourished, making it the perfect donor plant.
2. The plant material is washed with liquid detergent (*see* **Note 1**) for 10 min.
3. Introduced into a Captan® 3 g/L solution for 45 min.
4. Immersed in a 70% ethanol solution for 30 s.
5. Immersed in a 30% sodium hypochlorite solution for 10 min.
6. Wash three or four times with sterile distilled water to remove residues of disinfectant solutions (*see* **Note 3**).

7. Establishment of plant material in the culture medium (*see* **Note 5**).
8. The in vitro plants, free of microorganisms and more vigorous, were selected for the multiplication of the material.

3.2 Shoos Proliferation

1. The shoots with a size of 15 mm are established in a semi-solid medium (*see* Table 1), with an incubation period of 30 days.
2. The shoots are incubated at a temperature of 24 ± 2°C and a photoperiod of 16 h of light and 8 h of darkness with a light intensity of 2000 lux.
3. After 30 days of incubation, new shoots are obtained per explant, which are separated and used in subsequent processes.

3.3 Cryopreservation

1. Apical meristems measuring approximately 1.5 mm are used (*see* **Note 6**).
2. The shoot tips are placed in Petri dishes (90 × 20 mm) in a pre-culture medium (see Table 1), and incubated in a room with a temperature of 24 ± 2°C under darkness for 24 h.
3. Set the cryo-plates on the Petri dishes.
4. Place a drop (3 μL) of the sodium alginate solution (see Table 1) on the wells of the aluminium plate using a micropipette. Place the pre-cultivated shoot tips in each well with the tip of a scalpel blade. Another drop of sodium alginate (see Table 1) is then placed on the shoot tips. The entire plate is coated with calcium chloride (see Table 1). Let it incubate for 15 min.
5. Once the time has passed, the excess calcium chloride is removed and placed in the loading solution (see Table 1) for 30 min for osmoprotection.
6. After this time, they are placed in the PVS3 (see Table 1) solution for 55 min for vitrification. The cryo-plates are transferred to uncapped 2 mL cryotubes, which are help on cryo-canes, and directly in liquid nitrogen (LN), where they remained for at least 30 minutes.
7. The cryotubes are removed from the liquid nitrogen and the cryo-plates with shoot tips are transferred from the cryotubes and immersed in the regrowth solution (see Table 1) for 15 min at room temperature. The shoot tips are transferred on regrowth medium (see Table 1) and placed in a room at 25°C for 4 weeks so that the shoot tips that develop can be subsequently evaluated to obtain a seedling (Fig. 1).

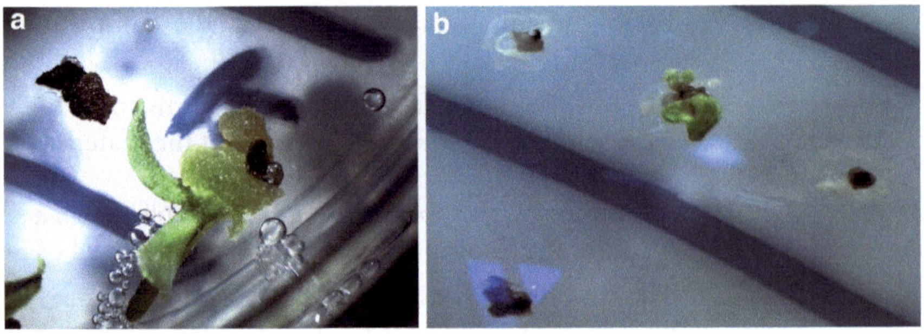

Fig. 1 Shoot tips cryopreserved. (**a**) and (**b**) Shoot tips regenerating after cryopreservation

4 Notes

1. The detergent is a commercial brand for washing dishes.
2. Disinfectant solutions are prepared with distilled water.
3. The water is sterilized in an autoclave at 1.5 kg/cm^2 pressure at a temperature of 121°C for 15 min.
4. Captan® is a commercial fungicide.
5. The disinfection protocol using the 70% ethanol solution is carried out inside the laminar flow hood.
6. The tissues are dissected, with a size of 15 mm, within the laminar flow hood and established in the culture medium (see Table 1).

Acknowledgments

This work was supported by the National Genetic Resources Center of National Institute of Forestry, Agriculture and Livestock Research.

References

1. Comisión Nacional Forestal (CONAFOR) (2018) *Swietenia macrophylla*. SIRE: CONAFOR CONABIO. Paquetes tecnológicos. México. Available at: http://www.conafor.gob.mx:8080/documentos/docs/13/1005Swietenia%20macrophylla.pdf. Accessed 20 Jun 2025
2. Carranza M, Reyes H, Mora W, Cevallos O, Escobar A, Cadme M, Nieto J, Morante J (2013) Propagación clonal *in vitro* de *Swietenia macrophylla* King (CAOBA). Ciencia y Tecnología 6(2):1–8
3. Uribe M, Delaveau C, Garcés M, Escobar R (2008) Efecto de asepsia y fitohormonas en el establecimiento *in vitro* de Berberidopsis corallina, a partir de segmentos nodales. Bosque 29(1):58–64
4. Panis B, Lambardi M (2006) Status of cryopreservation of plant germplasm *in vitro*. In Vitro Cell Dev Biol - Plant 42(6):612–624
5. Li Z, Wang L, Huang S, Zhang W, Wei R (2022) Vitrification-based cryopreservation of plant germplasm: a comprehensive review. Plants 11(13):1668

6. Sakai A, Kobayashi S, Oiyama I (1990) Cryopreservation of nucellar cells of navel orange (*Citrus sinensis* Osb. var. brasiliensis Tanaka) by vitrification. Plant Cell Rep 9(1):30–33
7. González-Arnao MT, Engelmann F (eds) (2013) Crioconservación de plantas en América Latina y el Caribe. IICA, San José
8. Sakai A, Kawamura Y, Uragami A (1991) Cryopreservation of protoplasts by vitrification. Plant Physiol 97(2):465–470
9. Murashige T, Skoog F (1962) A revised medium for rapid growth and bio assays with tobacco tissue cultures. Physiol Plant 15(3):473–497
10. Lloyd G, McCown B (1981) Woody plant medium: a mineral nutrient formulation for microculture of woody plant species. HortScience 16:453
11. Niino T, Yamamoto S-I, Maki S, Tanaka D, Engelmann F (eds) (2017) Manual of cryopreservation methods using cryo-plate. Plant Tissue Culture and Cryopreservation Group, p 185

Chapter 29

Biotechnological Tools for *Agave* Conservation

María de Lourdes Delgado-Aceves, José Juvencio Castañeda-Nava, José Manuel Rodríguez-Domínguez, Fernando Santacruz-Ruvalcaba, Liberato Portillo and Antonia Gutiérrez-Mora

Abstract

Nowadays, several *Agave* species have been described worldwide; they are used for ornamental, agroindustrial, ethnobotanical, and other purposes. Due to excessive use, there is the possibility of losing the germplasm of some species, mainly those that are not cultivated and are listed as endangered. These factors make it necessary to develop strategies to conserve germplasm. Plant biotechnology offers a number of techniques, such as in vitro culture of plant cells and tissues, that can be incorporated into conservation schemes for cultivated, overexploited, threatened, and/or endangered species. Other important strategies for creating in vitro germplasm banks that are complementary to cell and tissue culture are the indefinite conservation of living tissues under controlled conditions of light and temperature. This chapter will mention biotechnological tools for the conservation of *Agave* species, such as micropropagation, cryopreservation, and seed germination. These strategies have been developed in different species of *Agave*; in all species, it is necessary to start with micropropagation and later cryopreservation. On the other hand, methodologies are mentioned to improve seed germination and recover populations from the wild or to have a gene pool with commercial interests.

Keywords Cryopreservation, Genetic resources, Germplasm, Micropropagation

1 Introduction

The genus *Agave* of the Asparagaceae family, has about 75% of its species in Mexico, out of 55% are endemic [1]. Although most species are not domesticated, since pre-Hispanic times they have been used as food, construction material, ornament, production of fermented and distilled alcoholic beverages, cordage, pharmaceutical compounds, and biofuels. In addition, it is a highly efficient plant in biomass production with low water consumption [2, 3], consolidating itself as a valuable and versatile resource in various productive sectors. Some *Agave* species produce seeds, but this process can take many years, as the plant must reach maturity, which can take between 5 and 40 years, depending on the species. In addition, being semelparous

Marco A. Ramírez-Mosqueda et al. (eds.), *Conservation of Plant Genetic Resources*, Methods in Molecular Biology, vol. 3011, https://doi.org/10.1007/978-1-0716-5130-8_29,

plants, once the seeds mature in the inflorescence, the plant dies [4–6]. Most *Agave* species are propagated by rhizome shoots that form around the mother plant, and some by bulbils on the axillary buds of the inflorescence when seed production is limited [7].

Recent research by Alducin-Martínez et al. [8] in the *Agave* genus indicates a priority of conservation of species classified by subgenus *Agave* and *Littaea*. The use of *Agave* subgenus to produce distilled beverages stands out, while the *Littaea* subgenus is used for ornamental purposes. However, it is clear that it is essential to develop adequate conservation and management strategies for numerous species of the genus, independent the use due to the study reveals three categories:

(1) Severely endangered: species facing the possibility of extinction and that should be subjected to an urgent rescue and recovery program, considering the possibility of stopping harvest from the wild altogether.
(2) High risk of extinction: species need a recovery program, and their harvest should be subjected to a strong conservation program.
(3) Threatened species: species management and recovery plants must be prepared and implemented for their harvest.

In vitro plant propagation, also known as micropropagation, replicates many fundamentals of conventional propagation carried out in nurseries or greenhouses, offering several advantages. This technique allows the development of methods to multiply plants free of viruses, which improves crop quality. A small explant, a small portion of plant tissue, is sufficient to generate numerous propagules, such as shoots, embryos, or organs. Moreover, since it is carried out in aseptic environments, the plants produced are free of microorganisms, which guarantees their health [6]. Micropropagation uses slight adjustments in environmental conditions, such as light and temperature, as well as in the formulation of culture media and the use of growth regulators. This facilitates continuous propagation during any season of the year. It also requires minimal attention between subculturing times, as there is no need for watering, pest control, or spraying. There are also techniques for storing plant material for long periods using low consumption media and cryopreservation. Another interesting aspect is habituation, where plants develop novel characteristics, such as numerous sprouting, overproduction of stolons or rhizomatous stems, generating potential temporary advantages in their cultivation [9].

Plant cryopreservation is a scientific discipline aimed to study and develop techniques and methods to store in vitro plant tissues at ultra-low temperatures, typically in liquid nitrogen tanks, to halt all biological activity and preserve the material indefinitely [10, 11].

To determine a cryopreservation technique for storing a particular plant should be based on several factors such as origin and plant material starting. Effective cryopreservation procedures have been identified around the world in seeds, pollen, spores, protoplasts, protocorms, callus, suspension cell cultures, zygotic and somatic embryos, embryonic axes dormant buds, roots, nodal segments, shoot tips, and any in vitro derived explants [10, 12–14]. Also, it depends on the crop, available equipment, technical skills, budget, and project goals [15].

Generally, species candidates for cryopreservation are selected based on lack of complete control over many factors that influence the survival of plant materials and its genetic make-up in the wild [16], risk level based on classification of the International Union for Conservation of Nature (IUCN) Red List, anthropogenic use, commercial demand or genotype elite, their biological importance (endemism, knowledge of pollinator, floral visitor, and/or reproduction) [8, 13].

Agave is considered as a priority genus for conservation [8, 13, 17]. The rapid development of proper approaches for the long-term preservation of genetic resources seems to be necessary.

Ex situ seed conservation seeks to support the survival of species or varieties to recover populations from the wild or to have a gene pool with commercial interests, so it is important to have quality seed collections with a broad genetic base [18]. To be considered a successful conservation, seed storage must be longer than the natural phenological process of each species, from seed germination to seed production of the next generation [19].

2 Micropropagation

Plant tissue culture (micropropagation) is the propagation of plants in vitro by means of cells, tissues, or organs in sterile media, under aseptic and controlled conditions, usually in a laboratory, for their development in a specific artificial environment.

Micropropagation remains a key tool for the mass production of genetically faithful plants free of pests and diseases. This method, widely used commercially, makes it possible to maintain and enhance the economic value of outstanding agricultural species. In addition, it solves phytosanitary problems that facilitate access and expansion of high-quality plants for growers in various regions of the world. Although costs are higher compared to traditional vegetative propagation methods, micropropagated plants offer significant advantages: they are disease-free, show better initial development, and guarantee high genetic and physiological quality [20]. Also, cell and tissue culture addresses problems such as recalcitrant seed germination, seeds with short viability, and the rescue of zygotic embryos, contributing to the advancement of agricultural biotechnology [21].

It is advisable to promote sustainability and conservation of the plant genetic resource, preserving species, varieties, and cultivars [3, 22, 23]. There are three main methods for micropropagation: (1) shoot proliferation from apical or axillary buds, (2) induction of adventitious buds for rooting, and (3) induction of somatic embryo formation and germination. The efficiency of each method depends on plant regeneration, culture medium, and management conditions.

Micropropagation systems are well documented for several *Agave* species, the most common of them in all available publications is shoot proliferation; however, shoot regeneration is also verified by organogenesis through the multicellular event when new meristems (de novo) are formed. This new groups of meristematic cells are tissues resembling meristems known as meristemoids, which can be generated directly from the explant tissues or indirectly from disorganized tissue cells (callus). In some papers, the organogenetic structures are misunderstood and are considered somatic embryos. This is very common when the unicellular event to regenerate new embryos (de novo or embryoids) is not well known, and then the confusion arises [24]. Embryoids must be anatomically identical to those obtained from seeds, since the same genetic expression governs the process; if the morphology is not the same, then the next option is to consider organogenetic shoots.

2.1 Axillary Shoot Proliferation

In plant propagation by means of axillary shoot proliferation, there are two main techniques: (a) shoot culture and (b) node culture, which can be single or multiple. In shoot culture, the aim is to initiate crops from axillary buds as explants, with the objective of multiplying shoots by stimulating lateral branching, generally using cytokinins as a growth regulator. The length of axillary buds produced varies considerably depending on the type of plant [9].

Successful meristem stimulation and axillary shoot growth depend on overcoming the dominance of the apical growing point and activating the meristem so that lateral branching occurs. Subsequently, these shoots are separated and individualized to bring them to rooting, thus completing the formation of a plant.

The culture is initiated from a lateral or apical bud, or a group of them, which are isolated and established in vitro after a disinfection process. Generally, shoot tips or nodal segments, which contain at least one meristem, are used. To ensure success, it is essential to eliminate or inhibit apex dominance, which favors shoot formation in significant numbers. The growth of lateral branching is enhanced by the addition of cytokinins to the culture medium (chemical pruning effect). The shoots produced are used as mini-shoots to root and develop new complete plants [25]. This method allows obtaining genetically stable plants, which ensures genetic fidelity between the mother tissue and the offspring.

The first report of micropropagation in *Agaves* is presented at the end of the 70s [26], after this work there are already dozens of protocol developments for different species and in the great majority alternatives are sought for the preservation of wild materials, using in the first instance a method that allows a high probability of producing offspring from genetic fidelity of the mother material to the multiplied material. For this purpose, the axillary shoot proliferation technique is suggested. Work has been done on micropropagation by means of axillary shoot proliferation in *Agave* used for the production of alcoholic beverages, where the aim is to reduce the impact on wild populations due to unplanned collections: such is the case of *A. marmorata* [27, 28]; in *A. angustifolia* the same happens in different regions of Mexico [29–31]; for the species *A. salmiana* from which the spirit drink "pulque" is obtained, irrational collecting is carried out [22, 23, 32, 33]; *A. potatorum*, *A. karwinskii*, *A. cupreata*, and *A. durangensis*, endemic species used in mezcal production protocols are implemented [34–38]. Work has also been carried out to conserve and multiply elite plant lines in significant quantities, especially in the species reported as the most cultivated: in *A. tequilana* [2, 39, 40]; in *A. fourcroydes* (henequen) [41], and *A. sisalana* [42], the latter two species also used in the production of fine cordage. And also by the same methodology of in vitro multiplication, protocols were generated in ornamental plants in which looting has affected wild populations: *A. arizonica* [43], *A. victoria-reginae* [35], and *A. parrasana* [44]. Figure 1a shows an example of the proliferation of axillary buds in *A. tequilana*, and Fig. 1b shows an example of the proliferation of axillary buds in *A. maximiliana*.

It is important to highlight the work of ex situ conservation through the proliferation of buds, in which the creation of a germplasm bank is seen as an effective alternative, since several species are stored in these banks. One of the most widely used strategies is the conservation of plant tissues in low-consumption culture media, which seek to reduce the growth rate of plant cells, tissues, or organs. For this purpose, osmotic potential regulating agents are used in the culture medium to reduce the availability of water for plants. In the case of *A. potatorum*, a protocol for its conservation has been generated [45], and the same has been done for seven other threatened and three wild *Agave* species [3]. In addition, cryopreservation of *A. peacockii*, micropropagated through shoot proliferation, stimulating buds, and using the apical meristems of plants developed in vitro, has already been reported. These meristems were flash-frozen in liquid nitrogen for cryopreservation [46].

2.2 Organogenesis

Organogenesis is the process by which plant cells and tissues undergo changes that lead to the production of a unipolar primordium whose vascular system is generally connected to the parental tissues that gave rise to it [47], the primordia can produce adventitious shoots (caulogenesis) or adventitious roots (rhizogenesis) so that at least

Fig. 1 Micropropagation of *Agave*. (**a**) Proliferation of axillary buds in *A. tequilana*. Bar = 1 cm; (**b**) Proliferation of axillary buds in *A. maximiliana*. Bar = 1 cm; (**c**) Indirect organogenesis in *A. tequilana*. Bar = 1 mm; (**d**) Direct organogenesis in *A. tequilana*. Bar = 3 mm; (**e**) Indirect somatic embryogenesis in *A. tequilana*. Bar = 0.5 mm; (**f**) Indirect somatic embryogenesis in *A. tequilana*. Bar = 1.5 mm

two different culture media are needed to generate a complete plant, one to induce shoot formation and another to produce adventitious roots in the shoots generated [9, 48]. Regarding the origin of the shoots, it has been reported that this can be unicellular [49] or multicellular [50]; however, it is clear that meristemoids are multicellular structures.

Micropropagation of plants of the genus *Agave* by organogenesis has been carried out in several species, such as *A. fourcroydes* [41], *A.*

victoriae-reginae [51], *A. tequilana* [52], *A. americana* [53, 54], *A. angustifolia* [55], *A. marmorata* [56], *A. potatorum* [57], among others. Figures 1c and d show examples of indirect organogenesis and direct organogenesis in *Agave tequilana*.

Micropropagation is a useful tool for the conservation of germplasm of *Agave* plants, especially those that do not produce offshoots, take many years to produce seeds, or produce very few viable seeds. Out of the different methods of micropropagation, organogenesis has the characteristic of producing somaclonal variation with which the new shoots may offer tolerance to pests and diseases; in addition, organogenesis produces large numbers of shoots quickly, being less complex than somatic embryogenesis and more efficient in terms of cost and time. For conservation purposes, organogenesis has been carried out in different species of *Agave*, highlighting the work done by Martínez-Palacios et al. [51] in *A. victoriae-reginae*, where some specimens produce seeds until the plant has reached 40 years of age.

2.3 Somatic Embryogenesis

The ability of the plant cell to acquire competence and reflect itself in a new form of development, influenced by the modulation of growth conditions, is fundamental to the establishment of a regeneration system in plants. Regeneration is a physiological process that refers to the ability of a specific cell or group of cells to regenerate a new tissue, organ, or even an entire individual. The ultimate expression of cell totipotency is somatic embryogenesis, in which a plant cell can give rise to a new individual from a single event [58]. Traditionally, the most commonly described embryogenesis is zygotic embryogenesis, which involves the fusion of female and male gametes. However, recent research has shown that somatic cells are also capable of forming embryos that can germinate and develop; these embryos are referred to as somatic embryos or embryoids. In vitro somatic embryogenesis is a useful technique in various applications within plant sciences [25]. Somatic embryogenesis can occur by two routes: directly from a cell in organized tissues cultured in vitro, and indirectly, by a cell from disorganized tissue (callus). Both pathways depend on induction through the management of specific conditions and the formulation of the culture medium, with the use of growth regulators that favor the expression of embryogenic competence [59].

Over time, the somatic embryogenesis technique has had several applications: (1) micropropagation of species; (2) production of synthetic seeds; (3) regeneration in the genetic transformation of species; (4) cryopreservation of plant germplasm; (5) elimination of viruses in plants; (6) production of somatic hybrids and polyploid plants [60]. Synthetic seeds are somatic embryos encapsulated in calcium alginates, which are kept hydrated. This technique has several advantages, such as the possibility of prolonged conservation, easy handling, storage in large quantities, and automated production [25].

The somatic embryogenesis technique has been used as a method of regeneration in *Agave*, applying both direct and indirect routes. In both, strategies are used to conserve the species in vitro conditions, being cryopreservation one of the most outstanding. The first somatic embryogenesis protocol in *Agave* species was presented directly in *A. victoriae-reginae*, a species listed as threatened in CITES Appendix II. This protocol sought an efficient alternative of in vitro multiplication to implement a reforestation program for the species [4]. Subsequently, in *A. fourcroydes*, thirteen clonal lines with different direct embryogenic capacities were identified, using explants from stems cut in thin layers of tissue, which allowed the conservation of germplasm [61]. Several cases have already been reported by the indirect route from disorganized tissues (callus): In *A. tequilana* cultivar "Azul" was obtained from leaf segments, and calli are feasible for cryopreservation [62]. In *A. angustifolia*, somatic embryogenesis occurs from zygotic embryos after the induction of embryogenic callus. Conservation can be achieved by reducing pressure on wild populations through micropropagation of seeds or selected plants and cryopreservation of both embryos and portions of embryogenic callus [63, 64]. For *A. sisalana*, a somatic embryogenesis protocol was implemented in which the calli were taken to bioreactors in the search to reduce costs in plant micropropagation; well-directed mass multiplication allows the long-term conservation of the species and outstanding individuals [5]. An indirect somatic embryogenesis protocol was developed in *A. tequilana* cultivar 'Chato', the cultivar was used in the production of distilled alcoholic beverages in the past, and its wild populations have decreased considerably at present. The cultivar represents an important genetic reservoir in the species for genetic improvement works. This first protocol has important contributions in the subject for *Agave* cryopreservation of somatic embryos [17]. Figures 1e and f show an example of indirect somatic embryogenesis in *A. tequilana*.

3 Cryopreservation

The development of simple and reliable methods for cryopreservation is crucial for advancing various scientific fields. These methods not only enhance the viability and integrity of preserved cells but also reduce costs and time associated with complex protocols.

Often, more than a single method and techniques are suitable for preservation of a selected species (Table 1).

3.1 Stages in Cryopreservation in Agave Studies

Depending on the chosen technique, cryopreservation consists of several stages, including preparation and excision of the explant, preculture, cryoprotection, vitrification or dehydration, rapid cooling in liquid nitrogen, rewarming, removal of cryoprotectants, survival or regrowth, and plant cultivation [10, 11, 16] (Fig. 2).

3.2 Droplet-Vitrification

This method involves the rapid cooling of plant tissues in small droplets of a vitrification solution, which prevents the formation of

Table 1
Recommendations to select the appropriate method and technique (*) for different initial plant materials of *Agave* species for cryopreservation

Method/Technique	Seeds	Callus and cell supension	Embryos zygotic/ somatic	Shoot tips	Pollen
Controlled-rate cooling		*			*
Vitrification/Droplet-vitrification			*	*	
Vitrification/V-cryoplate			*	*	
Dehydration/air-desiccated or silica gel	*				*
Dehydration/D-cryoplate			*		
Direct immersion into liquid nitrogen	*				*

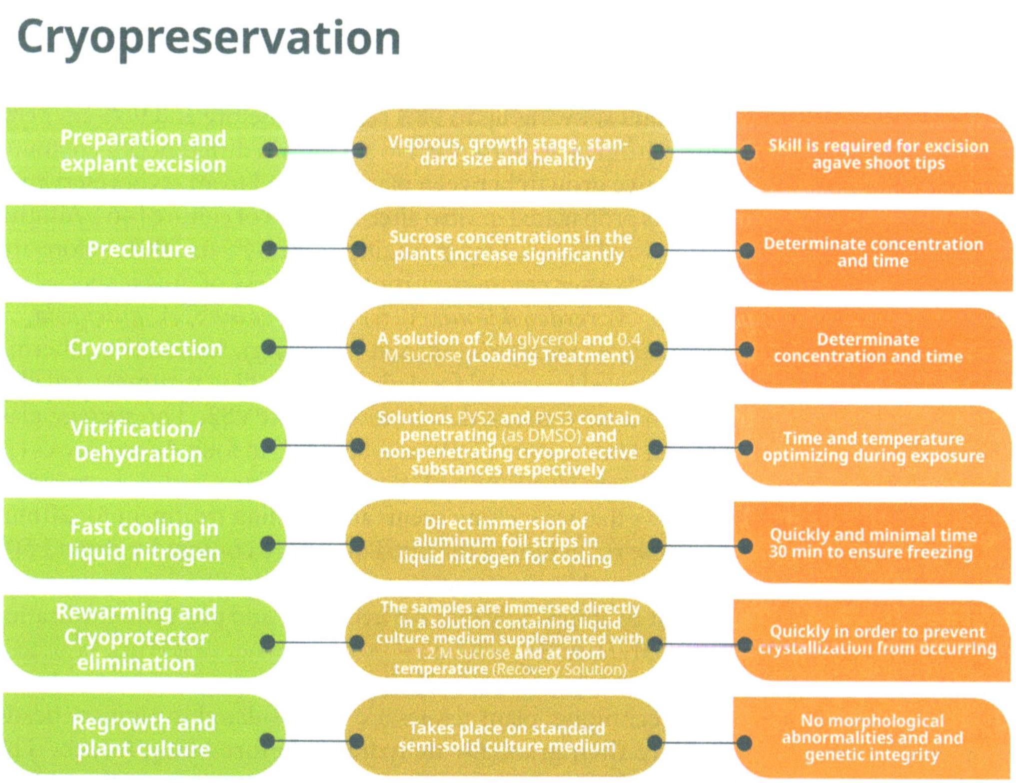

Fig. 2 Main stages in the process for cryopreservation of *Agave* plant species. Stages (green), specifications (yellow), recommendations (orange)

ice crystals that can damage cells. In monocots, meristem dissection is an acquired skill that requires time and dedication; therefore, it is advisable to practice the technique optimally and confirm survival before proceeding to cryopreservation [46, 65]. For *Agave*, shoot tips are commonly utilized as the preferred explant. Shoot tips are pre-cultured on a medium with 0.3 M sucrose, loaded in a solution with 0.4 M sucrose and 1.6 M glycerol, and then exposed to Plant Vitrification Solution 2 (PVS2) for 20 min before being immersed in liquid nitrogen. PVS 2 consists of 30% glycerol (w/v), 15% ethylene glycol (w/v), and 15% dimethylsulfoxide (DMSO; w/v) in basal culture medium MS [66] containing 0.4 M sucrose (pH 5.8) [67].

Advantages: versatility is highlighted by vitrification-based methods. This method has been applied to different tissues of over 800 plant species from temperate and tropical origins, and it is considered the most widely applied for plant cryo-biologists and applied routinely is increasing [68, 69]. As a process that does not require large quantities of cryoprotectants or expensive equipment, it can be more accessible to laboratories with limited budgets, contributing to the reduction of operating costs. Vitrification is a fast process, which minimizes the time exposure to extreme conditions and risk of toxicity and facilitates the recovery of cells or tissues after the freezing process.

Outcome: High survival and regrowth rates have been reported, with studies showing up to 98% regrowth before and 96% after cryopreservation in *A. peacockii*. No significant differences were found in vegetative growth between plants derived from cryopreservation and plants propagated in vitro after 120 days of culture [46]. Another study showed survival rates ranging from 84% to 100% before and 76% to 97% after cryopreservation regardless of *Agave* species (*A. peacockii*, *A. rzedowskiana*, *A. victoriae-reginae*, *A. guiengola*, *A. karwinskii*, *A. maximiliana*, and *A. lurida*). In addition, thermal analysis showed ice crystal formation after loading treatment and glass transition after osmotic dehydration with PVS2. The average glass transition temperature (Tg) was −55.44°C in four *Agave* species (*A. peacockii*, *A. rzedowskiana*, *A. victoriae-reginae*, and *A. karwinskii*). This shows that ice melting events after loading treatment for 20 min, despite the reduction of the initial moisture content by around 50% and suggests that the same exposure time to PVS2 was apparently sufficient to completely remove free water and promote vitrification events [70] (Fig. 3).

3.3 Encapsulation-Vitrification

In this technique, plant tissues are encapsulated in alginate beads, which are then gradually dehydrated before being frozen. This method helps protect the tissues from ice crystal formation during freezing. Cryopreservation was performed using V-cryoplate technique in somatic embryos of *A. tequilana* 'Chato' with vitrification solutions (PVS2 and PVS3) at different times. Plant vitrification solution 3 (PVS3) consists of 40% glycerol (w/v) and 40% sucrose

Fig. 3 Droplet-vitrification process. (**a**) Shoot tip dissection; (**b**) preculture; (**c**) loading solution exposed; (**d**) shoot tips in strip with plant vitrification solution 2; (**e**) liquid nitrogen immersion; (**f**) shoot tip recovery after 15 days

(w/v) in basal medium [71]. Warming was carried out at room temperature in basal medium with 1.2 M sucrose for 15 min, then calcium alginate gel from the explants was carefully removed before the recovery culture. Post-cryopreservation recovery involved transferring the embryos to a semisolid MS medium supplemented with 0.3 M sucrose for 1 week in darkness. The somatic embryos were recultured on semisolid basal MS medium with a photoperiod 16/8h light/dark at 25°C.

Advantages: This approach is cost-effective and handled during the procedure and high regrowth rates after cryopreservation, making it suitable for large-scale conservation efforts [72]. A common feature between V-cryoplate and D-cryoplate is that both methodologies benefit from the facilities provided by the handling of abundant material immobilized in calcium alginate, unlike the Droplet-Vitrification method, which involves direct handling of each tissue. Another benefit is that the duration of the entire cryogenic protocol is shortened compared to that required for encapsulation-dehydration [11, 73].

Outcome: The V-cryoplate method proved to be practical and efficient for the handling and manipulation of somatic embryos (SEs) in *A. tequilana* cultivar 'Chato'. Cryopreservation was achieved by testing two Plant Vitrification Solutions (PVS2 or PVS3). The highest post-cryopreservation regrowth (83%) and plant conversion rate (about 70%) were achieved with PVS2 at 0°C for 15 min. This work represents significant advances in somatic embryogenesis in *Agave*

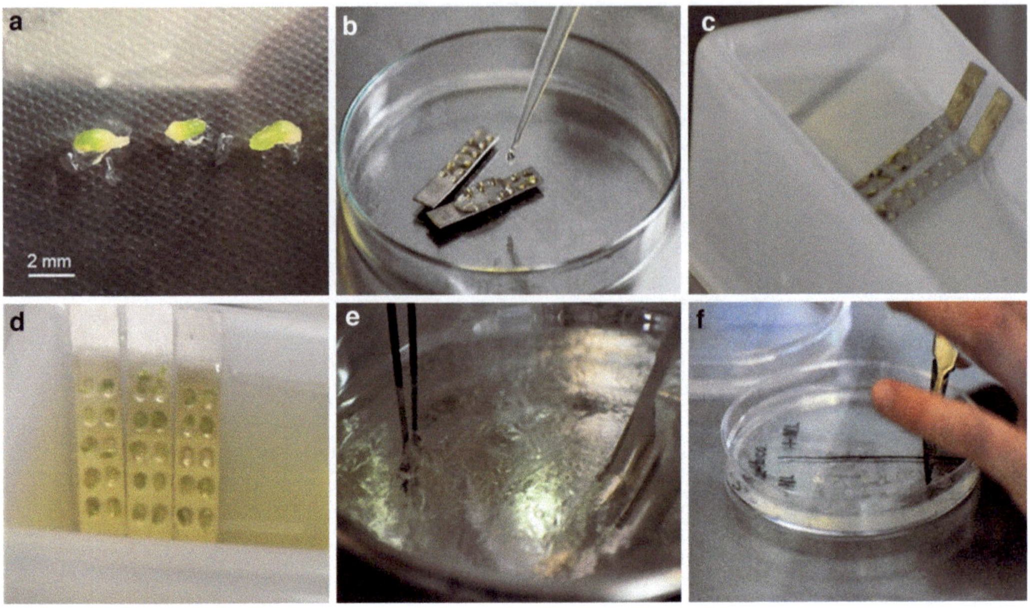

Fig. 4 Encapsulation-Vitrification in the cryoplate process. (**a**) Somatic embryos selection and preculture; (**b**) encapsulation with calcium alginate; (**c**) loading solution expose; (**d**) Somatic embryos in cryoplate with Plant Vitrification Solution 2; (**e**) Liquid nitrogen immersion; (**f**) Somatic embryo in semisolid medium for regrowth

and presents the first results on cryopreservation of somatic embryos of this species [17] (Fig. 4).

3.4 Direct Immersion

This involves a slow pre-freezing step to allow for controlled dehydration of the tissues, followed by rapid immersion in liquid nitrogen. This method is particularly useful for tissues that are sensitive to rapid cooling.

Advantages: simplified approach for long-term storage. Less handling of biological material. Direct immersion facilitates recovery much better than equilibrium hydration of partially dehydrated explants [74]. In some cases, direct immersion in liquid nitrogen promotes the seeds germination and seedling growth [75, 76].

Outcome: The initial application of cryopreservation for *Agave* seeds was reported in *A. victoriae-reginae* was successfully tested using liquid nitrogen immersion [77]. Subsequently, successful results were achieved with both nitrogen gas storage at −100°C and liquid nitrogen immersion at −196°C, utilizing seeds from *A. salmiana* and *A. lechuguilla* [78]. Recently, seeds have been evaluated through direct immersion in liquid nitrogen for the cryopreservation of species in this genus, such as *A. maximiliana* and *A. valenciana*, where a high germination rate has been observed (data not shown) (Fig. 4). These bioassays are promising for long-term conservation

Fig. 5 Direct immersion process. (**a**) Selection and drying of seeds; (**b**) seeds in cryotubes for direct immersion in liquid nitrogen; (**c**) seed germination after cryopreservation (15 days); (**d**) cryopreserved seedlings at 60 days

and represent a viable alternative for laboratories, producers from local communities, and germplasm banks, which serve as distribution bridges (Fig. 5).

4 *Agave* Seeds

In the world, flowering plants represent the dominant species, which are necessary for human beings with different approaches [79]. Agaves are monocarpic plants, they emit their flower axis at the center of the rosette, along this are the paniculate inflorescence (Fig. 6a) with its flowers grouped in umbels (Figs. 6a-c), the flowers are protandric (Figs. 6b and c) [80], the androecium consists of six stamens and a single stigma (Figs. 6b and c), *Agave* ovary is characterized by being tricarpellar and trilocular (Fig. 6d), each of them multiloculated (Figs. 6d and e), the fruit is a dry capsule with seeds in three lobules, when they are in development they are green in color (Fig. 6f), as they mature they change to a light to

Fig. 6 Reproductive phenology of *Agave*. (**a**) Panicle with flower buds; (**b**) Flower, I: Androecium, II: Gyneceum, III: Ovary; (**c**) Panicle with open flowers; (**d**) Cross section of ovary; (**a–d**) *A. guadalajarana*; (**e**) Longitudinal section of green fruit; (**f**) Green fruits; (**e and f**) *A. angustifolia*; (**g**) Dry fruit with its three openings for seed dispersal; (**h**) Empty seed; (**i**) Viable seed, arrow position of the embryo; (**j**) Embryo, under confocal microscopy; (**g and h**) *A. maximiliana*

dark brown color (Fig. 6g) [81]. Seed (Figs. 6h and i) has a tear-shaped and semi-flat morphology; sides have irregular surfaces [82]. The seed has a shiny black or lustrous black testa color, irregular, and rough (Fig. 6i). In seeds, the embryo (Fig. 6j) is located in an indentation in the center of the seed that extends in the direction of the micropyle (Fig. 6i), this is located at the widest part of the seed with marginal position [83]. Unviable seeds can be found in *Agave* fruits (Fig. 6h) and viable (Fig. 6i) percentages of each of them are influenced by different factors [84, 85]. The seed by nature is the strategy used by plants to reproduce and conserve germplasm, the seed remains dormant for a long time until the conditions are suitable for its development, it is also the way in which plant species disperse to colonize new areas [86].

For seed conservation, storage conditions should provide environmental conditions that minimize seed deterioration while maintaining the most homogeneous environmental conditions that will maximize the quality of seed available for future use [19]. Seed drying is recommended for seed conservation, which will extend the longevity of the seeds during storage. This will allow interaction with the environmental conditions recommended for seed conservation, which are 5–20°C and 10–25% relative humidity [87]. Among the few studies on the conservation of *Agave* seeds, Jimenez-Torres et al. [88] conducted a study with seeds stored for 10 years in hermetically sealed glass jars at room temperature, the average temperature during the storage period was 14°C. Under these conditions, germination percentages higher than 95% were observed. Hay and Probert [87] mentioned that the ideal temperature for storage ranges between 5 and 20°C, the temperature was found to be within the adequate parameters for conservation, considering the high percentage of germination at 10 years considering that *Agave* take approximately 10 years to reach maturity, this can be an alternative for conservation.

5 Conclusion

Habitat loss due to agricultural and urban expansion, as well as climate change, are some of the main concerns. In addition, overexploitation of *Agave* plants to meet the growing demand for derived products also poses a major challenge.

The use of micropropagation in *Agaves* is useful for mass propagation for commercial purposes and to obtain material for genetic improvement of individuals with outstanding characteristics. Also, micropropagation of plants is a biotechnological tool to produce large quantities of plants to generate in vitro germplasm banks, as well as to reforest areas that have had overexploitation of species with agro-industrial, ornamental, and other uses.

Recently, cryopreservation techniques have been performed as an essential technique for the long-term conservation of *Agave* species, especially those that are endangered or have significant economic and cultural value.

Cryopreservation offers a robust and reliable method for the conservation of *Agave* species. By utilizing techniques such as droplet-vitrification, encapsulation-dehydration, and direct immersion it is possible to preserve genetic diversity and ensure the survival of these important species for future generations.

The seed by nature is the strategy used by plants to reproduce and conserve germplasm; it is also the way in which plant species disperse to colonize new areas.

The successful results reported in this section prove to be a support for the choice of one or more techniques for the conservation of the *Agave* genus. In this sense, these conservation proposals are not infallible, but they are a quantitative alternative that can be decisive in protecting a species. Overall, these studies contribute to the knowledge and conservation of *Agave* species, which are crucial both for biodiversity and for their use in the production of various *Agave*-derived products.

References

1. García-Mendoza AJ, Franco IS, Sandoval D (2019) Four new species of *Agave* (Asparagaceae, Agavoideae) from southern Mexico. Acta Bot Mex 126:1–18. https://doi.org/10.21829/abm126.2019.1461
2. Barreto R, Nieto-Sotelo J, Cassab GI (2010) Influence of plant growth regulators and water stress on ramet induction, rosette engrossment, and fructan accumulation in *Agave tequilana* Weber var. Azul. Plant Cell Tiss Organ Cult 103:93–101. https://doi.org/10.1007/s11240-010-9758-9
3. Pérez-Molphe EB, Esparza-Araiza MJ, Pérez-Reyes ME (2012) *In vitro* conservation of *Agave* spp. germplasm under slow growth conditions. Rev Fitotec Mex 35:279–287. https://doi.org/10.35196/rfm.2012.4.279
4. Rodríguez-Garay B, Gutiérrez-Mora A, Acosta-Dueñas B (1996) Somatic embryogenesis of *Agave victoria-reginae* Moore. Plant Cell Tiss Org Cult 46:85–87. https://doi.org/10.1007/BF00039700
5. Nikam TD, Bansude GM, Kumar KCA (2003) Somatic embryogenesis in sisal (*Agave sisalana* Perr. ex. Engelm.). Plant Cell Rep 22: 188–194. https://doi.org/10.1007/s00299-003-0675-9
6. Bautista-Montes E, Hernández-Soriano L, Simpson J (2022) Advances in the micropropagation and genetic transformation of *Agave* species. Plants 11:1–12. https://doi.org/10.3390/plants11131757
7. Rodríguez-Garay B, Rodríguez-Domínguez JM (2018) Micropropagation of *Agave* species. In: Loyola-Vargas VM, Ochoa-Alejo N (Eds.) Plant cell cultures protocols. Methods in molecular biology, vol 1815. Springer Nature, USA, pp 151–160 https://doi.org/10.1007/978-1-4939-8594-4_8
8. Alducin-Martínez C, Ruiz-Mondragón KY, Jiménez-Barrón O, Aguirre-Planter E, Gasca-Pineda J, Eguiarte LE, Medellin RA (2023) Uses, knowledge and extinction risk faced by *Agave* species in Mexico. Plants 12:124. https://doi.org/10.3390/plants12010124
9. George EF, Hall MA, DeKlerk GJ (2008) Plant propagation by tissue culture. Vol. 1: The background. Springer, Netherlands, p 501
10. Reed BM (2008) Plant cryopreservation: a practical guide. Springer Science. p 513
11. González-Arnao M, Engelmann F (2013) Consideraciones teóricas y prácticas para la crioconservación de germoplasma vegetal 26-33. In: Crioconservación de plantas en América Latina y el Caribe San José, C.R. IICA, XII p 204
12. Chaudhury R, SK Malik (2017) Cryopreservation of plant species: practical approaches from handling to cryobanking. Indian Council of Agricultural Research (ICAR)-National Bureau of Plant Genetic Resources, Pusa Campus, New Delhi, p 52
13. Kaviani B, Kulus D (2022) Cryopreservation of endangered ornamental plants and fruit crops from tropical and subtropical regions. Biology (Basel) 11(6):847. https://doi.org/10.3390/biology11060847
14. Bettoni JC, Chen K, Volk GM (2024) Considerations when implementing shoot tip cryopreservation. In: Training in plant genetic resources: cryopreservation of clonal propagules. Colorado State University Plant and Food Research, Fort Collins, Colorado, https://colostate.pressbooks.pub/clonalcryopreservation/frontmatter/introduction/
15. Bettoni JC, Bonnart R, Volk GM (2021) Challenges in implementing plant shoot tip cryopreservation technologies. Plant Cell Tiss Org Cult 144:21–34. https://doi.org/10.1007/s11240-020-01846-x
16. Acheampong-Amankwaah V, Naa-Ashiokai Prempeh R, Dorcas-Quain M (2022) Plant cryopreservation importance, approaches and

future trends. IntechOpen, p 12. https://doi.org/10.5772/intechopen.108806

17. Delgado-Aceves L, González-Arnao MT, Santacruz-Ruvalcaba F, Folgado R, Portillo L (2021) Indirect somatic embryogenesis and cryopreservation of *Agave* tequilana Weber cultivar 'Chato'. Plants 10:249. https://doi.org/10.3390/plants10020249
18. Cochrane JA, Crawford AD, Monks LT (2007) The significance of *ex-situ* seed conservation to reintroduction of threatened plants. Aust J Bot 55:356–361
19. Bonner FT (1990) Storage of seeds: potential and limitations for germplasm conservation. For Ecol Manag 35:35–43. https://doi.org/10.1016/0378-1127(90)90230-9
20. Cardoso JC, Sheng-Gerald LT, Texeira da Silva JA (2018) Micropropagation in the twenty-first century. In: Loyola-Vargas VM, Ochoa-Alejo N (Eds) Plant cell cultures protocols. Methods in molecular biology, vol 1815. Springer Nature, USA, pp 3–16 https://doi.org/10.1007/978-1-4939-8594-4_2
21. Jain MS, Ochatt SJ (2010) Protocols for in vitro propagation of ornamental plants, In: Methods in molecular biology. Springer Nature, USA, p. 400
22. Silos-Espino G, González-Córtes N, Carrillo-López A, Guevaralara F, Valverde-González ME, Paredes-López O (2007) Chemical composition and *in vitro* propagation of *Agave salmiana* Gentry. J Hortic Sci Biotech 82:355–359. https://doi.org/10.1080/14620316.2007.11512242
23. Puente-Garza CA, Gutiérrez-Mora A, García-Lara S (2015) Micropropagation of *Agave salmiana*: means to production of antioxidant and bioactive principles. Front Plant Sci 6:1026. https://doi.org/10.3389/fpls.2015.01026
24. Delgado-Aceves L, Palacios H, Romo-Paz F, Portillo L (2019) Micropropagation systems in *Agave* spp: common errors. In: Gutiérrez-Mora A, Rodríguez-Garay B, Estarrón-Espinosa M, Gschaedler-Mathis AG, Kirchmayr MR, Moreno-Terrazas RL, Camacho-Ruiz P, Ortiz-Basurto RM, Aguilar-Uscanga RI, Aguilar Juárez MG, Gallardo-Valdez O, Hernández-López JJ, Sánchez-Osorio E (eds) Integral and sustainable use of Agave. CONACYT/CIATEJ/Agared, Jalisco, Mexico, pp 51–54. ISBN: 978-607-8734-03-0
25. Bhojwani SS, Dantu PK (2013) Plant tissue culture: an introductory text. Springer Science, New Delhi, India, p 309
26. Groenewald EG, Wessels DCJ, Koeleman A (1977) Callus formation and subsequent plant regeneration from seed tissue of an *Agave* species (Agavaceae). Z Planzenphysiol 81:369–373. https://doi.org/10.1016/S0044-328X(77)80105-0
27. Aguilar-Jiménez D, Rodríguez de la O JL (2018) Micropropagación y aclimatización de Maguey Pitzometl (*Agave marmorata* Roezl) en la Mixteca Poblana. Rev Colomb Biotecnol 20:124–131. https://doi.org/10.15446/rev.colomb.biote.v20n2.77084
28. Martínez-Rodríguez A, Beltrán-García C, Valdés-Salas B, Santacruz-Ruvalcaba F, Di Mascio P, Beltrán-García MJ (2022) Micropropagation of seed-derived clonal lines of the endangered *Agave marmorata* Roezl and their compatibility with endophytes. Biology 11:1423. https://doi.org/10.3390/biology11101423
29. Monja-Mio KM, Barredo Pool F, Herrera-Herrera G, Esqueda-Valle M, Robert ML (2015) Development of the stomatal complex and leaf surface of *Agave angustifolia* Haw. "Bacanora" plantlets during the *in vitro* to *ex vitro* transition process. Sci Hortic 189:32–40. https://doi.org/10.1016/j.scienta.2015.03.032
30. Ríos-Ramírez SC, Enríquez-Del Valle JR, Rodríguez-Ortiz G, Ruiz-Luna J (2017) Benzylaminopurine and indol-3-acetic acid concentrations in *in vitro* proliferation of *Agave angustifolia* adventitious shoots. Cien Inv Agr 44:285–294. https://www.scielo.cl/pdf/ciagr/v44n3/0304-5609-ciagr-44-03-0285.pdf
31. Enríquez JR, Chávez-Cruz IL, Rodríguez-Ortiz G, Campos-Angeles GV (2023) Vitroplants of *Agave angustifolia* Haw. obtained in contrasting incubation environments, acclimated on different substrates. Rev Fitotec Mex 46:291–298. https://doi.org/10.35196/rfm.2023.3.298
32. Puente-Garza CA, García-Lara S, Gutiérrez-Uribe JA (2017) Enhancement of saponins and flavonols by micropropagation of *Agave salmiana*. Ind Crops Prod 105:225–230. https://doi.org/10.1016/j.indcrop.2017.05.014
33. Puente-Garza CA, Meza-Miranda C, Ochoa-Martínez D, García-Lara S (2017) Effect of *in vitro* drought stress on phenolic acids, flavonols, saponins, and antioxidant activity in *Agave salmiana*. Plant Physiol Bioch 115:400–407. https://doi.org/10.1016/j.plaphy.2017.04.012
34. Domínguez-Rosales MS, Alpuche-Solís AG, Vasco Méndez NL, Pérez Molphe Balch E (2008) Effect of cytokinins on the *in vitro* propagation of Mexican *Agaves*. . Rev Fitotec Mex 31:317–322. https://doi.org/10.35196/rfm.2008.4.317
35. Ramírez-Malagón R, Borodanenko A, Pérez-Moreno L, Salas-Araiza MD, Nuñez-Palenius HG, Ochoa-Alejo N (2008) *In vitro* propagation of three *Agave* species used for liquor distillation and three for landscape. Plant Cell Tiss Org Cult 94:201–207. https://doi.org/10.1007/s11240-008-9405-x

36. Luna-Luna S, Enríquez-del Valle JR, Rodríguez-Ortiz G, Carrillo-Rodríguez JC, Velasco-Velasco VA (2017) Anatomy and morphology of micropropagated-acclimatized plants of *Agave* potatorum Zucc. fertigated in nursery. Rev Fitotec Mex 40:491–494. https://www.revistafitotecniamexicana.org/documentos/40-4/13a.pdf
37. Dévora-Rodríguez DG, Pulido-Díaz C, Chávez-Simental JA, Ortiz-Sánchez IA, Loera-Gallegos HM, Prieto-Ruiz JA (2021) Growth regulators in the vegetative development of *Agave durangensis* Gentry vitroplants. Ecosis Recur Agropec 8:e2720. https://doi.org/10.19136/era.a8n1.2720
38. Téllez-Torres AG, Jiménez-Rodríguez JA, González-Caballero O, Juárez-Pérez WR, Martínez-Martínez S, Chávez-Avila VM (2023) *In vitro* culture of *Agave potatorum* a threatened species, endemic to Mexico. Bot Sci 101:883–894. https://doi.org/10.17129/botsci.3279
39. Santamaría JM, Herrera JL, Robert ML (1995) Stomatal physiology of a micropropagated CAM plant; *Agave tequilana* (Weber). Plant Growth Regul 16:211–214. https://doi.org/10.1007/BF00024776
40. Rescalvo-Morales A, Monja-Mio KM, Robert ML, Sánchez-Teyer LF (2019) Telomere length in *Agave tequilana* Weber plants during the *in vitro* to *ex vitro* transition. Plant Cell Tiss Org Cult 136:133–140. https://doi.org/10.1007/s11240-018-1499-1
41. Robert ML, Herrera JL, Contreras F, Scorer KN (1987) *In vitro* propagation of *Agave fourcroydes* Lem. (Henequen). Plant Cell Tiss Org Cult 8:37–48. https://doi.org/10.1007/BF00040731
42. Nikam TD, Mulye KV, Chambhare MR, Nikule HA, Ahire ML (2019) Reduction in hyperhydricity and improvement in *in vitro* propagation of commercial hard fibre and medicinal glycoside yielding *Agave sisalana* Perr. ex Engelm by NaCl and polyethylene glicol. Plant Cell Tiss Org Cult 138:67–78. https://doi.org/10.1007/s11240-019-01603-9
43. Powers DE, Backhaus RA (1989) *In vitro* propagation of *Agave arizonica* Gentry & Weber. Plant Cell Tiss Org Cult 16:57–60. https://doi.org/10.1007/BF00044072
44. Santacruz-Ruvalcaba F, Gutiérrez-Pulido H, Rodríguez-Garay B (1999) Efficient *in vitro* propagation of *Agave parrasana* Berger. Plant Cell Tiss Org Cult 56:163–167. https://doi.org/10.1023/A:1006232911778
45. Spinoso-Castillo JL, Pérez-Sato JA, Schettino-Salomón SS, Bello-Bello JJ (2022) An alternative method for medium-term *in vitro* conservation of different plant species trough gibberellin inhibitors. In Vitro Cell Dev Biol Plant 58:606–614. https://doi.org/10.1007/s11627-022-10263-y
46. Delgado-Aceves L, Portillo L, Folgado R, Romo-Paz FJ, González-Arnao MT (2022) New approaches for micropropagation and cryopreservation of *Agave peacockii*, an endagered species. Plant Cell Tiss Org Cult 150:85–95. https://doi.org/10.1007/s11240-022-02246-z
47. Thorpe TA (1994) Morphogenesis and regeneration. In: Vasil IK, Thorpe TA (eds) Plant cell and tissue culture. Kluwer Academic Publishers, Dordrecht, The Netherlands, pp 17–36
48. Phillips GC (2004) *In vitro* morphogenesis in plants-recent advances. In Vitro Cell Dev Biol Plant 40:342–345. https://doi.org/10.1079/IVP2004555
49. Broertjes C, Van Harten AM (1985) Single cell origin of adventitious buds. Euphytica 34:93–95. https://doi.org/10.1007/BF00022867
50. Norris R, Smith RH, Vaughn KC (1983) Plant chimeras used to establish *de novo* origin of shoots. Science 220:75–76. https://www.jstor.org/stable/1690556
51. Martínez-Palacios A, Ortega-Larrocea MP, Chávez VM, Bye R (2003) Somatic embryogenesis and organogenesis of *Agave victoriae-reginae*: considerations for its conservation. Plant Cell Tiss Organ Cult 74:135–142. https://doi.org/10.1023/A:1023933123131
52. Valenzuela-Sánchez KK, Juárez-Hernández RE, Cruz-Hernández A, Olalde-Portugal V, Valverde ME, Paredes-Lopez O (2006) Plant regeneration of *Agave tequilana* by indirect organogenesis. In Vitro Cell Dev Biol Plant 42:336–340. https://doi.org/10.1079/IVP2006788
53. Reyes-Zambrano SJ, Lecona-Guzmán CA, Barredo-Pool FA, Calderón JDA, Abud-Archila M, Rincón-Rosales R, Ruiz-Valdiviezo VM, Gutiérrez-Miceli FA (2016) Plant growth regulators optimization for maximize shoots number in *Agave americana* L. by indirect organogenesis. Gayana Bot 73:124–131. https://gayanabotanica.cl/index.php/gb/article/view/336
54. Lecona-Guzmán CA, Reyes-Zambrano S, Barredo-Pool FA, Abud-Archila M, Montes-Molina JA, Rincón-Rosales R, Gutierrez-Miceli FA (2017) *In vitro* propagation of *Agave americana* by indirect organogenesis. HortScience 52:996–999. https://doi.org/10.21273/HORTSCI10498-16
55. Ríos-Ramírez SC, Valle JRE, Rodríguez-Ortiz G, Ruíz-Luna J, Velasco-Velasco VA (2018) *In vitro* formation of adventitious shoots on caulinary tissue of physiologically contrasting *Agave angustifolia* plants. Emir J Food Agric

30:49–56. https://pdfs.semanticscholar.org/43f9/d8d16c0fee1acb611a8ea3d42e-a3229c7c2a.pdf
56. Martinez-Martinez SY, Arzate-Fernandez AM, Álvarez-Aragon C, Martínez-Velasco I, Norman-Mondragón TH (2021) Regeneration of *Agave marmorata* Roelz plants in temporary immersion systems, via organogenesis and somatic embryogenesis. Trop Subtrop Agroecosyt 24:84. https://doi.org/10.56369/tsaes.3472
57. Ramírez-Mosqueda MA, Cárcamo-Corona RG, Aguilar-Jiménez D, Bello-Bello JJ (2022) Micropropagation of *Agave* (*Agave potatorum* Zucc.) through direct organogenesis. Agrociencia 56:1–10. https://doi.org/10.47163/agrociencia.v56i6.2823
58. Rocha DI, Melo-Vieira L, Días-Koehler A, Campos-Otoní W (2018) Cellular and morpho-histological foundations of *in vitro* plant regeneration. In: Loyola-Vargas VM, Ochoa-Alejo N (Eds.) Plant cell cultures protocols. Methods in molecular biology, vol 1815. Springer Nature, USA, pp 47–68 https://doi.org/10.1007/978-1-4939-8594-4_3
59. Krikorian AD (2000) Historical insights into some contemporary problems in somatic embryogenesis. In: Jain SM, Gupta PK, Newton RJ (eds) Somatic embryogenesis in woody plants, vol 6. Kluwer Academic Publishers, Dordretch, Netherlands, pp 17–49
60. Lambardi M, Ozudogru EA, Benelli C (2008) Cryopreservation of embryogenic cultures. In: Reed BM (Ed.) Plant cryopreservation, a practical guide. Springer Science New York, USA, pp 177–210
61. Monja-Mio KM, Robert ML (2013) Direct somatic embryogenesis of *Agave fourcroydes* Lem. through thin cell layer culture. In Vitro Cell Dev Biol Plant 49:541–549. https://doi.org/10.1007/s11627-013-9535-7
62. Portillo L, Santacruz-Ruvalcaba F, Gutiérrez-Mora A, Rodríguez-Garay B (2007) Somatic embryogenesis in *Agave tequilana* Weber cultivar azul. In Vitro Cell Dev Biol-Plant 43:569–575. https://doi.org/10.1007/s11627-007-9046-5
63. Arzate-Fernández AM, Mejía-Franco R (2011) Embryogenic capacity of induced calli on zygotic axis of *Agave angustifolia* Haw. Rev Fitotec Mex 34:101–106
64. Reyes-Díaz JI, Arzate-Fernández AM, Piña-Escutia JL, Vázquez-García LM (2017) Media culture factors affecting somatic embryogenesis in *Agave angustifolia* Haw. Ind Crops and Prod 108:81–85. https://doi.org/10.1016/j.indcrop.2017.06.021
65. Panis B (2008) Cryopreservation of monocots. In: Reed BM (eds) Plant cryopreservation: a practical guide. Springer, New York, NY, pp 241–278
66. Murashige T, Skoog F (1962) A revised medium for rapid growth and bioassays with tobacco tissue cultures. Physiol Plant 15(3):473–497. https://doi.org/10.1111/j.1399-3054.1962.tb08052.x
67. Sakai A, Kobayashi S, Oiyama I (1990) Cryopreservation of nucellar cells of navel orange (*Citrus sinensis* Osb. var. *brasiliensis* Tanaka) by vitrification. Plant Cell Rep 9(1):30–33. https://doi.org/10.1007/BF00232130
68. Sakai A, Engelmann F (2007) Vitrification, encapsulation-vitrification and droplet-vitrification: a review. CryoLetters 28(3):151–172
69. Zamecnik J, Faltus M, Bilavcik A (2021) Vitrification solutions for plant cryopreservation: modification and properties. Plants 10:2623. https://doi.org/10.3390/plants10122623
70. Delgado-Aceves L, Corona S, Marin-Castro UR, Rascón-Díaz MP, Portillo L, Gutiérrez-Mora A, González-Arnao MT (2024) Comparative studies for cryopreservation of *Agave* shoot tips by droplet-vitrification. Plants 13:2609. https://doi.org/10.3390/plants13182609
71. Nishizawa S, Sakai A, Amano Y, Matsuzawa T (1993) Cryopreservation of asparagus (*Asparagus officinalis* L.) embryogenic suspension cells and subsequent plant regeneration by vitrification. Plant Sci 91:67–73. https://doi.org/10.1016/0168-9452(93)90189-7
72. Matsumoto T (2017) Cryopreservation of plant genetic resources: conventional and new methods. RAS 5:13–20. https://doi.org/10.7831/ras.5.13
73. Yamamoto S, Rafique T, Fukui K, Sekizawa K, Niino T (2012) V-Cryo-plate procedure as an effective protocol for cryobanks: case study of mint cryopreservation. CryoLetters 33:12–23
74. Perán R, Pammenter NW, Naicker J, Berjak P (2004) The influence of rehydration technique on the response of recalcitrant seed embryos to desiccation. Seed Sci Res 14:179–184. https://doi.org/10.1079/SSR2004167
75. Kaya E, Souza F, Gökdoğan EY, Ceylan M, Jenderek M (2017) Cryopreservation of citrus seed via dehydration followed by immersion in liquid nitrogen. Turk J Biol 41:1–25. https://doi.org/10.3906/biy-1603-92
76. Acosta Fernández Y, Fontes-Marrero D, Martínez-Montero M (2021) Liquid Nitrogen as promotor of seeds germination and seedling growth in tropical legumes. INGE CUC 17:1–10. https://hdl.handle.net/11323/10177
77. Bejarano, V, 2011. Preservación de semillas de Agave victoriae-reginae (AGAVACEAE) y Glandulicactus uncinatus (CACTACEAE), especies endémicas y en peligro de extinción.

Universidad Michoacana de San Nicolás de Hidalgo, Maestría en Ciencias Biológicas, pp 45–58
78. Cruz-Cárdenas C, Contreras-Corona M, Sandoval-Cancino G, González-Arnao M (2019) Effect of drying and cryogenic temperatures on the physiological response of stored seeds of *Agave salmiana* and *Agave lechuguilla*. Cryobiology 91:194
79. Taylor EL, Taylor TN, Krings M (2009) Flowering plants. In: Taylor EL, Taylor TN, Krings M (eds) Paleobotany, 2nd ed. Academic Press, Cambridge. https://doi.org/10.1016/B978-0-12-373972-8.00022-X
80. García-Mendoza AJ (2010) Revisión taxonómica del complejo *Agave potatorum* zucc. (Agavaceae): nuevos taxa y neotipificación. Act Bot Mex 91:71–93. https://doi.org/10.21829/abm91.2010.292
81. García-Mendoza AJ (2007) Los *Agaves* de México. Ciencias 87:14–23
82. Vázquez DE, García NJR, Peña VCB, Ramírez THM, Morales RV (2011) Tamaño de la semilla, emergencia y desarrollo de la plántula de maguey (*Agave salmiana* Otto ex Salm-Dyck). Rev Fitotec Mex 34(3):167–173. https://doi.org/10.35196/rfm.2011.3.167
83. Sánchez-Salas J, Flores J, Jurado E, Sáenz-Mata J, Orozco-Figueroa P, Muro G (2017) Hidrocoria en semillas de *Agave victoriae-reginae* T. Moore, especie en peligro de extinción: Morfología y anatomía como facilitadores de la hidro-dispersión y germinación. Gayana Bot 74(2):251–261. https://doi.org/10.4067/S0717-66432017000200251
84. Cuéllar-Martínez M, Galindo-González J, Andrade-Torres A (2024) Germination of *Agave obscura* seeds: effects of storage time and crossing systems. Plant Ecol 225:189–199. https://doi.org/10.1007/s11258-023-01389-z
85. Trejo L, Soriano D, Romano-Grande E, Sánchez-Carmona B, Dávila-Navarro DE (2024) Diversity of reproductive characters, seed set, and viability of *Agave* seeds used for pulque production and their wild relatives in Tlaxcala, Mexico. Genet Resour Crop Evol 71:2877–2903. https://doi.org/10.1007/s10722-023-01803-5
86. Nautiyal PC, Sivasubramaniam K, Dadlani M (2023) Seed dormancy and regulation of germination. In: Dadlani M, Yadava DK (Eds.) Seed science and technology. Springer, Singapore, https://doi.org/10.1007/978-981-19-5888-5_3
87. Hay FR, Probert RJ (2013) Advances in seed conservation of wild plant species: a review of recent research. Conserv Physiol 1(1):cot030. https://doi.org/10.1093/conphys/cot030
88. Jimenez-Torres JA, Monroy-Gonzalez Z, Juarez-Muñoz J (2022) Evaluation of seed morphology, seedling genetic variation, and components for seed storage of *Agave* landraces of commercial interest. Exp Res 3(e25):1–10. https://doi.org/10.1017/exp.2022.16

Chapter 30

Advances in Cryogenic Techniques for the In Vitro Preservation of Somatic Embryos and Shoot Tips of Plants

Lourdes Delgado-Aceves, Fabiola Hernández-Ramírez, Carlos Alberto Cruz-Cruz, Marco A. Ramírez-Mosqueda and María Teresa González-Arnao

Abstract

This chapter describes advances in cryopreservation techniques for the safe long-term storage of somatic embryos and shoot apical meristems of different plant species. Several examples are presented that illustrate the progress of vitrification-based procedures, analyzing, from a practical perspective, the modifications incorporated to better adapt cryogenic protocols to these organized tissues.

Cryopreservation of somatic embryos ensures the preservation of genetically stable and totipotent embryogenic cultures, essential for large-scale propagation, breeding programs, and the rescue of elite genotypes. Likewise, cryopreservation of shoot apical meristems plays a fundamental role in maintaining pathogen-free plants, protecting rare and endangered species, and preserving plants with high commercial and agricultural value. Therefore, cryopreservation of somatic embryos and shoot apical meristems constitutes an important biotechnological strategy.

Keywords Tissue culture, Cryopreservation, Dehydration, Vitrification, Encapsulation

1 Importance of Cryopreserving Shoot Tips and Somatic Embryos

Cryopreservation is the storage of any biological material at ultra-low temperatures, usually in liquid nitrogen (−196°C), where it can be preserved for long periods of time due to the total halt of cell division and metabolic processes [1–3]. The use of cryopreservation techniques facilitates the storage of plant germplasm in a pathogen-free aseptic environment and optimization of physical space for conservation [4].

Cryobiotechnology is a discipline that integrates plant biotechnology and cryogenics for the long-term preservation of biological material [5]. In vitro culture techniques are commonly used for shoot multiplication (proliferation), organogenesis (formation of plant organs), and somatic embryogenesis (generation of embryos), adapting to the specific needs of each species. These practices allow to produce plant

Marco A. Ramírez-Mosqueda et al. (eds.), *Conservation of Plant Genetic Resources*, Methods in Molecular Biology, vol. 3011, https://doi.org/10.1007/978-1-0716-5130-8_30,

material that can subsequently be cryopreserved, thus ensuring its storage without altering its biological and functional characteristics [6].

Existing knowledge of related genera within the same family or phylogenetic group facilitates the standardization of any technique and favors progress in the conservation of new species and genera. Several studies [3, 7–9] have identified key principles for the cryopreservation of various plant biological materials, which are based on the following aspects:

- Origin and taxonomic classification according to the specific characteristics of the species.
- Optimization of in vitro culture systems, ensuring that the formulation of culture media and growth conditions are standardized.
- Viability and quality of donor plants.
- Biophysical aspects of natural freezing tolerance (cold sensitivity).
- Maximal removal of intracellular freezable water content.
- Tissue tolerance to cryoprotective treatments.
- Explant size in relation to the selected technique.
- Exceptional species, including recalcitrant seeds, species that produce few or no seeds, or endangered species.

These general principles guide the selection of one or more cryopreservation techniques, which must be specifically adapted to the type of explant chosen to ensure their subsequent successful application.

This chapter describes and analyzes the technical advances achieved in the cryopreservation of somatic embryos and apical meristems of plants as a valuable biotechnological tool for the safe long-term conservation of plant germplasm in support of restoration, preservation of genetic biodiversity, breeding programs and the subsequent development of scientific research. Somatic embryo cryopreservation ensures the preservation of genetically stable and totipotent embryogenic cultures, which are essential for large-scale propagation and rescue of elite genotypes. Similarly, meristem conservation plays a critical role in keeping plant material free of pathogens, protecting rare and endangered species, and enabling rapid regeneration of plants with high commercial and agricultural value. The topics presented here demonstrate that cryopreservation provides a robust strategy for ensuring the long-term preservation of plant biodiversity and guaranteeing the availability of valuable genetic resources for current and future uses.

2 Biological and Physiological Characteristics of Somatic Embryos and Shoot Tips

Organized growth is essential for the formation and maintenance of well-defined plant structures. This process occurs when plant organs, such as apical meristems, leaf initials, roots, or embryos, are transferred to culture and continue to grow while preserving their original structure. Additionally, organized growth can also arise when organs are induced in vitro, either from a tissue explant or during the culture of previously unorganized tissues. The development of new organs through this process is referred to as morphogenesis, meaning "the development of form" [10].

In plant propagation, meristem and embryo cultures are key examples of organized growth:

- Meristem cultures involve the development of very small, excised shoot apices, typically consisting of the apical meristematic dome, with or without leaf primordia, to form a single shoot.
- Shoot tip cultures, derived from excised shoot tips or buds, often result in the production of multiple shoots.
- Embryo cultures involve the dissection of fertilized or unfertilized zygotic embryos from seeds or the isolation of somatic embryos, which are then grown in vitro until they develop into seedlings.

Both meristem and embryo cultures are essential for maintaining organized growth and supporting plant propagation [11].

According to the principles, an important starting point for understanding the intrinsic characteristics of embryos and meristems is to consider the botanical classification of plants. This classification into two large groups, monocotyledons and dicotyledons, is a relevant factor for cryopreservation.

Although botanical classification does not directly affect species survival during the cryogenic process, it plays a crucial role in the handling and manipulation of the initial explant. This distinction, involving ontogenetic and structural aspects of tissues, influences the selection of appropriate strategies and techniques for each plant type, as well as the expected developmental response of the explant after cryopreservation [12, 13].

2.1 Dicotyledonous Plant Species

Dicots are characterized by having two cotyledons, a vascular system arranged in a ring, a prominent primary root, and a reduced endosperm compared to monocotyledons [14].

Embryogenesis in dicots begins with the fertilization of the ovule, forming the zygote, which undergoes an asymmetric division, followed by several mitotic divisions. These divisions lead to the formation of a suspensor, which facilitates the transport of nutrients from the endosperm to the embryo. The suspensor also undergoes programmed cell death as the embryo matures, while the apical cell of the suspensor gives rise to the root meristem. During development, the embryo progresses through the following stages: (1) Globular Stage, the embryo is spherical, and internal organization begins. (2) Heart Stage, the cotyledons are established, which will become the first photosynthetic organs of the seedling. It is during this stage that the number and arrangement of the cotyledons are determined. (3) Torpedo Stage, the embryo begins to elongate, and the root structure starts to form. (4) Cotyledonary Stage, the development of the cotyledons, the hypocotyl–radicle transition zone, and the radicle (the root primordium) is completed [14, 15]. At this stage, the seed undergoes desiccation, and the embryo reaches maturity. Cotyledons, which are key organs, may store nutritional reserves such as starch and proteins that will be used during seedling development once germination occurs. A comparative overview of the indirect somatic embryogenesis pathways in *Carica papaya* and *Agave tequilana*, emphasizing the developmental differences between dicotyledonous and monocotyledonous species. The figure also indicates the position of the apical meristem in the mature plant (Fig. 1).

The shoot apical meristem is a crucial structure for plant growth, as it generates aerial organs. In dicotyledons, the shoot apical meristem is composed of three histogenic layers [16]:

- Layer L1: This layer gives rise to the epidermis of the plant and generally divides anticlinally (perpendicular to the growth axis).
- Layer L2: This forms the mesophyll of the leaves and mainly divides anticlinally, except in areas where leaf initiation is occurring, where periclinal division takes place.
- Layer L3 (corpus): This generates the inner parts of the plant, such as parenchyma and vascular bundles, as well as the primordia of the root and shoot.

The division capabilities of these layers in various directions are crucial for the formation of the stem and leaves. In this way, the apical meristem regulates the continuous growth of the plant by producing new cells that form vegetative organs. Dicots present exposed shoot buds, which makes their isolation easier and allows for a more straightforward collection of explants for cryopreservation [13].

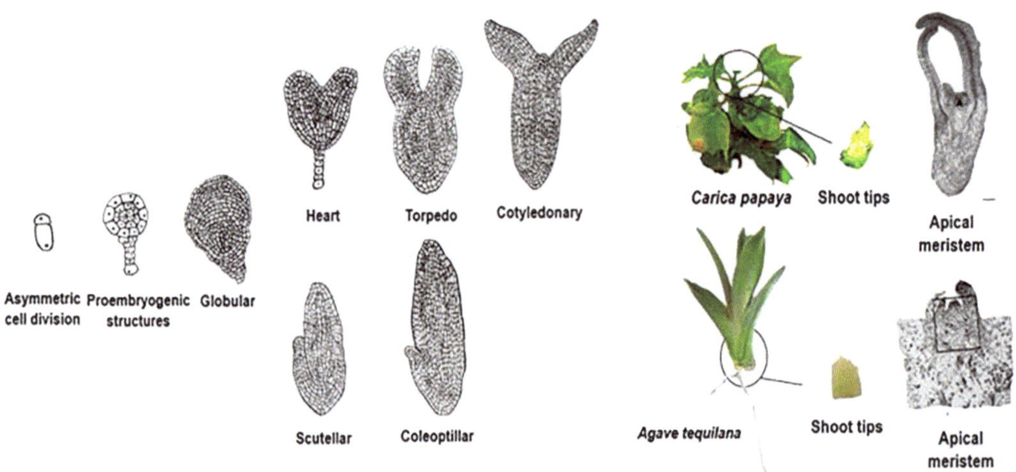

Fig. 1 Indirect somatic embryogenesis pathways in *Carica papaya* and *Agave tequilana*. The ideograms are based on the models proposed by Vale et al. (2018) for *C. papaya* and Delgado-Aceves et al. (2021) for *A. tequilana*. Drawing by Romo-Paz, 2025

2.2 Monocotyledonous Plant Species

Monocot plants are characterized by having one cotyledon, a dispersed vascular system, a greater amount of endosperm, and an adventitious root system [14, 17].

Embryogenesis in monocots begins with the fertilization of the ovules by the sperm cell, forming a zygote. This zygote undergoes an asymmetric division, giving rise to a basal cell that does not divide, and a terminal cell that forms the suspensor. Subsequently, the axial cell leads to the first cellular changes, creating the embryo and the suspensor.

The embryo initially has a globular shape and, with further cell divisions, elongates into a club shape with radial symmetry. It then develops bilateral symmetry when structures such as the scutellum and coleoptile differentiate, covering the first leaf of the seedling. The scutellum is a modified cotyledon that aids in nutrient absorption from the endosperm.

As the embryo matures, the coleoptile stage establishes the embryonic axis and the suspensor. Later, during the leaf stage, the shoot (aerial part) and root meristems are defined, and leaf primordia differentiate [14, 18]. Cell development halts during embryo maturation, which desiccates and enters a dormant state. During this drying process, the embryo becomes quiescent, and in orthodox species, the seed can survive under low moisture conditions. However, in some recalcitrant species, seeds do not survive in a dry state [19].

Unlike dicots, the embryo in monocots is generally smaller and more compact, with a single, elongated cotyledon that is not as voluminous. The seedling does not rely on the cotyledons for storing nutrients; instead, it depends on the scutellum to absorb nutrients from the endosperm.

The shoot apical meristem in monocots is essential for continued plant growth. The shoot apical meristem may consist of two or three cell layers [20] and the orientation of the cell divisions in these layers ensures that the cells do not mix during morphogenesis, allowing proper tissue differentiation.

In monocots, shoot apical meristems are usually located in the inner basal part of the shoot or bulb, which makes them less visible and complicates their dissection compared to dicots, where the meristematic structures are more superficial [13].

Within the shoot apical meristem, the meristematic dome serves as an active region composed of totipotent cells with "embryogenic" characteristics. These cells are distinguished by their isodiametric shape, small size, thin cell walls, frequent mitotic divisions, and a high nucleus-to-cytoplasm ratio, lacking large central vacuoles [16, 21–23]. The population of meristematic cells typically ranges from a few hundred to over a thousand, depending on the species [16]. Beyond cell number, meristem plasticity is a key factor in determining its structure and function, as well as its ability to adapt to heterogeneous environmental conditions. The tunica-corpus organization is a fundamental structural pattern in the apical meristem (Fig. 2). The corpus consists of a group of cells where divisions occur in various planes, giving rise

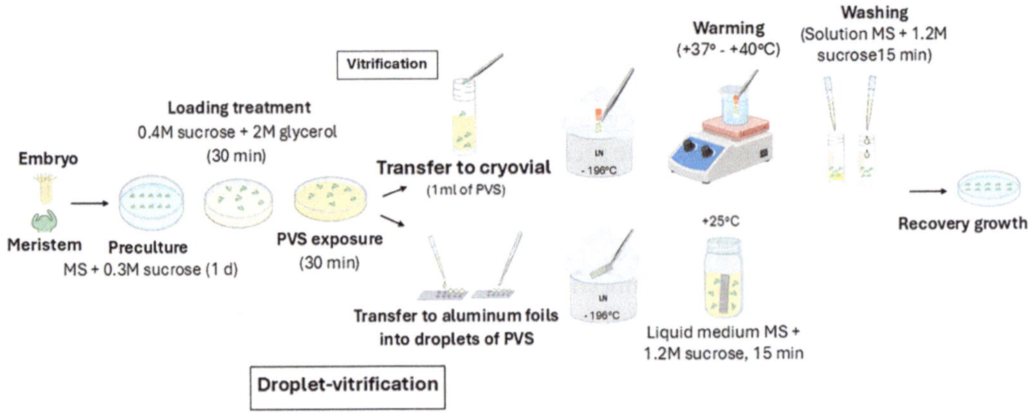

Fig. 2 Schematic diagram of the cytohistological zonation of the shoot apical meristem (SAM). The diagram illustrates the spatial organization of different tissue regions within the SAM. It highlights the corpus, which consists of the initial apical tunic and corpus cells, along with the transition zone (yellow), peripheral zone (green), and rib meristem (gray). The tunica layers are depicted surrounding the meristematic dome, while the leaf primordium is shown protecting the meristematic dome. Illustration courtesy of Morales-Juárez, 2025

to multiple tunica layers. In dicotyledonous plants, the tunica typically comprises three layers (L1–L3), whereas in monocotyledons, only two layers are present. The outermost layers primarily undergo anticlinal or periclinal divisions, contributing to surface expansion and the maintenance of meristematic structure [20]. Several cytohistological zones can be distinguished within the apical meristem. The central zone is composed of a cluster of relatively large cells located at the core of the meristem. Surrounding this, the peripheral zone contains smaller, actively dividing cells that proliferate at a higher rate than those in the central zone. These cells differentiate to form lateral organs and fundamental tissues, playing a key role in organogenesis [23]. Variations in mitotic activity rates, division planes, cell size, and cytoplasmic density across different regions of the meristem collectively define the structural and functional dynamics of meristems, influencing its capacity for growth and regeneration [16].

3 Technological Aspects for the Development of a Plant Cryopreservation Protocol

The first requirement for establishing a cryopreservation protocol is to define the in vitro culture conditions that guarantee the regeneration, multiplication, and recovery of biological material, both before and after cryopreservation [24]. Furthermore, in vitro culture is the only possible strategy for generating new biotechnological products, including somatic embryos. Therefore, tissue culture techniques contribute to diversifying and increasing the availability of materials that can potentially be cryopreserved and to ensuring the long-term storage of new sources of plant genetic information [25].

Several aspects must be considered to optimize the use of in vitro culture techniques and improve the biological response to any cryopreservation process. The nutritional composition of the culture medium is a key factor that can vary depending on the plant species, the type of culture, and the stage in which it is used (initial or final) of the cryogenic protocol [26, 27]. The most common practice is to supplement the basal medium formulation with growth regulators such as auxins and cytokinins. However, other useful alternatives have also been reported to improve the regeneration of cryopreserved tissues [28]. These include: the total elimination or reduction of the concentration of ammonium salts in the in vitro regrowth medium [29]; the gradual decrease in sucrose concentration to avoid osmotic shock during the first days of culture; the addition of exogenous antioxidants; and the use of nanoparticles in the encapsulation medium [28, 30]. Table 1 presents some examples of culture media, which can alternatively be used for the regrowth of somatic embryos and shoot tips after cryopreservation.

Modification of physical and environmental conditions is also an important factor for the in vitro recovery of cryopreserved explants. After cryopreservation, samples are usually incubated in the dark for at least the first 7 days to avoid photooxidation stress [25, 41].

Table 1

Examples of alternative culture media for the regrowth of somatic embryos (SE) and shoot tips after cryopreservation

Plant species	Explant		Reference
Allium spp.	Shoot-tips	PGR: IAA 0.3 + 2-iP 2.0 mg L^{-1} (66%)	[31]
Castanea sativa	SE	PGR: NAA 0.1 + BA 0.1 mg L^{-1} (68%)	[32]
Citrus limon L. Burm. F	Shoot-tips	Decrease to 1/4 NH_4NO_3 (70%)	[33]
Coriandrum sativum	SE	PGR: 2,4-D 1.0 mg L^{-1} (98%)	[34]
Lamprocapnos spectabilis L. "Gold Heart"	Shoot-tips	5 ppm AgNP enc (60%), 15 ppm ZnONP enc (71%), added to the alginate pearl matrix	[35]
Lamprocapnos spectabilis L. "Valentine"	Shoot-tips	10 ppm AuNP (70%), added to the alginate pearl matrix	[36]
Manihot esculenta	Shoot-tips	Regrow in medium with 0.3 M sucrose for 2 days, followed by 0.058 M for 1 month (70%–75%)	[37]
	SE	One-day exposure to 0.3 M sucrose (75%)	[38]
Nicotiana tabacum and *Hypericum perforatum*	Shoot-tips	Antioxidant action 0.1–0.5 μM Melatonin: Improved regrowth by 30%–40%	[39]
Vitis spp.	Shoot-tips	Regrow in medium with 0.6 M sucrose, followed by a medium with standard (3%) sucrose concentration (43%)	[40]

As mentioned above, plant germplasm cryopreservation relies on the use of ultra-low storage temperatures and, consequently, the success of any cryogenic process depends on avoiding the ice crystal formation in the intracellular medium [2]. Vitrification is the only mechanism by which the freezing of water can be inhibited at low temperatures and is defined as the phase transition from a viscous liquid to an amorphous solid at the glass transition temperature (Tg) [42]. Water removal is essential for inducing the glass transition phase, preventing freezing damage, and maintaining viability after cryopreservation. Following classical protocols, intracellular water content reduction is achieved through a freeze-dehydration process using a slow freezing regime [4]. However, these procedures were not sufficiently effective for the homogeneous dehydration of organized biological systems such as somatic embryos and shoot tips. Nevertheless, they have been useful for the cryopreservation of apical meristems of some plant species from temperate and subtropical climates [43].

A classical cryopreservation protocol involves the use of a slow cooling rate to an intermediate temperature (in most cases −40°C) before rapid immersion of the samples in liquid nitrogen. The use of a slow rate is strategically necessary to promote freeze-dehydration of cells in a small volume (approximately 1 ml) of a cryoprotective solution during the gradual decrease in temperature. Thawing must be rapidly performed in a water bath at +37°C–+40°C for 1–2 min and the cryoprotective solution is removed using a washing solution composed of liquid culture medium [44].

Since 1990, with the development of new technological approaches to dehydration, cryogenic procedures have evolved and diversified, giving rise to so-called vitrification-based techniques. The key modification of these methods was to replace the freeze–dehydration process characteristic of classical protocols with the application of drastic dehydration treatments prior to cooling. For this purpose, highly concentrated solutions known as plant vitrification solutions (PVS) or air drying have been successfully used. The implementation of dehydration strategies at non-freezing temperatures allowed the removal of a greater fraction of freezable water from cells, thus improving the tolerance of organized structures to cryopreservation [44]. Among the first vitrification-based procedures are the vitrification, encapsulation–dehydration, and encapsulation–vitrification methods [45].

The cryogenic procedure called vitrification (Vitr) was first applied to nucellar cells of navel orange [46]. This methodology, later extended to meristems and somatic embryos, includes a first loading treatment generally performed at room temperature for 20 min in a solution containing sucrose and glycerol. After loading, the samples are osmotically dehydrated using a highly concentrated plant vitrification solution (PVS) and then rapidly transferred to the LN contained in cryovials with approximately 1 mL of fresh PVS solution. Rewarming is quickly performed in a water bath at +37°C to +40°C for 1–2 min, and the vitrification solution is removed using a washing solution composed of liquid culture medium supplemented with 1.2 M sucrose. This basic procedure led to the subsequent definition of new cryoprotective approaches that resulted in the development of other cryogenic techniques such as encapsulation–vitrification, droplet–vitrification and cryoplates V and D [45].

Encapsulation/dehydration (E/D) is a technique in which shoot meristems and/or somatic embryos are encapsulated in spherical calcium alginate beads before being submitted to a cryogenic process. This approach was first reported by Fabre and Dereuddre [47] using potato meristems. The main advantage of this method is that encapsulation allows it to withstand drastic desiccation processes that would otherwise be lethal for non-encapsulated samples [48].

The basic protocol of E/D consists of encapsulating the biological material in calcium alginate, pre-culturing the coated explants in liquid medium supplemented with a high concentration of sucrose, drying the microspheres exposed to air in a laminar flow hood or using silica gel and then, after placing them in cryovials, the samples are rapidly immersed in liquid nitrogen. Rewarming is usually performed at room temperature, and it is not always necessary to remove the alginate coating to achieve explant recovery during reculturing [49, 50].

The encapsulation–vitrification (E/Vitr) technique combines the same process of encapsulating explants in calcium alginate beads, with the use of a PVS solution to dehydrate and cool the samples placed in cryovials before rapidly immersing them in liquid nitrogen. Rewarming is performed rapidly in a water bath at +37°C to +40°C for 1–2 min. The main advantage of this method is that encapsulation in the calcium alginate matrix mitigates the effect of osmotic dehydration caused by exposure to PVS, when this treatment proved to be very drastic and lethal for non-encapsulated explants [45].

Years later, additional modifications to cryopreservation protocols were implemented to increase cooling and warming rates from rapid to ultrarapid. This was achieved by replacing the use of cryovials with aluminum foil strips or aluminum cryoplates to immerse the samples directly in liquid nitrogen. Furthermore, the volume of PVS and the encapsulating gel layer used to cryopreserve the explants were significantly reduced. Cryogenic procedures that incorporated these modifications represented a new generation of vitrification-based techniques and are known as droplet–vitrification [51, 52] and the V and D cryoplate methods [53].

Droplet–vitrification (D–V) technique involves the same cryoprotective treatments used in the vitrification procedure, i.e, the biological material is first treated with the loading and vitrification solutions as described above, but then, the samples are placed in small drops of fresh PVS on strips of aluminum foil that are quickly immersed in liquid nitrogen [4, 45]. Rewarming is performed at room temperature for about 15 min, rapidly transferring the cryoplates, the Cryo-plate that contained the samples, to an unloading solution composed of a liquid medium supplemented with 1.2 M sucrose. This procedure was described by Panis et al. [52] using banana meristems.

The V-cryoplate (V-Cp) and D-cryoplate (D-Cp) methods are the most recent cryogenic procedures applied to shoot apical meristems and somatic embryos of various plant species. In fact, they innovatively summarize the strategic advances that have proven effective for cryoprotection. The V-Cp technique [53] involves the use of PVS for explant dehydration, while the D-Cp technique [54] is based on air dehydration. However, both methods implement the encapsulation of explants in small wells of a cryoplate using calcium alginate. After encapsulation, the samples adhered to the cryoplates undergo loading treatment and are subsequently dehydrated according to the specific methodology of each technique. Rewarming is performed rapidly at room temperature for about 15 min, transferring the cryoplates to an unloading solution composed of a liquid medium supplemented with 1 M sucrose.

The common feature of the protocols derived from the D-V, V-Cp, and D-Cp techniques is their high cooling and warming rates [25]. Furthermore, the V-Cp and D-Cp methods, which involve encapsulation of biological material directly in cryoplates, offer additional advantages such as ease of simultaneous manipulation of numerous tissues, lower risk of material loss in later stages of the procedure and, most importantly, improved recovery of larger tissues after cryopreservation [54].

Thus, over the years, vitrification-based protocols have significantly contributed to the establishment of successful methods for the cryopreservation of complex organs, especially those derived from tropical plant species [4]. The characteristic steps of the most widely applied vitrification-based methodologies are graphically illustrated in Figs. 3–5.

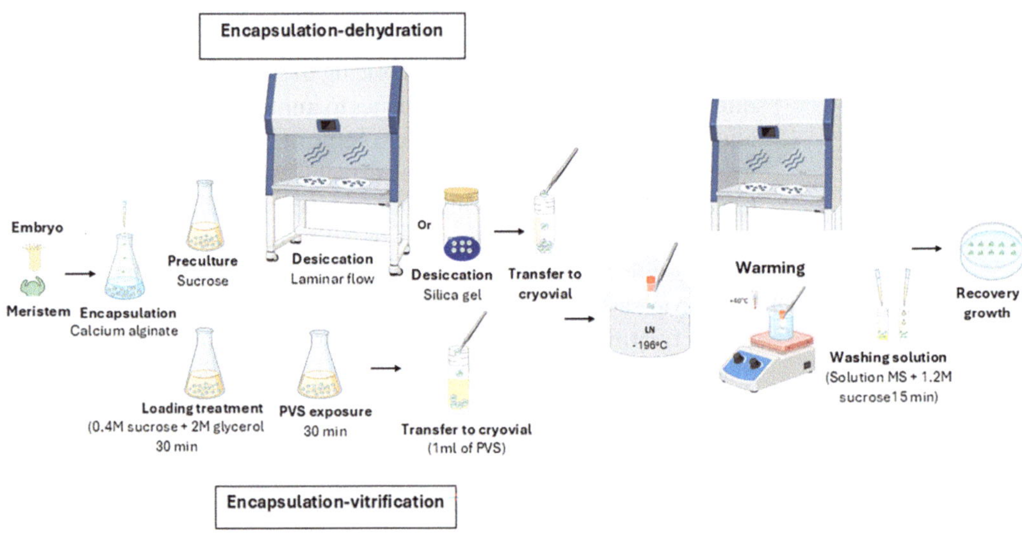

Fig. 3 Graphical representation of the successive steps of vitrification (above) and droplet–vitrification (down) procedures

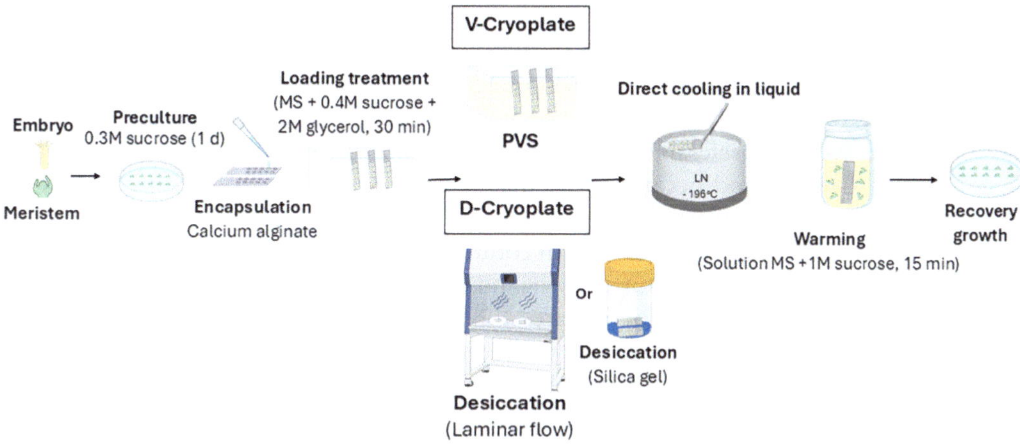

Fig. 4 Graphical representation of the successive steps of encapsulation–dehydration (above) and encapsulation–vitrification (down) procedures

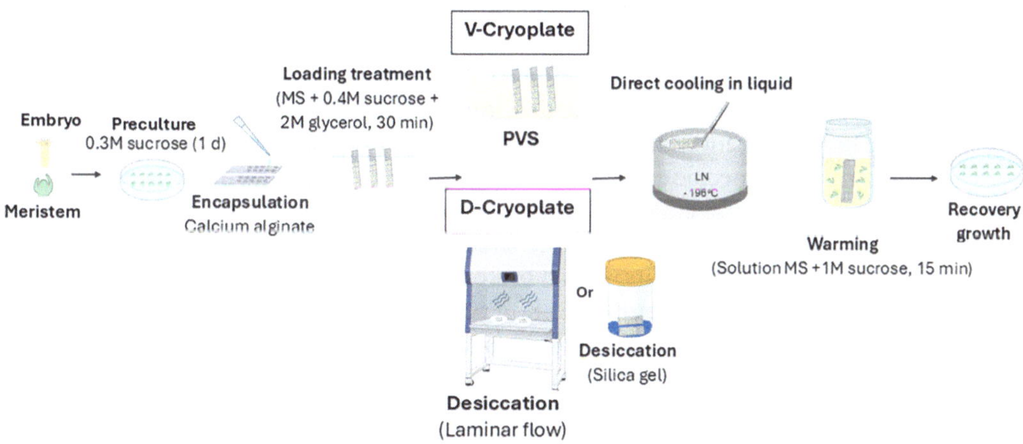

Fig. 5 Graphical representation of the successive steps of V-cryoplate (above) and D-cryoplate (down) procedures

4 Cryopreservation of Somatic Embryos and Apical Meristems

Based on the practical approaches reviewed previously, the first important step for the cryopreservation of somatic embryos and meristems is the establishment of a reproducible and efficient tissue culture protocol that ensures the availability of abundant regenerable material, both before and after cryopreservation. Under strict selection criteria, the material must be as homogeneous as possible in both size and phenological state, as these factors decisively influence survival.

Regarding the size of the explant, the meristems usually range between 0.5 and 2 mm [13]. However, this size correlates with the growth stage and positional origin within the plant (apical or axillary). Recent studies in potato suggested that axillary meristems are preferable to apical meristems due to higher post-thaw regeneration, and because more material can be excised from a single plant (six axillary meristems from a 6-week-old plant versus only one apical meristem) [55]. In contrast, in cassava, significantly higher viability rates, shoot growth, and plant regeneration were observed in apical meristems compared to lateral meristems [56].

On the other hand, somatic embryos tend to be larger, with sizes ranging between 1 and 5 mm. In this context, the embryonic stage is crucial for determining the most suitable cryopreservation technique. Larger explants contain greater amounts of potentially extractable water, requiring more drastic dehydration treatments and faster cooling and rewarming rates are required to achieve and stabilize a vitrified intracellular state and prevent the formation of lethal ice crystals during a cryopreservation process. Therefore, explant size is a key factor, and its optimal range varies by plant species [44, 57].

The quantity and quality of regenerated somatic embryos often exhibit asynchronous development, meaning that somatic embryos exist in different developmental stages simultaneously. Therefore, standardizing a cryogenic protocol may require the selection of a specific embryogenic stage to homogenize the material used in terms of size, water content, and histological composition [25, 58].

Early embryogenic stages are significantly more resistant to cryopreservation than more differentiated stages, such as cotyledonary somatic embryos; however, they are more difficult to manipulate. The greater sensitivity of cotyledonary embryos is likely due to their higher levels of vacuolation and cellular differentiation, while the active cell division of the upper layers of the proembryogenic masses or globular embryos, being less structurally developed, are more tolerant to cryopreservation [28].

According to Ballesteros et al [59], the main challenges in somatic embryo cryopreservation include:

- Developing protocols for embryos at more advanced developmental stages, such as cotyledonary embryos.
- Expanding current procedures to conserve a larger number of embryogenic lines.
- Addressing species-specific responses to cryopreservation techniques, as different species exhibit varying tolerances to cryogenic conditions.

The cryopreservation of the shoot apical meristem has been successfully applied across a wide range of plant taxonomic groups, including species from both tropical and temperate ecosystems. However, the success of cryopreservation is not solely determined by post-thaw

Table 2
Cryopreservation of somatic embryos of different plant species using vitrification-based techniques

Plant species	Plant type/uses	Technique	Survival (%)	Reference
Aesculus hippocastanum L.	Medicinal species	Vitrification	94	[62]
Ammocharis coranica	Used as an adhesive	Vitrification	30	[63]
Anadenanthera colubrina (Vell.) Brenan	Medicinal plant/ ornamental species	Vitrification	65	[64]
Agave peacockii	Food and beverage production	Droplet–vitrification	96	[65]
Agave spp.	Food and beverage production/ ornamental	Droplet–vitrification	76-97	[66]
Agave tequilana Weber Cultivar "Chato"	Food and beverage production	Vitrification	83	[18]
Araucaria angustifolia (Bertol.) Kuntze	Dioecious perennial conifer native	Vitrification	100	[67]
Atropa belladonna	Medicinal plant	Vitrification	75	[68]
Brunsvigia gregaria	Ornamental species	Vitrification	60	[68]
Brunsvigia orientalis	Ornamental species	Vitrification	55	[68]
Castanea sativa	Timber tree/edible seeds	Vitrification	60	[69]
Castanea sativa	Timber tree/edible seeds	Encapsulation–vitrification	70	[70]
Citrus indica	Medicinal fruit	Encapsulation–dehydration	85	[71]
Citrus indica	Medicinal fruit	Vitrification	80	[71]
Citrus sinensis (L.) Osbeck	Edible fruits	V-cryoplate	75	[72]
Citrus spp.	Edible fruits	Encapsulation–dehydration	76-100	[44]
Cupressus sempervirens L.	Ornamental species/timber tree	Vitrification	75	[57]
Elaeis guineensis Jacq.	Edible seeds/oil production	Droplet vitrification	68	[73]
Haemanthus humulus humulus	Ornamental species	Vitrification	70	[68]
Larix spp.	Timber tree	Vitrification	80	[74]

Table 2
(continued)

Plant species	Plant type/uses	Technique	Survival (%)	Reference
Malus spp.	Edible fruits	Encapsulation–vitrification	76	[75]
Musa acuminata		Droplet–vitrification	77	[76]
Olea europaea L.		Vitrification	38	[77]
		Droplet–vitrification	60	[78]
		Droplet–vitrification	90	[79]
Persea americana Mill.		Droplet–vitrification	77-100	[80]
Persea americana Mill. "A10"		Droplet–vitrification	100	[81]
		Vitrification	91	
Persea americana Mill. "Duke-7"		Vitrification	80	
Persea americana Mill. "Reed"		Droplet–vitrification	85	
		Vitrification	73	
Persea americana Mill. "Velvick"		Droplet–vitrification	93	
		Vitrification	86	
Phoenix dactylifera L.	Ornamental species/edible fruits	D-cryoplate	92-96	[82]
Phoenix dactylifera L. cv. Sagai		Vitrification	87	[83]
Picea abies L. Karst	Ornamental species/timber tree	Vitrification	87-100	[84]
Picea glauca × *engelmannii*	Timber tree	Vitrification	97	[85]
Pinus pinaster	Medicinal and timber tree	Vitrification	100	[86]
Podophyllum hexandrum	Medicinal plant	V-cryoplate	90	[87]
Quercus ilex L. genotype Q8	Cattle feed	Vitrification	80	[88]
Quercus suber L.	Cork production	Encapsulation–dehydration	90	[89]
Robinia pseudoacacia	Ornamental and timber tree/ Acacia honey production	Vitrification	78	[90]
Strumaria discifera	Ornamental species	Vitrification	65	[68]
Syzygium maire	Ornamental tree/ edible fruits	Encapsulation–dehydration	37	[91]

Table 2
(continued)

Plant species	Plant type/uses	Technique	Survival (%)	Reference
Theobroma cacao L.	Edible fruits/ chocolate production	Vitrification	75	[92]
Theobroma cacao L. genotype AMAZ12		Encapsulation–dehydration	72	[93]
Theobroma cacao L. genotype EET272		Encapsulation–dehydration	25	
Theobroma cacao L. genotype IMC14		Encapsulation–dehydration	40	
Thuja koraiensis	Ornamental tree	Vitrification	28	[94]
Tsuga ulleungensis	Endemic conifer species	Vitrification	33	[95]
Vitis "Albariño" and "Tempranillo"	Edible fruits/wine production	Encapsulation–dehydration	100	[96]

viability; it also depends on the physiological and functional integrity of meristematic cells. These cells must regain their capacity for division to enable plant regeneration. During the cooling and warming processes, meristematic tissues may experience various types of damage, including alterations in cell membranes, loss of intracellular compartmentalization, oxidative stress, and even apoptosis. Identifying the morphological and physiological changes that meristems undergo after storage in liquid nitrogen is therefore essential for optimizing protocols and improving whole-plant recovery rates [60].

Plant recovery after cryopreservation of apical meristems depends on the survival of a viable number of meristematic cells. The apical meristems contain both stem cells and initial cells, which drive plant tissue growth and differentiation. Successful regeneration does not require the survival of all meristematic cells; however, the preservation of a functional core of viable and totipotent cells is essential. The specific characteristics of meristematic cells, such as their small size, reduced vacuole volume, and prominent nucleus, help minimize the formation of ice crystals within them during freezing. The surviving meristematic cells retain their capacity for self-renewal, division, differentiation, and regeneration, enabling them to repopulate damaged tissues [22]. This principle is the basis of cryotherapy, a useful biotechnological approach for obtaining pathogen-free plants through cryopreservation. Cells located farther from the apical dome are more likely to be infected by viruses or phytoplasmas and are also more sensitive to the effects of any cryogenic procedure. Therefore, cryopreservation of shoot apical meristems offers additional benefits to germplasm conservation, as it also contributes to pathogen

Table 3
Cryopreservation of shoot apical meristems of different plant species using vitrification-based techniques

Plant species	Plants type/uses	Technique	Survival (%)	Reference
Acampe rigida	Epiphytic orchid/ornamental	D-cryoplate	74	[97]
Actinidia spp.	Native trees of temperate eastern Asia/edible fruits	Encapsulation–dehydration	22-56	[98]
Allium chinense G. Don, "Shima-rakkyo"	Type of bulbous plant/use as food	V-cryoplate	100	[99]
Allium wakegi		Vitrification	80	[100]
Amelanchier alnifolia (Nutt.) M. Roem.	Native shrub of North America/edible fruits	V-cryoplate	50	[101]
		D-cryoplate	20	
Castilleja levisecta	Rare species of flowering plant	Droplet–vitrification	66	[102]
Centaurea ultreiae	Perennial plant with underground stolons in danger of extinction	Vitrification	96	[103]
Ceratopetalum gummiferum	Ornamental shrub from New South Wales	Encapsulation–dehydration	61	[104]
Chrysanthemum × *grandiflorum*	Perennial suffruticose plant/medicinal use	Droplet–vitrification	80	[105]
		Droplet–vitrification	85	[106]
		Encapsulation–dehydration	85	[107]
Clinopodium odorum	Plant with branchy, aromatic subshrubs/medicinal use	V-cryoplate	71	[1]
		D-cryoplate	29	
Cosmos atrosanguineus	Perennial herbaceous plant native to Mexico/ornamental use	Encapsulation–dehydration	100	[108]

Table 3
(continued)

Plant species	Plants type/uses	Technique	Survival (%)	Reference
Dendranthema grandiflorum Kitam	Ornamental plant	Vitrification	74	[109]
		Encapsulation–dehydration	60	
Dioscorea alata	Species of true yam/edible use	Encapsulation–dehydration	19	[110]
Dioscorea bulbifera		Encapsulation–dehydration	60	
Dioscorea deltoidea	Nepalese Yam/edible use	Vitrification	30-54	[111]
		Droplet–vitrification	21-51	
Dioscorea Prazeri	Species of yam native to the central Himalayas to the Malaysian peninsula/edible use	Vitrification	92	[112]
		Encapsulation–dehydration	75	
		Encapsulation–vitrification	38	
Dioscorea spp.	Species of true yam/edible use	Encapsulation–dehydration	100	[113]

Table 3
(continued)

Plant species	Plants type/uses	Technique	Survival (%)	Reference
Fragaria × *ananassa*	Stoloniferous creeping plants/edible fruits	V-cryoplate	70-97	[52]
Fragaria × ananassa "Cley"		V-cryoplate	63	[101]
		D-cryoplate	83	
Fragaria × *ananassa* Duch.		Encapsulation–vitrification	90	[114]
Fragaria × *ananassa* Elsanta.		Encapsulation–dehydration	86	[115]
Gentiana kurroo	Critically endangered perennial herbs/medicinal use of the root	Vitrification	35	[116]
		Droplet–vitrification	60	
		Vitrification	49	
		Droplet–vitrification	72	[117]
		Encapsulation–vitrification	38	
		Encapsulation–dehydration	31	
Hladnikia pastinacifolia	Herbaceous plant/medicinal use	Encapsulation–dehydration	53	[118]
		Encapsulation–vitrification	64	
		D-cryoplate	90	

Table 3
(continued)

Plant species	Plants type/uses	Technique	Survival (%)	Reference
Lamprocapnos spectabilis	Monotypic perennial rhizomatous plant, native to eastern Asia/ornamental use	Encapsulation–vitrification	73	[119]
		Vitrification	26	
		Droplet–vitrification	52	
Lilium lancifolium Thunb	Flowering plant native to China, Japan, Korea/ornamental use	Droplet–vitrification	84	[120]
Lilium spp.		Vitrification	80	[121]
		Vitrification	60-90	[122]
Photinia × fraseri	Hybrid shrub between *Photinia glabra* and *Photinia serratifolia*/ornamental use	Droplet–vitrification	27	[123]
		Encapsulation–dehydration	13	[124]
Pogostemon yatabeanus	Aquatic plant/ornamental use	Droplet–vitrification	92	[125]
Saccharum officinarum	Large perennial plant/used for sugar production	V-cryoplate	100	[126]
Saccharum officinarum N-1		D-cryoplate	98	[127]
		Vitrification	100	
Solanum tuberosum "Sayaka"		D-cryoplate	93	[128]
Stevia rebaudiana Bertoni	Evergreen shrub/use of leaves as a sweetener	Droplet–vitrification	80	[5]
		V-cryoplate	93	
Tarenaya spinosa Jacq.	Perennial/ornamental use	D-cryoplate	76	[129]
Taxus baccata	Dioecious conifer/ornamental horticulture	Encapsulation–dehydration	100	[130]

Table 3
(continued)

Plant species	Plants type/uses	Technique	Survival (%)	Reference
Vaccinium corymbosum	Plant of the genus *Vaccinium*/edible fruits	V-cryoplate	62-81	[101]
		D-cryoplate	89	
Vanilla planifolia Andrews.	Monocotyledonous orchid with climbing habit/ extraction of vanillin from its fruits	Droplet–vitrification	30	[131]
Vanilla planifolia Jacks.		D-cryoplate	33	[132]
Viola stagnina	Violet native to central and northern Europe and northern Asia/ornamental use	Encapsulation–dehydration	48	[133]
		Encapsulation–vitrification	28	
		Encapsulation–vitrification	38	

eradication. The lethal cell injury that occurs during cryopreservation also results in the destruction of infected tissues [61].

In Tables 2 and 3 are summarized several cryopreservation protocols applied to somatic embryos and shoot apical meristems of different plant species.

5 Conclusions

To date, cryopreservation has been successfully applied to somatic embryos and shoot tips of numerous plant species, including those from temperate and tropical regions. The most significant advances for these organized tissues have been achieved using vitrification-based methods, which have proven to be particularly efficient and versatile. Over the years, some genebanks have also incorporated the application of these protocols for the long-term storage of genetic resources, especially of vegetatively propagated species.

This chapter briefly reviews several theoretical and practical aspects that support the selection and optimization of these protocols. The objective is to share the practical knowledge acquired through scientific research and specialized collaboration in the field of plant cryobiology.

References

1. Engelmann-Sylvestre I, Engelmann F (2015) Cryopreservation of in vitro-grown shoot tips of *Clinopodium odorum* using aluminium cryo-plates. In Vitro Cell Dev Biol Plant 51:185–191
2. Benson EE (2008) Cryopreservation of phytodiversity: a critical appraisal of theory & practice. CRC Crit Rev Plant Sci 27:141–219
3. Bettoni CJ, Bonnart R, Volk GM (2021) Challenges in implementing plant shoot tip cryopreservation technologies. Plant Cell Tissue Organ Cult 144(1):21–34
4. Cruz-Cruz CA, Gonzalez-Arnao MT, Engelmann F (2013) Biotechnology and conservation of plant biodiversity. Resources 2:73–95
5. Benelli C, Carvalho SL, Merzougui FS, Petruccelli R (2021) Two advanced cryogenic procedures for improving *Stevia rebaudiana* (Bertoni) cryopreservation. Plants 10(2):277
6. Martínez-Montero ME, Mora N, Quiñones J, González-Arnao MT, Engelmann F, Lorenzo JC (2002) Effect of cryopreservation on the structural and functional integrity of cell membranes of sugarcane (Saccharum sp.) embryogenic calluses. Cryoletters 23(4):237-244
7. Reed BM (2008) Plant cryopreservation: a practical guide. Springer, New York
8. González-Arnao M, Engelmann F (2013) Consideraciones teóricas y prácticas para la crioconservación de germoplasma vegetal. Crioconservación de Plantas en América Latina y el Caribe. IICA, San José, pp 26–33
9. Nagel M, Pence V, Ballesteros D, Lambardi M, Popova E, Panis B (2024) Plant cryopreservation: principles, applications, and challenges of banking plant diversity at ultralow temperatures. Annu Rev Plant Biol 75:797–824. https://doi.org/10.1146/annurev-arplant-070623-103551
10. Loyola-Vargas VM, Ochoa-Alejo N (eds) (2018) Plant cell culture protocols (4th ed). Methods in molecular biology, Vol 1815. Humana Press. https://doi.org/10.1007/978-1-4939-8594-4
11. George EF, Debergh CP (2008) Micropropagation: uses and methods. In: George EF, Hal MA, de Klerk GJ (eds). Plant propagation by tissue culture. 3rd ed. Springer, Dordrecht, 29–64 https://doi.org/10.1007/978-1-4020-5005-3_2
12. Panis B (2008) Cryopreservation of monocots. In: Reed BM (ed). Plant cryopreservation: a practical guide. Springer, New York, 241–278
13. Keller ERJ, Senula A, Kaczmarczyk A (2008) Cryopreservation of herbaceous dicots. In:

Reed BM (ed). Plant cryopreservation: a practical guide. Springer, New York, pp 281–332
14. Von A (2008) Somatic embryogenesis. In: George EF, Hall MA, de Klerk GJ (eds). Plant propagation by tissue culture. 3rd ed. Springer, Dordrecht, 29–64 https://doi.org/10.1007/978-1-4020-5005-3_2
15. Yuan YH, Kagale S, Ferrie RAM (2024) Multifaceted roles of transcription factors during plant embryogenesis. Front Plant Sci 14:1322728. https://doi.org/10.3389/fpls.2023.1322728
16. Fosket DE (1994) Apical meristems and the formation of the plant body. In: Plant growth and development. Academic Press, pp 459–516. https://doi.org/10.1016/B978-0-12-262430-8.50013-9
17. Samuels FMD, Stich DG, Bonnart R, Volk GM, Levinger NE (2021) Non-uniform distribution of cryoprotecting agents in rice culture cells measured by CARS microscopy. Plants 10:589. https://doi.org/10.3390/plants10030589
18. Delgado-Aceves L, González-Arnao MT, Santacruz-Ruvalcaba F, Folgado R, Portillo L (2021) Indirect somatic embryogenesis and cryopreservation of *Agave tequilana* Weber cultivar 'Chato'. Plants 10(2):249
19. Sliwinska E, Bewley DJ (2014) Overview of seed development, anatomy, and morphology. In: Seeds: the ecology of regeneration in plant communities. 2nd ed. CABI, Wallingford, UK. https://doi.org/10.1079/9781780641836.0001
20. Chriqui D (2008) Developmental biology. In: George EF, Hall MA, de Klerk GJ (eds). Plant propagation by tissue culture. 3rd ed. Springer, Dordrecht, 283–334
21. Engelmann F (2004) Plant cryopreservation: progress and prospects. In Vitro Cell Dev Biol Plant 40:427–433
22. Jiroutová P, Sedlák J (2020) Cryobiotechnology of plants: a hot topic not only for gene banks. Appl Sci 10(13):4677. https://doi.org/10.3390/app10134677
23. Kean-Galeno T, Lopez-Arredondo D, Herrera-Estrella L (2024) The shoot apical meristem: an evolutionary molding of higher plants. Int J Mol Sci 25(3):1519. https://doi.org/10.3390/ijms25031519
24. Engelmann F (2011) Cryopreservation of embryos: an overview. Methods Mol Biol 710:155–184. https://doi.org/10.1007/978-1-61737-988-8_13
25. Gonzalez-Arnao MT, Martinez-Montero EM, Cruz-Cruz AC, Engelmann F (2014) Advances in cryogenic techniques for the long-term preservation of plant biodiversity. In: Biotechnology and Biodiversity, pp 129–170
26. Bettoni CJ, Chen K, Volk GM (2024) Considerations when implementing shoot tip cryopreservation. Training in plant genetic resources: cryopreservation of clonal propagules. Colorado State University Plant and Food Research, Fort Collins, Colorado, pp 1–28
27. Bettoni CJ, Wang RM, Wang CQ (2024) In vitro regeneration, micropropagation and germplasm conservation of horticultural plants. Horticulturae 10(1):45
28. Popova E, Kulichenko I, Kim HH (2023) Critical role of regrowth conditions in post-cryopreservation of in vitro plant germplasm. Biology 12(4):542
29. Kuriyama A, Watanabe K, Ueno S, Mitsuda H (1989) Inhibitory effect of ammonium ion on recovery of cryopreserved rice cells. Plant Sci 64:231–235
30. El Merzougui S, Benelli C, El Boullani R, Serghini MA (2023) The cryopreservation of medicinal and ornamental geophytes: application and challenges. Plants 12(11):2143
31. Kim HH, Popova E, Shin DJ, Yi JY, Kim CH, Lee JS, Yoon MK, Engelmann F (2012) Cryobanking of Korean Allium germplasm collections: results from a 10-year experience. CryoLetters 33:45–57
32. Corredoira E, San-José MC, Ballester A, Vieitez AM (2004) Cryopreservation of zygotic embryo axes and somatic embryos of European chestnut. CryoLetters 25:33–42
33. Yi JY, Balaraju K, Baek HJ, Yoon MS, Kim HH, Lee YY (2018) Cryopreservation of *Citrus limon* (L.) Burm. f. shoot tips using a droplet-vitrification method. Korean. J Plant Res 31:684–694
34. Popova E, Kim HH, Paek KY (2010) Cryopreservation of coriander (*Coriandrum sativum* L.) somatic embryos using sucrose preculture and air desiccation. Sci Hortic (Amsterdam) 124:522–528
35. Kulus D, Tymoszuk A, Kulpińska A, Wojnarowicz J, Szałaj U (2024) Nanoparticle-mediated enhancement of plant cryopreservation: cultivar-specific insights into morphogenesis and biochemical responses in *Lamprocapnos spectabilis* (L.) Fukuhara 'Gold Heart' and 'Valentine'. PLoS One 19(5):e0304586
36. Kulus D, Tymoszuk A (2021) Gold nanoparticles affect the cryopreservation efficiency of in vitro-derived shoot tips of bleeding heart. Plant Cell Tissue Organ Cult 146:297–311
37. Escobar RH, Muñoz L, Rios A, Núñez A, Tohme J (2014) Using a droplet-vitrification method to partially overcome the recalcitrance of cassava to cryostorage. Acta Hortic 1039:227–232

38. Danso KE, Ford-Lloyd BV (2004) Cryopreservation of embryogenic calli of cassava using sucrose cryoprotection and air desiccation. Plant Cell Rep 22:623–631
39. Uchendu EE, Shukla MR, Reed BM, Saxena PK (2014) An efficient method for cryopreservation of St John's Wort and tobacco: role of melatonin. Acta Hortic 1039:233–243
40. Bettoni JC, Kretzschmar AA, Bonnart R, Shepherd A, Volk GM (2019) Cryopreservation of 12 Vitis species using apical shoot tips derived from plants grown in vitro. HortScience 54:976–981
41. Normah MN, Sulong N, Reed BM (2019) Cryopreservation of shoot tips of recalcitrant and tropical species: advances and strategies. Cryobiology 8:1–14
42. Fahy GM, MacFarlane DR, Angell CA, Meryman HT (1984) Vitrification as an approach to cryopreservation. Cryobiology 21(4):407–426
43. Reed BM (2001) Implementing cryogenic storage of clonally propagated plants. CryoLetters 22(2):97–104
44. Gonzalez-Arnao MT, Panta A, Roca WM, Escobar RH, Engelmann F (2008) Development and large scale application of cryopreservation techniques for shoot and somatic embryo cultures of tropical crops. Plant Cell Tissue Organ Cult 92:1–13
45. Sakai A, Engelmann F (2007) Vitrification, encapsulation-vitrification and droplet-vitrification: a review. CryoLetters 28(3):151–172
46. Sakai A, Kobayashi S, Oiyama I (1990) Cryopreservation of nucellar cells of navel orange (*Citrus sinensis* Osb. var. brasiliensis Tanaka) by vitrification. Plant Cell Rep 9(1):30–33
47. Fabre J, Dereuddre J (1990) Encapsulation-dehydration: a new approach to cryopreservation of Solanum shoot tips. CryoLetters 11:413–426
48. Gonzalez-Arnao MT, Panta A, Roca MW, Escobar HR, Engelmann F (2008) Development and large scale application of cryopreservation techniques for shoot and somatic embryo cultures of tropical crops. Plant Cell Tissue Organ Cult 92:1–13
49. Engelmann F, Arnao MTG, Wu Y, Escobar R (2008) Development of encapsulation dehydration. In: Reed BM (ed). Plant cryopreservation: a practical guide. Springer, New York, 59–75
50. Sherlock G, Block W, Benson EE (2005) Thermal analysis of the plant encapsulation-dehydration cryopreservation protocol using silica gel as the desiccant. CryoLetters 26(1):45–54
51. Keller ERJ, Dreiling M (2002) Potato cryopreservation in Germany – using the droplet method for the establishment of a new large collection. In: XXVI International Horticultural Congress: Plant Genetic Resources, The Fabric of Horticulture's Future, Acta Horticulturae 623:193–200
52. Panis B, Piette BMAG, Swennen R (2005) Droplet vitrification of apical meristems: a cryopreservation protocol applicable to all Musaceae. Plant Sci 168(1):45–55
53. Yamamoto IS, Fukui K, Rafique T, Khan IN, Martinez CCR, Sekizawa K, Niino TM (2012) Cryopreservation of in vitro-grown shoot tips of strawberry by the vitrification method using aluminium cryo-plates. Plant Genetic Resour 10(1):14–19
54. Niino T, Yamamoto IS, Fukui K, Martínez CCR, Arizaga VM, Matsumoto T, Engelmann F (2013) Dehydration improves cryopreservation of mat rush (Juncus decipiens Nakai) basal stem buds on cryo-plates. CryoLetters 34(6):549–560
55. Wilms H, Sleziak FN, van der Auweraer M et al (2020) Development of a fast and user-friendly cryopreservation protocol for sweet potato genetic resources. Sci Rep 10:14674. https://doi.org/10.1038/s41598-020-70869-3
56. Smets F, Panis B (2016) Cryopreservation of cassava (Manihot esculenta) by droplet vitrification: influence of plant age, meristem position, and used protocol. Proceedings of the Cryopreservation Symposium, May 30, 2016
57. Lambardi M, Ozudogru AE, Barberini S, Danti R (2018) Strategies for fast multiplication and conservation of forest trees by somatic embryogenesis and cryopreservation: a case study with cypress (*Cupressus sempervirens* L.). Not Bot Hortic Agrobot Cluj Napoca 46(1):32–38
58. Corredoira E, Toribio M, Vieitez MA (2014) Clonal propagation via somatic embryogenesis in Quercus spp. In: Ramawat KG, Mérillon J-M, Ahuja MR (eds). Tree biotechnology. CRC Press, Boca Raton, pp 262–302
59. Ballesteros D, Pence V, Nagel M, Lambardi M, Popova E, Panis B (2024) Principles, applications, and challenges of banking plant diversity at ultralow temperatures. Annu Rev Plant Biol 75:797–824. https://doi.org/10.1146/annurev-arplant-070623-103551
60. Barraco G, Sylvestre I, Collin M et al (2014) Histocytological analysis of yam (*Dioscorea alata*) shoot tips cryopreserved by encapsulation–dehydration. Protoplasma 251:177–189. https://doi.org/10.1007/s00709-013-0536-5
61. Wang QC, Panis B, Engelmann F, Lambardi M, Valkonen JPT (2009) Cryotherapy of shoot tips: a technique for pathogen eradication to produce healthy planting materials

and prepare healthy plant genetic resources for cryopreservation. Ann Appl Biol 154:351–363
62. Lambardi M, De Carlo A, Capuana M (2005) Cryopreservation of embryogenic callus of *Aesculus hippocastanum* L. by vitrification/ one-step freezing. CryoLetters 26(3):185–192
63. Nomali Z, Ngobese S, Sershen N, Berjak P, Pammenter NW (2014) Effects of some cryopreservation procedures on recalcitrant zygotic embryos of *Ammocharis coranica.* CryoLetters 35(4):299–307
64. Nery FC, Paiva R, Campos ACAL, Nogueira GF, Stein VC, Alvarenga AA (2011) Cryopreservation of *Anadenanthera colubrina* (Vell.) Brenan embryonic axes. Acta Hortic 908:227–232
65. Delgado-Aceves L, Portillo L, Folgado R, Romo-Paz FJ, González-Arnao MT (2022) New approaches for micropropagation and cryopreservation of *Agave peacockii*, an endangered species. Plant Cell Tissue Organ Cult 150(1):85–95
66. Delgado-Aceves L, Corona S, Marin-Castro RU, Rascón-Díaz PM, Portillo L, Mora-González A, González-Arnao MT (2024) Comparative studies for cryopreservation of Agave shoot tips by droplet-vitrification. Plants 13(18):2609
67. Fraga PH, Vieira NL, Puttkammer CC, Silva MJ, Anjos GK, Oliveira ME, Guerra PM (2016) High-efficiency cryopreservation of *Araucaria angustifolia* (Bertol.) Kuntze embryogenic cultures: ultrastructural characterization and morpho-physiological features. Plant Cell Tissue Organ Cult 124:307–318
68. Sershen NP, Berjak P, Wesley-Smith J (2007) Cryopreservation of embryonic axes of selected amaryllid species. CryoLetters 28(5):387–399
69. Vieitez MA, San-José CM, Corredoira E (2011) Cryopreservation of zygotic embryonic axes and somatic embryos of European chestnut. In: Plant embryo culture: methods and protocols, pp 201–213
70. Gaidamashvili M, Khurtsidze E, Kutchava T, Lambardi M, Benelli C (2021) Efficient protocol for improving the development of cryopreserved embryonic axes of chestnut (*Castanea sativa* Mill.) by encapsulation–vitrification. Plants 10(2):231
71. Malik KS, Kaur S, Choudhary R, Chaudhury R, Pritchard WH (2023) Comparative cryopreservation of Indian wild orange (*Citrus indica* Tanaka) embryonic axes. CryoLetters 44(3):142–150
72. Souza DFV, Kaya E, Vieira DJL, Souza DSA, Carvalho DSJMD, Santos BE, Ellis AAACD (2017) Cryopreservation of Hamilin sweet orange [*Citrus sinensis* (L.) Osbeck] embryogenic calli using a modified aluminum cryo-plate technique. Sci Hortic (Amsterdam) 224:302–305
73. Gantait S, Sinniah RU, Suranthran P, Palanyandy RS, Subramaniam S (2015) Improved cryopreservation of oil palm (*Elaeis guineensis* Jacq.) polyembryoids using droplet vitrification approach and assessment of genetic fidelity. Protoplasma 252:89–101
74. Ma M, Wang X, Zhang C, Pak S, Wu H, Yang J, Li C (2023) Enhancing the cryopreservation system of larch embryogenic culture by optimizing pre-culture, osmoprotectants, and rapid thawing. Forests 14(8):1621
75. Dereuddre J, Bertrand-Desbrunais A, Brison M, Boucaud D, Engelmann F, Paul H, Scottez SBS et al (1994) Cryoconservation d'apex et d'embryons somatiques d'arbres fruitiers. Genet Sel Evol 26(Suppl):279s–290s
76. Roostika I, Rahayu DME, Prawestri DA, Poerba SY, Handayani T, Ahmad F, Herlina MD, Sudarmonowati LDE (2024) Cryopreservation of zygotic embryos derived from BBTV resistant accessions of wild banana *Musa acuminata*. South Afr J Bot 173:255–265
77. Benelli C, De Carlo A, Lambardi TP, Lynch M (2001) Vitrification of shoot tips, nodal segments and embryogenic tissue of olive (*Olea europaea* L.) for germplasm cryopreservation. In: IV International Symposium on In Vitro Culture and Horticultural Breeding. Acta Hortic 560:137–140
78. Bradaï F, Almagro-Bastante J, Sánchez-Romero C (2017) Cryopreservation of olive somatic embryos using the droplet-vitrification method: The importance of explant culture conditions. Sci Hortic (Amsterdam) 218:14–22
79. Bradaï F, Almagro-Bastante J, Sánchez-Romero C (2023) Effect of sucrose preculture and culture inoculum density on cryopreservation of olive somatic embryos. Sci Hortic (Amsterdam) 322:112385
80. Guzmán-García E, Bradaï F, Sánchez-Romero C (2013) Cryopreservation of avocado embryogenic cultures using the droplet-vitrification method. Acta Physiol Plant 35:183–193
81. O'Brien C, Constantin M, Walia A, Yiing LYJ, Mitter N (2016) Cryopreservation of somatic embryos for avocado germplasm conservation. Sci Hortic (Amsterdam) 211:328–335. https://doi.org/10.1016/j.scienta.2016.09.008
82. Salma M, Fki L, Engelmann-Sylvestre I, Niino T, Engelmann F (2014) Comparison of droplet-vitrification and D-cryoplate for cryopreservation of date palm (*Phoenix*

dactylifera L.) polyembryonic masses. Sci Hortic (Amsterdam) 179:91–97

83. Alansi S, Al-Qurainy F, Nadeem M, Khan S, Tarroum M, Alshameri A, Gaafar ZAR (2019) Cryopreservation: a tool to conserve date palm in Saudi Arabia. Saudi J Biol Sci 26(7):1896–1902
84. Varis S, Ahola S, Jaakola L, Aronen T (2017) Reliable and practical methods for cryopreservation of embryogenic cultures and cold storage of somatic embryos of Norway spruce. Cryobiology 76:8–17
85. Cyr RD, Lazaroff RW, Grimes MS, Quan G, Bethune DT, Dunstan ID, Roberts RD (1994) Cryopreservation of interior spruce (*Picea glauca* × engelmannii complex) embryogenic cultures. Plant Cell Rep 13:574–577
86. Álvarez MJ, Cortizo M, Ordás JR (2012) Cryopreservation of somatic embryogenic cultures of *Pinus pinaster*: effects on regrowth and embryo maturation. CryoLetters 33(6):476–484
87. Parasher K, Sharma S, Mukherjee P, Qazi HP (2023) Cryopreservation of zygotic embryos of *Podophyllum hexandrum* Royle, an endangered medicinal plant, by vitrification and V cryo-plate techniques. CryoLetters 44(4):219–228
88. Barra-Jiménez A, Aronen ST, Alegre J, Toribio M (2015) Cryopreservation of embryogenic tissues from mature holm oak trees. Cryobiology 70(3):217 225
89. Fernandez P, Rodriguez E, Pinto G, Roldan-Ruiz I, Loose MD, Santos C (2008) Cryopreservation of *Quercus suber* somatic embryos by encapsulation-dehydration and evaluation of genetic stability. Tree Physiol 28(12):1841–1850
90. Verleysen H, Fernandes P, Pinto SI, Bockstaele VE, Debergh P (2005) Cryopreservation of *Robinia pseudoacacia*. Plant Cell Tissue Organ Cult 81:193–202
91. Nadarajan J, Walt K. van der, Lehnebach AC, Saeiahagh H, Pathirana R (2021) Integrated ex situ conservation strategies for endangered New Zealand Myrtaceae species. N Z J Bot 59(1):72–89
92. Adu-Gyamfi R, Wetten A (2012) Cryopreservation of cocoa (*Theobroma cacao* L.) somatic embryos by vitrification. CryoLetters 33(6):494–505
93. Fang YJ, Wetten A, Hadley P (2004) Cryopreservation of cocoa (*Theobroma cacao* L.) somatic embryos for long-term germplasm storage. Plant Sci 166(3):669–675
94. Ahn HC, Heo K, Park SH, Choi EY (2019) In vitro propagation and cryopreservation of *Thuja koraiensis* Nakai via somatic embryogenesis. In Vitro Cell Dev Biol Plant 55:605–614
95. Ahn HC, Han YJ, Kim SY, Choi EY (2018) Propagation and cryopreservation of Ulleungdo hemlock (*Tsuga ulleungensis*) via somatic embryogenesis. Trees 32:1801–1808
96. Gonzalez-Benito ME, Martin C, Vidal JR (2009) Cryopreservation of embryogenic cell suspensions of the Spanish grapevine cultivars 'Albariño'and 'Tempranillo'. Vitis 48(3):131–136
97. Imsomboon T, Thammasiri K, Kosiyajinda P, Chuenboonngarm N, Panvisavas N (2018) Cryopreservation of non-precultured protocorms of Acampe rigida using V and D cryo-plate methods. In: III International Symposium on Plant Cryopreservation. Acta Horticult 1234, pp 269–278
98. Wu Y, Zhao Y, Engelmann F, Zhou M (2001) Cryopreservation of kiwi shoot tips. CryoLetters 22(5):277–284
99. Tanaka D, Sakuma Y, Yamamoto S, Matsumoto T, Niino T (2018). Development of the V cryo-plate method for cryopreservation of in vitro rakkyo (Allium chinense G. Don). In: Acta Horticulturae, 1234, 287–292. Proceedings of the symposium, Bangkok, Thailand, 26–28 March 2018
100. Kohmura H, Ikeda Y, Sakai A (1994) Cryopreservation of apical meristems of Japanese shallot (*Allium wakegi* A.) by vitrification and subsequent high plant regeneration. CryoLetters 15:289 298
101. Vujović T, Anđelić T, Marković ZGAHJC, Strawberry and Blueberry Breeding Group (2024) Methods of micropropagation and cryopreservation of small fruits. In Vitro Cell Dev Biol Plant 60:85–97
102. Salama A, Popova E, Jones MSMRFNS SPKC (2018) [Incomplete reference in source]
103. Mallón R, Bunn E, Turner RS, González LM (2008) Cryopreservation of *Centaurea ultreiae* (Compositae), a critically endangered species from Galicia (Spain). CryoLetters 29(5):363–370
104. Shatnawi AM, Johnson AK (2004) Cryopreservation by encapsulation-dehydration of 'Christmas bush' (*Ceratopetalum gummiferum*) shoot tips. In Vitro Cell Dev Biol Plant 40:239 244
105. Halmagyi A, Fischer-Klüver G, Mix-Wagner G, Schumacher HM (2004) Cryopreservation of *Chrysanthemum morifolium* (Dendranthema grandiflora Ramat.) using different approaches. Plant Cell Rep 22:371–375
106. Lee YG, Popova E, Cui HY, Kim HH, Park SU, Bae CH et al (2011) Improved cryopreservation of chrysanthemum (*Chrysanthemum morifolium*) using droplet-vitrification. CryoLetters 32(6):487–497

107. Sakai A, Matsumoto T, Hirai D, Niino T (2000) Newly developed encapsulation-dehydration protocol for plant cryopreservation. CryoLetters 21(1):53–62
108. Wilkinson MTI, Wetten A, Prychid C, Fay FM (2003) Suitability of cryopreservation for the long-term storage of rare and endangered plant species: a case history for *Cosmos atrosanguineus.* Ann Bot (Lond) 91(1):65–74
109. Osorio-Saenz OA, Gallardo MOJ, López MC, González-Arnao TM (2011) Cryopreservation of apices of chrysanthemum (*Dendranthema grandiflorum* Kitam.) by encapsulation-dehydration and vitrification. Revista Chapingo. Serie Horticult 17(SPE2):33–43
110. Malaurie B, Trouslot FM, Engelmann F, Chabrillange N (1998) Effect of pretreatment conditions on the cryopreservation of in vitro–cultured yam (Dioscorea alata 'Brazo fuerte' and D. bulbifera 'Nouméa imboro') shoot apices by encapsulation–dehydration. CryoLetters 19:15–26
111. Sharma N, Malhotra VE, Chandra R, Gowthami R, Sultan MS, Bansal S, Agrawal A (2022) Cryopreservation and genetic stability assessment of regenerants of the critically endangered medicinal plant *Dioscorea deltoidea* Wall. ex Griseb. for cryobanking of germplasm. In Vitro Cell Dev Biol Plant 58(4):521–529
112. Thankappan SS, Morawala-Patell V (2021) Germplasm conservation techniques and assessment of physiological, biochemical and molecular integrity of indigenous, endangered *Dioscorea prazeri.* CryoLetters 42(3):168–177
113. Mandal BB, Dixit-Sharma S (2007) Cryopreservation of in vitro shoot tips of *Dioscorea deltoidea* Wall, an endangered medicinal plant: effect of cryogenic procedure and storage duration. CryoLetters 28(6):461–470
114. Hirai D, Shirai K, Shirai S, Sakai A (1998) Cryopreservation of in vitro-grown meristems of strawberry (Fragaria × ananassa Duch.) by encapsulation-vitrification. Euphytica 101(1):109–115
115. Navatel JC, Capron M (1997) Cryopreservation of alginate-coated strawberry axillary buds. Acta Hortic 439:659–661
116. Sharma N, Gowthami R, Devi VS, Malhotra VE, Pandey R, Agrawal A (2021) Cryopreservation of shoot tips of *Gentiana kurroo* Royle—a critically endangered medicinal plant of India. Plant Cell Tissue Organ Cult 144(1):67–72
117. Pence CV, Philpott M, Culley MT, Plair B, Yorke RS, Lindsey K, Ballesteros D (2017) Survival and genetic stability of shoot tips of Hedeoma todsenii R.S. Irving after long-term cryostorage. In Vitro Cell Dev Biol Plant 53:328–338
118. Ciringer T, Martín C, Šajna N, Kaligarič M, Ambrožič-Dolinšek-Dolinšek J (2018) Cryopreservation of an endangered *Hladnikia pastinacifolia* Rchb. by shoot tip encapsulation-dehydration and encapsulation-vitrification. In Vitro Cell Dev Biol Plant 54:565–575
119. Kulus D (2020) Effect of bead composition, PVS type, and recovery medium in cryopreservation of bleeding heart 'Valentine'—preliminary Study. Agronomy 10(6):891
120. Chen LX, Li HJ, Xin X, Zhang EZ, Xin PP, Lu XX (2011) Cryopreservation of in vitro-grown apical meristems of Lilium by droplet-vitrification. South Afr J Bot 77(2):397–403
121. Matsumoto T, Sakai A, Yamada K (1995) Cryopreservation of in vitro-grown apical meristems of lily by vitrification. Plant Cell Tissue Organ Cult 41:237–241
122. Bouman H, Tiekstra A, Petutschnig E, Homan M, Schreurs R (2003, August) Cryopreservation of Lilium species and cultivars. In: XXI International Eucarpia Symposium on Classical versus Molecular Breeding of Ornamentals—Part I. Acta Horticult 612, 147–154
123. Akdemir H, Ozden Tokatli Y (2011) Long-term storage of Fraser Photinia shoot tips via conventional and droplet vitrification techniques. In: I International Symposium on Cryopreservation in Horticultural Species. Acta Horticult 908, 297–302
124. Ozden Tokatli Y, Ozudogru EA, Akdemir H, Gumusel F, De Carlo A (2008) Application of cryopreservation for pistachio germplasm conservation. In: I International Symposium on Biotechnology of Fruit Species: BIOTECHFRUIT2008. Acta Horticult 839, 245–251
125. Lee H, Kim H (2022) Vigorous growing of donor plantlets by liquid overlay in subcultures is the key to cryopreservation of endangered species *Pogostemon yatabeanus.* Plants 11:3127
126. Rafique T, Yamamoto I.S, Fukui K, Tanaka D, Arizaga V.M, Abbas M, Niino TM (2016). Cryopreservation of shoot-tips from different sugarcane varieties using D cryo-plate technique. Pak J Agric Sci 53(1): 151-158
127. Rafique T, Yamamoto S, Fukui K, Mahmood Z, Niino T (2015) Cryopreservation of sugarcane using the V cryo-plate technique. CryoLetters 36:51–59
128. Yamamoto IS, Wunna Rafique T, Arizaga VM, Fukui K, Gutierrez CEJ, Niino TCMCRWK (2015) The aluminum cryo-plate increases efficiency of cryopreservation protocols for potato shoot tips. Am J Potato Res 92:250–257

129. Vilardo MAFR, Lima SGDD, Cordeiro SLD, Albarello N, Simões-Gurgel C (2023) Application of D cryo-plate technique for the cryopreservation of in vitro-grown shoot tips of *Tarenaya spinosa* (Cleomaceae). CryoLetters 44(6):333–342
130. Moghaddam SA, Kaviani B, Torkashvand MA, Abdossi V, Eslami RA (2024) In vitro propagation, cold preservation and cryopreservation of Taxus baccata L, an endangered medicinal and ornamental shrub. Acta Sci Pol Hortorum Cultus 23(1):13–28
131. Gonzalez-Arnao MT, Lazaro-Vallejo CE, Engelmann F, Gamez-Pastrana R, Martinez-Ocampo YM, Pastelin-Solano MC, Diaz-Ramos C (2009) Multiplication and cryopreservation of *Vanilla planifolia* 'Andrews'. In Vitro Cell Dev Biol Plant 45:574–582
132. Hernández-Ramírez F, Dolce N, Flores-Castaños O, Rascón PM, Ángeles-Álvarez G, Folgado R, González-Arnao MT (2020) Advances in cryopreservation of *Vanilla planifolia* Jacks. shoot-tips: assessment of new biotechnological and cryogenic factors. In Vitro Cell Dev Biol Plant 56:236–246
133. Żabicki P, Mikuła A, Sliwinska E, Migdałek, Nobis A, Żabicka J, Kuta E (2021) Cryopreservation and post-thaw genetic integrity of Viola stagnina Kit, an endangered species of wet habitats—a useful tool in ex situ conservation. Sci Hortic (Amsterdam) 284:110056

Chapter 31

Application of Nanomaterials in the Conservation of Plant Genetic Resources

José Luis Aguirre-Noyola, Marco A. Ramírez-Mosqueda,
Jorge David Cadena-Zamudio, Juan Ramos-Garza
and Gustavo Cuaxinque-Flores

Abstract

The preservation of plant genetic resources is crucial to safeguard food security. These resources encompass a wide array of plant material, including cells, seeds, tissues, and their genetic material, derived from both wild and domesticated species. They serve as the fundamental raw material for crop improvement. The ex situ conservation of these resources in germplasm banks poses significant challenges, which can be addressed through the application of nanobiotechnology. This field utilizes nanomaterials (10–100 nm) that exhibit a wide range of shapes and chemical compositions, along with remarkable optical, electronic, and mechanical properties. These materials are capable of catalyzing a multitude of biochemical reactions and processes, making them highly effective tools for agricultural research and development. Nanobiotechnology approaches in plant genetic resources conservation can range from the stimulation of seed germination and the improvement of in vitro culture methods, to cryopreservation assisted by nanomaterials and the conservation of nucleic acids using nanostructures. The use of nanomaterials opens a window for innovation in germplasm biobanks; however, it must be considered that the ideal dose and type of nanomaterial for each purpose must be determined in order to ensure the genomic integrity and reactivation of the plant cells.

Keywords Nanotechnology, Nanobiotechnology, Nanoparticles, Nanopriming, Nanogardening

1 Introduction

The term "Plant Genetic Resources" is understood to encompass all plant material that harbors actual or potential value for food and agriculture, or for environmental sustainability [1]. Plant genetic resources include a wide range of plant material, including seeds, pollen, cells, tissues, whole plants, and their genetic material, from a variety of genotypes, ranging from traditional and modern varieties to closely related wild species [2]. The sustainable use of genetic resources has a wide range of implications for agriculture and the food security of present and future generations. These include increased

Marco A. Ramírez-Mosqueda et al. (eds.), *Conservation of Plant Genetic Resources*, Methods in Molecular Biology, vol. 3011,
https://doi.org/10.1007/978-1-0716-5130-8_31,

crop productivity, enhanced resistance to pests and diseases, and facilitated adaptation to environmental fluctuations, including climate change, drought, and salinity [3]. In recent years, an integrated approach to biodiversity, including agricultural biodiversity, conservation, sustainable use and development, is increasingly being recognized. Novel approaches are needed to ensure the effective use and conservation of plant genetic resources, and nanobiotechnology offers a number of methodologies for achieving this goal.

Nanobiotechnology is a multidisciplinary field that integrates the technological expertise of nanotechnology with the chemical and biological components of biotechnology [4, 5]. The field of nanotechnology is a cutting-edge scientific domain that involves the design and utilization of materials and devices with dimensions ranging from 1 to 100 nm, exhibiting novel optical, electronic, mechanical and chemical properties [6]. Conversely, biotechnology employs biological instruments and expertise to regulate cellular and molecular operations, thereby yielding valuable products and services in domains such as medicine, environment and agriculture [7]. Consequently, nanobiotechnology facilitates the generation of nanoscale materials with distinctive functionalities, capable of catalyzing, enhancing, or innovating biological, physical, and chemical processes [4].

Nanomaterials are characterized by their high surface-to-volume ratio and quantum effects, which can result in enhanced reactivity, altered electrical conductivity, and unique absorption and emission optical properties that exhibit different properties and behavior than materials at larger scales [8, 9]. The structure of nanomaterials is diverse, ranging from quantum dots, spherical nanoparticles, nanotubes, nanowires, nanofibers, nanoplatelets, and nanoliposomes (Fig. 1). Each has unique structural properties and potential applications in different areas of industry and scientific research.

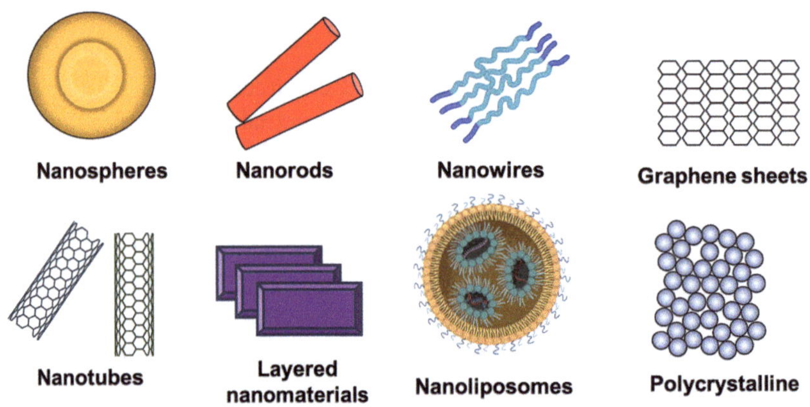

Fig. 1 Examples of some nanomaterial structures and compositions. The diagram was created with BioRender.com

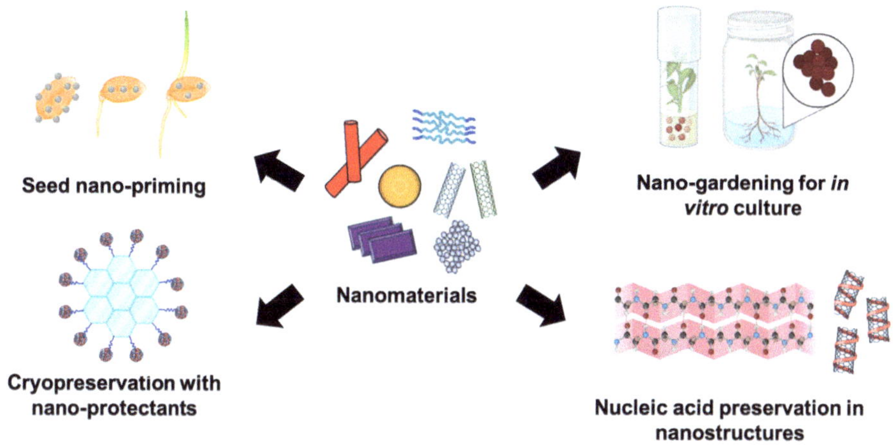

Fig. 2 Examples of some applications of nanomaterials to plant genetic resources conservation. The diagram was created with BioRender.com

But how are nanomaterials made? There are two main approaches to their synthesis: the top-down approach and the bottom-up approach [8]. In top-down methods, bulk materials are subdivided to produce nanostructured materials. Mechanical milling, laser ablation, etching, sputtering, and electrospinning are some examples of this approach. In contrast, bottom-up methods such as pyrolysis, spinning, gel polymerization, green synthesis, and chemical vapor deposition employ chemical, physical, or biological forces to assemble basic units into larger structures [9].

Nanobiotechnology has become a fundamental tool for the conservation of genetic resources, since it allows progress in the innovation of methods and materials to prolong the long-term stability of the material stored in germplasm banks [10]. Applications include enhancing in vitro culture by adding antimicrobial, antioxidant, and phytostimulant nanomaterials; nanopriming for seeds; nanoprotectant-assisted cryopreservation; and DNA nanocomplexes for stabilization against environmental changes (Fig. 2).

These innovations facilitate the conservation of biodiversity and the preservation of key genetic resources for nutritional food security, environmental sustainability, and scientific research. Here, we review the main approaches that nanobiotechnology has used to support the conservation of plant genetic resources.

2 Seed Nanopriming and Biofortification

The plant seed is fundamental to inheritance because it contains the genetic material that is passed from one generation to the next. Through sexual reproduction, plant DNA recombines, resulting in seeds with new genetic combinations that contribute to the biological variation essential for adaptation and evolution [11]. Plant genetic information determines key traits such as size, color, phenology, physiology, disease resistance, and environmental adaptability. Therefore, seed storage and management enable the conservation and generation of varieties, facilitating sustainable agriculture and the protection of species from climate change [2]. Seed conservation faces several challenges that threaten the viability and genetic stability of plant genetic resources. Major problems include deterioration and loss of viability due to factors such as moisture, inadequate temperatures and exposure to light. There are also problems associated with attack by pathogens such as fungi, bacteria and insects, which can affect the physical integrity of seeds. Together, these factors limit the longevity of seeds in biobanks. The high costs associated with germplasm conservation are also a major challenge [12]. Seed nanopriming is an innovative tool to overcome these challenges.

Nanopriming technology is an innovative technique involving treatment of seeds with nanomaterials to enhance germination, seedling yield and tolerance to different types of stresses [13]. This process typically involves exposing seeds to nanoscale solutions, such as nanoparticles, nanocoatings, or nanocarriers, which can be absorbed by the seed or adhere to the surface layer. The nanoparticles penetrate the seed and allow water to enter, triggering molecular and biochemical responses associated with breaking dormancy and promoting germination [14]. These mechanisms include the production of reactive oxygen species and gibberellic acid (GA3) and the expression of genes encoding amylases [15].

The benefits of seed nanopriming are many and varied. It has been successfully applied to cereal seeds such as corn, wheat, rice and barley [16–18] and legumes such as bean, faba bean, soybean, and pea [19–22]. Further details on the types of nanoparticles and their biological effects can be found in Table 1.

In addition to promoting faster and more uniform germination, nanopriming stimulates early growth and root development, resulting in healthier and more vigorous plants. Several studies show increased germination rates under adverse environmental conditions (drought, cold, salinity, and metal stress) by using nanomaterial-treated seeds [34, 35]. Nanopriming also improves nutrient uptake (biofortification), allowing more efficient uptake of essential elements and promoting the controlled release of growth regulators. This leads to improved seedling development and a corresponding increase in crop yield [36].

Table 1
Examples of seed nanopriming and its biological effects

Nanomaterial	Plant species	Physiological effects	Reference
PVP-Ag NPs	Barley (*Hordeum vulgare* L.)	Reduction of salinity stress by decreasing ROS generation.	[23]
ZnO NPs	Faba bean (*Vicia faba* L.)	Improved antioxidant system and copper stress tolerance	[24]
Ag NPs	Cucumber (*Cucumis sativus* L.),	Seedling development	[25]
$CoFe_2O_4$ NPs	Common bean (*Phaseolus vulgaris* L.)	Improvement of seed mineral composition	[26]
FeON Ps	Rice (*Oryza sativa* L.)	Seed germination	[27]
Se NPs	Tomato (*Solanum lycopersicum* L.)	Induced systemic resistance	[28]
ZnO NPs	Maize (*Zea mays* L.)	Protection against cobalt stress	[29]
Se and ZnO NPs	Rapeseed (*Brassica napus* L.)	Induction of salinity tolerance and protection against oxidative damage	[30]
Mn_2O_3 NPs	Watermelon (*Citrullus lanatus* Thunb. Matsum. & Nakai)	Modification of chlorophyll and antioxidant activity	[31]
Si NPs	Wheat (Triticum aestivum)	Reduction of Cd translocation and oxidative stress	[32]
Ag NPs	Chinese cabbage (*Brassica pekinensis*)	Increased germination and plant development	[33]

3 Nanogardening: Biotechnological Innovations for In Vitro Culture Systems

In vitro culture methods involve the growth of plant cells, tissues or organs in a controlled and sterile environment on culture media, outside their natural habitat (ex situ). Consequently, these methods are widely used for propagation, conservation and study of plant genetic resources [37]. There are many challenges to in vitro culture that can hinder its efficiency and success. Bacterial, fungal, or viral contamination is a common problem that often spreads rapidly in culture media and negatively affects plant growth. Some of these microbes colonize and live naturally in plant tissues and are known as endophytes. The endophyte population can increase when nutrients and moisture are increased, as occurs during micropropagation [38]. To mitigate this risk, strict aseptic conditions are required, which translates into higher costs, skilled labor and technical preparation.

Another recurring problem is phenolic oxidation or phenolization. This process is triggered by oxidative enzymes such as polyphenol oxidases, peroxidases, or tyrosinases that are released during injury or tissue stress. These enzymes produce highly reactive quinones from compounds with aromatic functional groups such as phenols and catechols, which negatively affect plant growth, causing browning, reduced viability and ultimately cell death [39, 40]. To counteract this, antioxidants or sorbents can be added to the culture medium, such as activated charcoal and organic acids (such as ascorbic and citric acids), in addition to careful handling to minimize the risk of tissue damage [41]. Phenolic oxidation management is essential to improve the success rate of in vitro establishment.

Recently, nanotechnology has been resorted to using nanomaterials and nanodevices to overcome all the challenges associated with in vitro culture. This multidisciplinary approach is called nanogardening [42]. Specifically, nanogardening is the use of nanomaterials to promote the growth, development and overall success of plants during in vitro culture. Nanoparticles can also be used to disinfect explants (plant tissue fragments for in vitro propagation), which reduces the contamination rate and growth of endophytic microorganisms [43]. Nanoparticles, such as silver (AgNPs), gold (AuNPs), silica (SiNPs), zinc oxide (ZnO-NPs), and carbon nanotubes improve plant growth rates, stress tolerance, and microbial contamination.

Nanogardening begins with the selection of the type, size, and chemical composition of the nanomaterial based on the desired effect. For example, silver nanoparticles are characterized by their antimicrobial properties, while silicon nanoparticles are effective in inducing developmental and systemic abiotic stress tolerance responses [44, 45]. Nanomaterials are incorporated into the culture medium at concentrations optimized for the plant species. It is advisable to perform dose-response experiments beforehand to avoid nanotoxicity [46]. Then, plant explants, such as nodes, leaves, or stems, are placed in nanoenriched media and incubated under appropriate nutrient, temperature, and light conditions. Key parameters of response to nanomaterials include growth measurements (number and length of shoots and roots, biomass, number of leaves), stress resistance tests (e.g., exposure to salinity or drought), and biochemical analyses of chlorophyll, carotenes and secondary metabolites [47].

The use of nanomaterials as antioxidants to reduce phenolization and plant death is another interesting application of nanotechnology. Certain types, including those made as metal oxides, exhibit significant antioxidant activity [48].

Metal nanoparticles can neutralize reactive oxygen species (ROS) through various mechanisms, such as electron donation or activation of antioxidant systems [49]. For example, cerium oxide nanoparticles (CeO_2–NPs) have demonstrated the unique ability to mimic the enzymatic activity of superoxide dismutase (SOD) and catalase (CAT), both of which are critical for mitigating oxidative stress [50].

The consequences of ROS include lipoperoxidation and DNA damage; when they occur at high frequency, cell membrane rupture and genome mutations are generated [51]. Similarly, selenium nanoparticles (SeNPs) act as potent antioxidants by increasing glutathione peroxidase (GSH-Px) activity [50, 52]. GSH-Px protects cells from the degrading effect of hydroperoxides [53]. In summary, the use of nanomaterials during in vitro culture opens a window of opportunity for the conservation and genetic stability of plant genetic resources.

4 Nanomaterial-Assisted Cryopreservation

Cryopreservation is the process of preserving the structure and function of plant material (cells, organs, tissues, or organisms) by reducing their metabolism and enzymatic activity while maintaining them at extremely low temperatures [54]. Cryoprotectants such as dimethyl sulfoxide, glycerol, and ethylene glycol are cheap and commonly used, but they can induce cytotoxicity when used at high concentrations or in prolonged exposures [55]. Cryoprotectants have reactive functional groups (such as hydroxyls groups) that can generate free radicals and then oxidative stress. Several effects have been reported at the biochemical level as a result of meristem cryopreservation in tomato (*Solanum lycopersicum* L.) seedlings, including increased levels of peroxidases and phenolic compounds. Damage to the plant wall infrastructure was also reported [56].

As a solution to these problems, nanoprotectants have been developed that would reduce phytoxicity and increase the cell reactivation success rate. Nanoparticle-based cryoprotectants interact with plant cells and protect them from oxidative stress due to their high surface area contact, cell penetration ability, and high thermal conductivity [57]. They may also stabilize cell membranes, proteins and nucleic acids during freezing and thawing processes [58]. For example, graphene oxide and carbon nanotubes have high thermal conductivity [59, 60]. This would allow biological systems to achieve uniform cooling and heating rates during the cryopreservation approach. Nanoparticles, such as those coated with cerium dioxide or gold, can inhibit ice nucleation, which is essential for maintaining cell integrity during freezing [61–63] . Overall, it has been proposed to combine conventional cryopreservation methods with new nanomaterials to increase and maintain the viability of cells, tissues and even plant organs. A successful example was described in the Asian bleeding heart (*Lamprocapnos spectabilis*) plants by Kulus and Tymoszuk [64]. The authors demonstrated that the addition of gold nanoparticles during the pre-cooling steps (encapsulation) at cryopreservation protocol, improves the recovery potential of derived-explants by 20 %, without significantly altering the DNA integrity.

5 Conservation of Nucleic Acids by Nanobiotechnology

DNA carries the genetic information necessary for the development and functioning of each species, and the set of DNA of an organism is called the genome. Variation in genome sequence and expression is responsible for the diversity of phenotypes observed among varieties of the plant species. DNA analysis is essential for genotypic characterization of plant species, whether wild, domesticated, or improved [65]. Understanding the genome allows for more rapid assessment of the metabolic and adaptive potential of plant genetic resources, and enables the identification of relevant genes for crop genetic improvement [66, 67]. In plants, we find genetic material associated with the nucleus, but also with chloroplasts and mitochondria, each with its own characteristics and functions. The conservation of DNA from crop plants and their wild relatives is essential for genomic characterization and bioprospecting in the short, medium and long term, without the need for the whole organism [65]. New technologies make it possible to clone and express genetic material from a plant species in another, and even to modify the nucleotide sequence to produce gain or loss of function of a specific region of the genome [68]. The most widely used technologies are CRISPR (clustered regularly interspaced short palindromic repeats), TALE (transcription activator-like effector nucleases), and zinc finger nucleases [69, 70].

Once isolated, DNA is commonly stored at −80°C or under refrigeration for its preservation. While these methods are practical for small-scale storage of nucleic acids in laboratories, large-scale storage in biobanks requires ultra-cold freezers, which are usually very expensive. In addition to the initial costs of purchasing and installing ultrafreezers, there are ongoing expenses associated with maintenance, repair, and electricity. Consequently, there is a growing interest in exploring the use of DNA storage at room temperature. New DNA preservation methods, particularly those involving nanomaterials, have been developed that provide protection against environmental factors (e.g., pH, temperature, and salinity) and improve storage stability at room temperature [71–73]. These methods also reduce damage to the physical integrity of DNA during continuous freeze–thaw cycles.

A clear example of the use of nanobiotechnology for nucleic acid preservation is magnetite nanoparticles, which are composed of iron oxide (Fe_3O_4). Iron oxide nanoparticles have superparamagnetic properties, meaning that they exhibit magnetism only in the presence of an external magnetic field [74, 75]. DNA molecules are negatively charged due to the phosphate groups in the outer structure of their double helix. Depending on surface modifications, iron oxide nanoparticles can be positively charged or coated with positively charged molecules. The electrostatic attraction between the negatively charged DNA and the positively charged nanoparticles makes it easier for them to bind together. Van der Waals forces are weak and non-specific interactions that enhance the overall binding of DNA to magnetite. Therefore, nanomaterials have potential as carriers for the delivery of genetic material into cells, in DNA purification methods and in nucleic acid preservation [71, 76].

6 Conclusions and Future Directions

The use of nanomaterials for the conservation of plant genetic resources offers remarkable advantages, particularly to improve the efficiency of storage and preservation protocols. Some nanomaterials, such as those made of silver and copper, exhibit strong antimicrobial properties that can significantly reduce microbial growth in seeds and in vitro plants. Others, such as those made of gold, silica, and zinc oxide, induce plant growth and abiotic stress tolerance. On the other hand, cerium oxide and selenium nanoparticles are potent antioxidants. Nanoparticles have also been shown to protect nucleic acids from fluctuations in temperature and pH, helping to prevent degradation during prolonged storage at room temperature. Similarly, in cryopreservation, nanomaterials with high thermal conductivity allow for much more controlled freezing and thawing, reducing cell damage and increasing the survival rate of preserved plant tissues.

Despite these benefits, there are concerns and uncertainties associated with using nanotechnology to conserve plant genetic resources. Nanomaterials may have cytotoxic and genotoxic effects at high concentrations. The long-term ecological and health effects of nanomaterials remain uncertain, raising concerns about their safe disposal and environmental impact. It is essential, therefore, to carefully evaluate and optimize the concentration, shape, size, and chemical composition of nanomaterials to maximize their benefits.

References

1. Kindie B, Abdala T (2024) Review of plant genetic resources, utilization and conservation of genetic resources. Blue Eyes Intelligence Engineering and Sciences Publication (BEIESP)
2. Salgotra RK, Chauhan BS (2023) Genetic diversity, conservation, and utilization of plant genetic resources. Genes 14(1):174
3. Humphreys MO (2003) Utilization of plant genetic resources in breeding for sustainability. Plant Genet Resour 1(1):11–18
4. Iqbal A, Khan TF, Iqbal Y (2024) Nanobiotechnology. In: Handbook of nanomaterials, vol 2). Elsevier, pp 685–713
5. Balaji S (2019) Nanobiotechnology. MJP Publisher
6. Hulla JE, Sahu SC, Hayes AW (2015) Nanotechnology: history and future. Hum Exp Toxicol 34(12):1318–1321
7. Arundel AV, van Beuzekom B, Gillespie I (2007) Defining biotechnology–carefully. Trends Biotechnol 25(8):331–332
8. Baig N, Kammakakam I, Falath W (2021) Nanomaterials: a review of synthesis methods, properties, recent progress, and challenges. Mater Adv 2(6):1821–1871
9. Kolahalam LA, Viswanath IK, Diwakar BS, Govindh B, Reddy V, Murthy YLN (2019) Review on nanomaterials: synthesis and applications. Mater Today 18:2182–2190
10. Krishnamoorthi A (2024) Conservation of plantgenetic resources: in situ and ex situ

conservation. A monthly peer reviewed magazine for agriculture and allied sciences, vol 10
11. Waterworth WM, Bray CM, West CE (2019) Seeds and the art of genome maintenance. Front Plant Sci 10:706
12. Engelmann F (2020) Genetic resources: seeds conservation. In: Landscape andland capacity. CRC Press, pp 123-127
13. do Espirito Santo Pereira A, Caixeta Oliveira H, Fernandes Fraceto L, Santaella C (2021) Nanotechnology potential in seed priming for sustainable agriculture. Nanomaterials 11(2):267
14. Shelar A, Singh AV, Maharjan RS, Laux P, Luch A, Gemmati D, Patil R (2021) Sustainable agriculture through multidisciplinary seed nanopriming: prospects of opportunities and challenges. Cells 10(9):2428
15. Setty J, Samant SB, Yadav MK, Manjubala M, Pandurangam V (2023) Beneficial effects of bio-fabricated selenium nanoparticles as seed nanopriming agent on seed germination in rice (*Oryza sativa* L.). Sci Rep 13(1):22349
16. Shah T, Latif S, Saeed F, Ali I, Ullah S, Alsahli AA, Ahmad P (2021) Seed priming with titanium dioxide nanoparticles enhances seed vigor, leaf water status, and antioxidant enzyme activities in maize (*Zea mays* L.) under salinity stress. J King Saud Univ-Sci 33(1):101207
17. Rai-Kalal P, Tomar RS, Jajoo A (2021) Seed nanopriming by silicon oxide improves drought stress alleviation potential in wheat plants. Funct Plant Biol 48(9):905–915
18. Mahakham W, Sarmah AK, Maensiri S, Theerakulpisut P (2017) Nanopriming technology for enhancing germination and starch metabolism of aged rice seeds using phytosynthesized silver nanoparticles. Sci Rep 7(1):8263
19. Bayat N, Ghanbari AA, Bayramzade V (2020) Nanopriming a method for improving crop plants performance: a case study of red beans. J Plant Nutr 44(1):142–151
20. Younis ME, Abdel-Aziz HMM, Heikal YM (2019) Nanopriming technology enhances vigor and mitotic index of aged Vicia faba seeds using chemically synthesized silver nanoparticles. S Afr J Bot 125:393–401
21. Farajollahi Z, Eisvand HR, Nazarian-Firouzabadi F, Nasrollahi AH (2023) Nano-Fe nutrition improves soybean physiological characteristics, yield, root features and water productivity in different planting dates under drought stress conditions. Ind Crops Prod 198:116698
22. Tamindžić G, Azizbekian S, Miljaković D, Turan J, Nikolić Z, Ignjatov M, Vasiljević S (2023) Comprehensive Metal-Based Nanopriming for Improving Seed Germination and Initial Growth of Field Pea (*Pisum sativum* L.). Agronomy 13(12):2932
23. Cembrowska-Lech D, Rybak K (2023) Nanopriming of barley seeds—a shotgun approach to improve germination under salt stress conditions by regulating of reactive oxygen species. Plants 12(2):405
24. Abo-Shanab WA, Elsilk SE, Afifi SS, El-Shenody RA (2025) Ecofriendly biosynthesis of zinc oxide nanoparticles (ZnO NPs) from *Arthrospira platensis* and their assessment for antimicrobial, antibiofilm, anticancer potency and alleviation of copper stress in *Vicia faba* (L.) plant. J Soil Sci Plant Nutr 25:2742–2762. https://doi.org/10.1007/s42729-025-02295-1
25. Pintos B, de Diego H, Gomez-Garay A (2024) Nanopriming-induced enhancement of cucumber seedling development: exploring biochemical and physiological effects of silver nanoparticles. Agronomy 14(8):1866
26. Perea-Vélez YS, Carrillo-González R, González-Chávez MDCA, Vangronsveld J, Tapia Maruri D, López-Luna J (2024) Nanopriming of *Phaseolus vulgaris* OTI cultivar with cobalt ferrite nanoparticles enhances the mineral composition of progeny seeds. J Nanopart Res 26(9):205
27. Afzal S, Sharma D, Singh NK (2021) Eco-friendly synthesis of phytochemical-capped iron oxide nanoparticles as nano-priming agent for boosting seed germination in rice (*Oryza sativa* L.). Environ Sci Pollut Res 28:40275–40287
28. Joshi SM, De Britto S, Jogaiah S (2021) Myco-engineered selenium nanoparticles elicit resistance against tomato late blight disease by regulating differential expression of cellular, biochemical and defense responsive genes. J Biotechnol 325:196–206
29. Salam A, Khan AR, Liu L, Yang S, Azhar W, Ulhassan Z, Gan Y (2022) Seed priming with zinc oxide nanoparticles downplayed ultrastructural damage and improved photosynthetic apparatus in maize under cobalt stress. J Hazard Mater 423:127021
30. El-Badri AM, Batool M, Wang C, Hashem AM, Tabl KM, Nishawy E, Wang B (2021) Selenium and zinc oxide nanoparticles modulate the molecular and morpho-physiological processes during seed germination of Brassica napus under salt stress. Ecotoxicol Environ Saf 225:112695
31. Kasote DM, Lee JH, Jayaprakasha GK, Patil BS (2021) Manganese oxide nanoparticles as safer seed priming agent to improve chlorophyll and antioxidant profiles in watermelon seedlings. Nanomaterials 11(4):1016
32. Hussain A, Rizwan M, Ali Q, Ali S (2019) Seed priming with silicon nanoparticles improved the biomass and yield while reduced the oxidative stress and cadmium concentration in wheat grains. Environ Sci Pollut Res 26:7579–7588

33. Zhou X, Jia X, Zhang Z, Chen K, Wang L, Chen H, Zhao L (2022) AgNPs seed priming accelerated germination speed and altered nutritional profile of Chinese cabbage. Sci Total Environ 808:151896
34. Khan MN, Fu C, Li J, Tao Y, Li Y, Hu J, Li Z (2023) Seed nanopriming: how do nanomaterials improve seed tolerance to salinity and drought? Chemosphere 310:136911
35. Imtiaz H, Shiraz M, Mir AR, Siddiqui H, Hayat S (2023) Nano-priming techniques for plant physio-biochemistry and stress tolerance. J Plant Growth Regul 42(11):6870–6890
36. Shelar A, Nile SH, Singh AV, Rothenstein D, Bill J, Xiao J, Patil R (2023) Recent advances in nano-enabled seed treatment strategies for sustainable agriculture: challenges, risk assessment, and future perspectives. Nano Micro Lett 15(1):54
37. Kumar N, Reddy MP (2011) In vitro plant propagation: a review. J For Environ Sci 27(2):61–72
38. Bunn E, Tan B (2002) Microbial contaminants in plant tissue culture propagation. In Microorganisms in plant conservation and biodiversity. Springer Netherlands, Dordrecht, 307–335
39. Croguennec T (2016) Enzymatic browning. Handbook of food science and technology 1: food alteration and food quality. John Wiley & Sons, Inc., New York, pp. 159–181. https://doi.org/10.1002/9781119268659.ch6
40. Ahmad I, Hussain T, Ashraf I, Nafees M, Maryam RM, Iqbal M (2013) Lethal effects of secondary metabolites on plant tissue culture. Am Eurasian J Agric Environ Sci 13(4):539–547
41. Onuoha IC, Eze CJ, Unamba CI (2011) *In vitro* prevention of browning in planta in culture. Online J Biol Sci 11(1):13–17
42. Guha PS, Gupta SD, Saha N (2024) Nanogardening: harnessing metal nanoparticles for enhanced in vitro plant regeneration. Bionanoscience 14(3):3555–3571
43. Pastelín-Solano MC, Ramírez-Mosqueda MA, Bogdanchikova N, Castro-González CG, Bello-Bello JJ (2020) Las nanopartículas de plata afectan la micropropagación de vainilla (Vanilla planifolia Jacks. ex Andrews). Agrociencia 54(1):1–13
44. Durán N, Nakazato G, Seabra AB (2016) Antimicrobial activity of biogenic silver nanoparticles, and silver chloride nanoparticles: an overview and comments. Appl Microbiol Biotechnol 100:6555–6570
45. Wang L, Ning C, Pan T, Cai K (2022) Role of silica nanoparticles in abiotic and biotic stress tolerance in plants: a review. Int J Mol Sci 23(4):1947
46. Biswas A, Pal S (2024) Plant-nano interactions: a new insight of nano-phytotoxicity. Plant Physiol Biochem. https://doi.org/10.1016/j.plaphy.2024.108646
47. Prasad A, Sidhic J, Sarbadhikary P, Narayanankutty A, George S, George BP, Abrahamse H (2024) Role of metal nanoparticles in organogenesis, secondary metabolite production and genetic transformation of plants under in vitro condition: a comprehensive review. Plant Cell Tissue Organ Cult 158(2):33
48. Bedlovičová Z, Strapáč I, Baláž M, Salayová A (2020) A brief overview on antioxidant activity determination of silver nanoparticles. Molecules 25(14):3191
49. Valgimigli L, Baschieri A, Amorati R (2018) Antioxidant activity of nanomaterials. J Mater Chem B 6(14):2036–2051
50. Dutta D, Mukherjee R, Ghosh S, Patra M, Mukherjee M, Basu T (2022) Cerium oxide nanoparticles as antioxidant or pro-oxidant agents. ACS Appl Nano Mater 5(1):1690–1701
51. Jakubczyk K, Dec K, Kałduńska J, Kawczuga D, Kochman J, Janda K (2020) Reactive oxygen species-sources, functions, oxidative damage. Pol Merkuriusz Lek 48(284):124–127
52. Kondaparthi P, Flora SJS, Naqvi S (2019) Selenium nanoparticles: an insight on its Pro-oxidant and antioxidant properties. Front Nanosci Nanotechnol 6(1):5
53. Brigelius-Flohé R, Maiorino M (2013) Glutathione peroxidases. Biochim Biophys Acta 1830(5):3289–3303
54. Jaiswal AN, Vagga A (2022) Cryopreservation: a review article. Cureus 14(11):e31564. https://doi.org/10.7759/cureus.31564
55. Robertson SM, Lawrence AL, Neill WH, Arnold CR, McCarty G (1988) Toxicity of the cryoprotectants glycerol, dimethyl sulfoxide, ethylene glycol, methanol, sucrose, and sea salt solutions to the embryos of red drum. Prog Fish-Cult 50(3):148–154
56. Zevallos B, Cejas I, Rodríguez RC, Yabor L, Aragón C, González J, Lorenzo JC (2016) Biochemical characterization of Ecuadorian wild *Solanum lycopersicum* Mill. plants produced from non-cryopreserved and cryopreserved seeds. CryoLetters 37(4):413–421
57. Asadi Z, Safari-Faramani R, Aghaz F (2023) Effects of adding antioxidant nanoparticles on sperm parameters of non-human species after the freezing and thawing process: a systematic review and meta-analysis. Anim Reprod Sci https://doi.org/10.1016/j.anireprosci.2023.107323. Get rights and content
58. Susilowati S, Mustofa I, Suparyogi TW, Akintunde AO, Purwanto DA, Wurlina W, Mulyati S (2024) Adding chitosan nanoparticles of green tea extract in diluent and thawing temperatures ameliorate the post-thawed quality of Boer buck semen. Asian Pac J Reprod 13(1):42–48

59. Im H, Kim J (2012) Thermal conductivity of a graphene oxide–carbon nanotube hybrid/ epoxy composite. Carbon 50(15):5429–5440
60. Hu M, Wang D, Kokogiannakis G, Darkwa J, Li Y, Wang L, Su W (2024) Enhancement of thermal and mechanical properties of microencapsulated phase change materials with graphene oxide. Chem Eng J 479:147855
61. Impresari E, Peqini K, Hogervorst TP, Faustini A, Bodega F, Porta C, Pellegrino S (2024) Gold nanoparticles decorated with HPLC6-derived peptides as a platform for ice recrystallization inhibition. Biomacromolecules 26(1):715–725
62. Bobrova O, Falko O, Polyakova A, Klochkov V, Faltus M, Chizhevskiy V (2025) Nanocrystalline cerium dioxide reduces recrystallization in cryopreservation solutions. Cryobiology 118:105167
63. Zhang Y, Li B, Liu J, Han D, Rohani S, Gao Z, Gong J (2024) Inhibition of crystal nucleation and growth: a review. Cryst Growth Des 24(6):2645–2665
64. Kulus D, Tymoszuk A (2021) Gold nanoparticles affect the cryopreservation efficiency of *in vitro*-derived shoot tips of bleeding heart. Plant Cell Tissue Organ Cult 146(2):297–311
65. Govindaraj M, Vetriventhan M, Srinivasan M (2015) Importance of genetic diversity assessment in crop plants and its recent advances: an overview of its analytical perspectives. Genet Res Int 2015(1):431487
66. Petereit J, Bayer PE, Thomas WJ, Tay Fernandez CG, Amas J, Zhang Y, Edwards D (2022) Pangenomics and crop genome adaptation in a changing climate. Plants 11(15):1949
67. Zou J, Huang Y, Gao C, Wang K (2023) Unlocking crop diversity: enhancing variations through genome editing. Sci Bull 69:281–284
68. Chinnusamy V, Schepler-Luu V, Mangrauthia SK, Ramesh SV (2023) Genome editing in plants: a tool for precision breeding and functional genomics. J Plant Biochem Biotechnol 32(4):657–660
69. Cardi T, Murovec J, Bakhsh A, Boniecka J, Bruegmann T, Bull SE, Van Laere K (2023) CRISPR/Cas-mediated plant genome editing: outstanding challenges a decade after implementation. Trends Plant Sci 28(10):1144–1165
70. Bhushan K, Pratap D, Sharma PK (2016) Transcription activator-like effector nucleases (TALENs): an efficient tool for plant genome editing. Eng Life Sci 16(4):330–337
71. Organick L, Nguyen BH, McAmis R, Chen WD, Kohll AX, Ang SD, Strauss K (2021) An empirical comparison of preservation methods for synthetic DNA data storage. Small Methods 5(5):2001094
72. Mathur N, Singh N (2023) Melting of dsDNA attached with AuNPs. Eur Phys J E 46(7):58
73. Seferos DS, Prigodich AE, Giljohann DA, Patel PC, Mirkin CA (2009) Polyvalent DNA nanoparticle conjugates stabilize nucleic acids. Nano Lett 9(1):308–311
74. Sangaiya P, Jayaprakash R (2018) A review on iron oxide nanoparticles and their biomedical applications. J Supercond Novel Magn 31(11):3397–3413
75. Szymczyk A, Drozd M, Kamińska A, Matczuk M, Trzaskowski M, Mazurkiewicz-Pawlicka M, Malinowska E (2022) Comparative evaluation of different surface coatings of Fe3O4-based magnetic nano sorbent for applications in the nucleic acids extraction. Int J Mol Sci 23(16):8860
76. Samanta A, Medintz IL (2016) Nanoparticles and DNA—a powerful and growing functional combination in bionanotechnology. Nanoscale 8(17):9037–9095

Chapter 32

Omics Sciences: Driving the Conservation and Characterization of Plant Genetic Resources

María Isabel Iñiguez-Luna, Lorena Jacqueline Gómez-Godínez, Jorge David Cadena-Zamudio, Daniel Alejandro Cadena-Zamudio, José Luis Aguirre-Noyola and Luis A. Barrera-Guzmán

Abstract

Omics sciences have revolutionized the conservation and characterization of plant genetic resources by enabling a comprehensive understanding of genetic diversity, molecular mechanisms, and adaptive traits. Advances in genomics, transcriptomics, proteomics, metabolomics, and metagenomics have facilitated the identification of genes and metabolic pathways associated with stress tolerance, nutritional value, and agronomic performance. These technologies have enhanced the efficiency of germplasm banks by improving genetic resource characterization, optimizing conservation strategies, and accelerating breeding programs for climate-resilient crops. Additionally, omics approaches contribute to biodiversity conservation by revealing evolutionary relationships, ecosystem dynamics, and the functional roles of microbial communities in plant health. The integration of multi-omics data with bioinformatics and artificial intelligence further enhances predictive capabilities, enabling targeted conservation and breeding efforts. This review highlights the pivotal role of omics sciences in securing plant genetic resources for sustainable agriculture and global food security.

Keywords Omics sciences, Plant genetic resources, Conservation, Genomics, Transcriptomics, Metabolomics, Biodiversity

1 Introduction

The significance of omics sciences in agricultural and botanical research lies in their ability to provide comprehensive and detailed analyses of the molecular and genetic foundations of plants. The term "omics" originates from the Latin suffix oma, meaning "mass" or "abundance." Thus, omics sciences encompass a vast number of endpoint measurements [1]. In the biological context, this suffix is used to refer to the study of large sets of biological molecules [2]. The advancement of new technologies has enabled the simultaneous monitoring of hundreds or even thousands of macromolecules and

Marco A. Ramírez-Mosqueda et al. (eds.), *Conservation of Plant Genetic Resources*, Methods in Molecular Biology, vol. 3011, https://doi.org/10.1007/978-1-0716-5130-8_32,

small molecules, allowing for real-time functional analysis of multiple key cellular pathways. These approaches, known as "omics" technologies, are distinguished by their ability to comprehensively assess entire families of cellular molecules such as RNA, proteins, and intermediate metabolites in a single analysis [3]. Thanks to these innovations, it is now possible to obtain comprehensive evaluations of biochemical pathway activity while also identifying structural genetic differences between individuals and species. Today, fundamental biological processes can be studied with greater precision through the integrated application of various omics technologies genomics, transcriptomics, proteomics, metabolomics, and others on a single biological sample [4].

In the field of plant genetic resources (PGRs), these technologies have revolutionized research by enabling the identification of genes responsible for key traits such as drought resistance, pest tolerance, and high nutritional value. This accelerates the development of improved crop varieties and promotes sustainable agricultural practices [4]. Moreover, omics tools are essential for understanding plant biodiversity, ecosystem dynamics, and evolutionary relationships among species. This, in turn, paves the way for innovative solutions to pressing challenges such as food security and climate change. Thus, omics sciences have transformed the approach to agricultural and botanical challenges, providing powerful tools for the comprehensive analysis of complex biological systems [5]. Their ability to reveal molecular interactions and underlying mechanisms advances our understanding of fundamental processes and facilitates the practical application of this knowledge in crop improvement, biodiversity conservation, and agricultural sustainability. As these technologies continue to evolve, they are becoming indispensable pillars for addressing both current and future challenges in these fields.

1.1 Importance of Omics Sciences in Plant Genetic Resources Research

Omics sciences have revolutionized the study of plant genetic resources by providing advanced tools for a comprehensive analysis of molecular components and their relationship with phenotypic traits of interest. These technologies including genomics, transcriptomics, proteomics, and metabolomics have deepened our understanding of the genetic and biochemical mechanisms responsible for key plant characteristics, such as resistance to adverse environmental conditions and improved nutritional quality [6]. The integration of these technologies with bioinformatics tools has transformed biology into an information-rich discipline, enabling the large-scale generation and analysis of molecular and phenotypic data. This has optimized the characterization and utilization of plant genetic resources (PGR), enhancing efficiency in both their conservation and application [7]. A prominent example is their impact on germplasm banks, where genomic analysis allows for precise identification of genetic variations, tracking of desirable phenotypic traits, and their incorporation into breeding programs to address challenges such as climate change and the need for more resilient crops [8]. Multi-omics analysis has identified genes associated with tolerance to drought, salinity, and diseases, providing essential tools for developing crops better adapted to harsh conditions. This approach has been further strengthened by integrated data platforms and advanced techniques such as gene editing, which has accelerated the development of climate-smart crops, optimized resource use, and reducing dependence on external inputs. Additionally, tools such as machine learning have enhanced the prediction of phenotypic traits and variety selection, while genetic biofortification strategies have driven the development of more nutritious crops to combat malnutrition [9, 10]. Amid global population growth and the impacts of climate change, these technologies have emerged as innovative solutions to ensure food production, agricultural sustainability, and biodiversity conservation. By integrating multiple disciplines, omics sciences offer a detailed and multifaceted analysis of plant genetic resources, making a significant contribution to the development of more resilient and sustainable agricultural systems [5].

The above has profoundly transformed genetic resource research, providing advanced tools such as genomics, transcriptomics, proteomics, and metabolomics to analyze the molecular bases of key phenotypic traits. Genomics, as a cornerstone in this field, has enabled the unraveling of the evolutionary histories of domesticated plants and their wild relatives, facilitating the use of molecular markers and gene-assisted selection for the development of more resilient and productive crops [6, 11]. Through approaches such as genome-wide association studies (GWAS) and next-generation sequencing (NGS), genomics has identified critical genes for stress tolerance, yield, and disease resistance in crops such as rice, maize, and wheat [5, 12]. Meanwhile, transcriptomics and proteomics have provided essential insights into gene expression patterns and protein dynamics in response to adverse environmental conditions, revealing the molecular mechanisms underlying plant adaptation. Metabolomics complements this analysis by linking these genetic and molecular changes to key metabolites, which are fundamental for nutritional quality, pest resistance, and crop resilience [12, 13]. These tools have not only driven the development of transgenic and resistant crops but have also enhanced the conservation of genetic resources by identifying unique genomic variations in wild and cultivated species, which are essential for ensuring food security in the context of climate change and resource degradation [11]. Emerging technologies, such as gene editing through CRISPR/Cas, complement omics sciences by enabling precise modifications in plant genomes, accelerating the creation of varieties adapted to modern agricultural challenges, including drought, pest, and disease tolerance [5, 11]. Additionally, genomics has driven innovative approaches such as pangenomics, which captures the complete genetic diversity of a species and reveals key variants for adaptation to changing environments [14].

A critical dimension in the advancement of omics sciences is interdisciplinary collaboration and associated governance. The reliance on globally shared data highlights the importance of inclusive and transparent governance structures that foster trust among stakeholders such as farmers, researchers, companies, and governments. Access to open data platforms has proven crucial for overcoming institutional and geographical barriers, promoting innovative solutions in plant genetic resource research and management [8]. This allows us to conclude that omics sciences, together with cutting-edge technologies such as gene editing, represent a comprehensive and interdisciplinary approach to improving agricultural productivity, sustainability, and global food security. This set of tools provides a solid foundation for addressing present agricultural challenges and ensuring the resilience of agricultural systems in the future.

2 Genomics in the Conservation of Genetic Resources

The growing biodiversity crisis, exacerbated by climate change and human activities, poses a direct threat to ecosystem sustainability and global food security. Genetic resources, essential for developing resilient crops and ensuring agricultural stability, are facing rapid genetic erosion, jeopardizing their availability for future generations [15]. In this context, advancements in genomics have emerged as key tools to address these challenges by enabling more precise characterization and strategic utilization of the genetic diversity preserved in germplasm banks [16].

Germplasm banks play a crucial role in the ex situ conservation of the genetic diversity of both cultivated and wild species, serving as invaluable reservoirs for research and crop improvement [17]. The integration of genomic technologies, such as next-generation sequencing (NGS) and molecular markers, has revolutionized the management of these collections by facilitating the identification of genes associated with desirable traits, including drought, pest, and disease resistance, as well as enhanced yield and crop quality [18, 19]. These tools have also accelerated the selection of specific germplasm to mitigate the adverse effects of climate change and ensure global food security. In recent years, the implementation of molecular markers and marker-assisted selection (MAS) has significantly improved the precision of incorporating desirable genes into crop breeding programs [17]. Similarly, genome-wide association studies (GWAS) have enabled the identification of key loci linked to important agronomic traits. Moreover, reduced-representation genome (RGR) technologies, such as RAD-seq, have complemented these approaches by providing cost-effective genomic data for nonmodel species and traditional crops [20].

Given these advancements, the integration of cutting-edge genomic tolos MAS, GWAS, and RGR has strengthened both the conservation and utilization of genetic resources, underscoring their impact on germplasm bank management, food security, and global sustainability.

2.1 Molecular Markers

Molecular markers are fundamental tools in molecular biology and genetics, designed to identify DNA variations among organisms or individuals. These genetic variations, which can be found in both coding and noncoding regions of the genome, are detected through techniques such as polymerase chain reaction (PCR), hybridization, and DNA sequencing. Thanks to these technologies, molecular markers have revolutionized biodiversity studies, enabling detailed and precise analyses of genetic diversity within populations, the identification of genes associated with specific traits, and the advancement of crop improvement through marker-assisted selection (MAS) [21, 22]. Molecular markers are classified into two main categories based on their development methodology. Hybridization-based markers, such as Restriction Fragment Length Polymorphism (RFLP), were pioneers in this field, offering high precision and reproducibility. On the other hand, PCR-based markers including Random Amplified Polymorphic DNA (RAPD), Amplified Fragment Length Polymorphism (AFLP), Simple Sequence Repeats (SSR), and Single Nucleotide Polymorphism (SNP) are more recent and have been widely adopted due to their efficiency in analyzing large volumes of genetic data [23]. Each of these markers presents distinct advantages and limitations that determine their suitability for various research applications. For instance, RFLPs are highly precise and co-dominant, making them ideal for genetic mapping and gene identification. However, they require high-quality DNA, are costly, and involve technically demanding procedures. In contrast, RAPDs are fast and cost-effective, but their low reproducibility and dominant inheritance limit their use in detailed studies. SSRs, also known as microsatellites, stand out for their high polymorphism and reproducibility, making them valuable for kinship analysis and genetic characterization. Nevertheless, their application requires prior genomic information and can be expensive. Meanwhile, AFLPs and SNPs have proven to be versatile tools for advanced genomic studies, such as whole-genome mapping and genome-wide association studies (GWAS) [24, 25]. In genetic resource conservation, molecular markers have become indispensable. They enable the precise characterization of accessions in germplasm banks, ensuring the adequate representation of genetic diversity within collections. Additionally, they facilitate the identification of genetic duplicates and support the development of regeneration strategies to minimize genetic variability loss. In situ conservation, these markers are equally valuable, allowing for the monitoring of genetic structure in natural populations and assessing their adaptability to environmental changes, including those driven by climate change [26]. Such applications are essential for designing management strategies that preserve both genetic diversity and the functionality of agricultural and natural ecosystems. Beyond conservation, molecular markers play a crucial role in crop genetic improvement. They enable the identification and selection of genes responsible for key agronomic traits, such as disease resistance, drought tolerance, and yield enhancement. This is particularly relevant in the context of growing global food insecurity, where crop improvement must align with sustainability and resilience demands. Advances in technologies like SNPs and GWAS have further optimized these processes, allowing for the identification of loci associated with specific traits and accelerating the development of improved varieties [27]. Each marker possesses distinct characteristics that determine its suitability for specific applications (Table 1) [21]. Collectively, these markers form a powerful set of tools that, when integrated with advanced genomic approaches, enhance our understanding and management of genetic resources. They have transformed the way biodiversity is studied and preserved, providing essential solutions to global challenges such as climate change and food security.

2.2 Genomic Innovation: GBS, GWAS, and QTL as Key Tools for the Conservation of Genetic Resources and Food Security

Advancements in genomic technologies have revolutionized the study and management of genetic resources, enabling a deeper understanding of genetic diversity while enhancing their conservation and utilization. Among the most prominent tools in this field are Genotyping-by-Sequencing (GBS), genome-wide association studies (GWAS), and quantitative trait loci (QTL) mapping. These methodologies have been instrumental in addressing critical challenges such as genetic erosion, climate change, and global food security, providing a comprehensive framework for identifying and leveraging key genes in both cultivated and wild species [2, 28].

GBS is a next-generation sequencing-based technique that enables large-scale genotyping by analyzing specific regions of the genome. This technology has proven to be highly efficient and cost-effective, making it ideal for characterizing species with complex genomes or those lacking a reference genome. In the context of genetic resource conservation, GBS is used to detect genomic variations, generate high-density genetic maps, and identify alleles associated with traits such as resistance to abiotic and biotic stress. This allows for a detailed characterization of germplasm bank collections, optimizing their management and selection for breeding programs focused on sustainability [29, 30]. Meanwhile, GWAS are powerful tools for identifying genetic variants associated with complex traits. By utilizing diverse populations and dense genomic markers, GWAS pinpoint specific loci linked to key characteristics such as drought tolerance, pest resistance, and productivity under adverse conditions. This approach has facilitated the integration of highly adaptive genetic materials into breeding programs, ensuring the long-term sustainability of genetic resources stored in germplasm

Table 1
Comparative analysis of molecular markers: characteristics, applications, and considerations

Marker	Characteristics	Advantages	Disadvantages	Applications	When to use	When not to use	References
RFLP	Hybridization-based; high precision	High reproducibility, co-dominant inheritance	Requires large amounts of high-quality DNA; expensive and labor-intensive	Genetic mapping, identification of specific genes	Complex genetic diversity studies	When available DNA is limited or of low quality	[21]
RAPD	PCR-based with random primers	Simple, fast, and cost-effective	Low reproducibility, dominant inheritance	Rapid genetic diversity characterization	Preliminary diversity studies	For detailed genetic analysis or fine mapping	[23]
SSR	Microsatellite-based with short tandem repeats	Highly polymorphic, co-dominant, reproducible	Expensive; requires prior genome information	Diversity studies, parentage analysis, genetic improvement	Intraspecific diversity studies	In genomes without identified microsatellites	[22]
AFLP	PCR-based with DNA restriction and amplification	High sensitivity, genome-wide polymorphism detection	Technically complex and costly	Genome-wide genetic characterization, kinship studies	Whole-genome evaluations	When simple and rapid techniques are needed	[21–23]
SNP	Single-nucleotide polymorphism; genome-wide variations	High precision, abundant across genomes	Requires advanced equipment and high initial costs	GWAS, marker-assisted selection (MAS)	Advanced breeding programs	In preliminary studies with limited resources	[21, 23]

banks [2, 31]. Recent studies have demonstrated its effectiveness in crops such as wheat and maize, where critical genetic variants have been identified to address climate change challenges [32]. QTL analysis complements these technologies by mapping genomic regions that influence key phenotypic traits. This approach has significantly evolved with the incorporation of tools like GBS and GWAS, enhancing precision in identifying loci associated with important agronomic characteristics. QTLs enable the accumulation of desirable genes through marker-assisted selection (MAS), facilitating the combination of multiple beneficial traits in a single variety. This process has been instrumental in breeding resilient crops, particularly through the introgression of genes from wild relatives [32, 33]. The integration of GBS, GWAS, and QTL in genetic resource management has redefined the role of germplasm banks in global food security. These technologies not only allow for the precise characterization of genetic diversity but also help prioritize the conservation of accessions containing essential genes for adapting to extreme climate conditions, emerging pests, and increasing food demands. In the face of climate change, these tools provide innovative solutions by identifying and preserving highly adaptive genetic variants, which are essential for developing more resilient and sustainable crops. Additionally, QTL analysis facilitates the incorporation of tolerance and resistance genes into commercial varieties, ensuring that breeding programs can effectively respond to future agricultural challenges. This approach not only safeguards the conservation of genetic biodiversity within germplasm banks but also directly contributes to the stability of agricultural systems, which are crucial for ensuring global food security in a rapidly changing world [28, 30]. In conclusion, these technologies have revolutionized genetic resource conservation, providing advanced tools to explore, characterize, and efficiently utilize genetic diversity. By addressing global challenges related to climate change and food insecurity, they ensure the preservation of biodiversity for future generations, solidifying their role as fundamental pillars of modern biology and agriculture [2, 29].

3 Transcriptomics in the Conservation of Genetic Resources

Transcriptomics has emerged as a powerful tool in the era of omics technologies, enabling a detailed understanding of the molecular mechanisms that regulate plant and other organism responses to various environmental factors [34]. This approach, focused on analyzing transcribed RNA, provides insight into the dynamic behavior of genes under different conditions, which is crucial for addressing global challenges such as climate change, biodiversity loss, and food security [35, 36]. Unlike traditional genomic analyses, transcriptomics specifically examines genes that are active at given moments, offering a more functional perspective on biological processes [37].

The application of transcriptomics in genetic resource conservation has revolutionized the evaluation and management of ecologically and economically significant species [36, 38]. For instance, this technology has enabled the identification of key genes and metabolic pathways associated with resistance to abiotic stresses such as drought and high temperatures crucial traits for developing crops capable of withstanding extreme climatic conditions [39]. By unveiling the transcriptional responses of plants to such stressors, transcriptomics not only facilitates the identification of stress-tolerant genes but also aids in prioritizing accessions in germplasm banks for conservation and breeding programs [10, 12]. One of the most remarkable advantages of transcriptomics is its ability to study nonmodel species through de novo transcriptome assembly. This approach is particularly valuable for organisms whose genetic and adaptive diversity remains largely unexplored due to the absence of reference genomes [40]. Transcriptomic analyses enable the identification of functional genes and key metabolic pathways linked to essential traits, such as resistance to biotic and abiotic stress. Furthermore, transcriptomic data support the development of molecular tools such as microsatellite markers (SSRs) and single-nucleotide polymorphisms (SNPs), which are instrumental in assessing genetic diversity and population dynamics in conservation contexts [40]. These markers not only facilitate detailed monitoring of natural populations but also play a crucial role in identifying genetically diverse accessions in germplasm banks, ensuring the preservation of genetic resources with the highest adaptive potential for changing environmental conditions [40, 41]. The integration of transcriptomics with other omics technologies, such as genomics, metabolomics, and proteomics has significantly expanded its impact on agricultural research and genetic resource conservation [41]. These multidisciplinary approaches enable a deeper analysis of the factors regulating phenotypic plasticity, stress responses, and the molecular mechanisms underlying complex biological processes [42]. For example, combining transcriptomic and metabolomic data has revealed critical interactions between metabolic pathways and gene regulation in crops subjected to drought stress, leading to the identification of key candidate genes for crop improvement [42, 43]. Additionally, these tools facilitate the modeling of regulatory networks that explain how organisms coordinate their responses to multiple stressors simultaneously, which is particularly valuable for designing crop varieties that integrate both stress resistance and high yields. The ability to perform integrated analyses accelerates the identification of specific genes and metabolic pathways that can be leveraged to develop crops more resilient to extreme climate conditions [10, 12, 35]. Beyond its applications in research and breeding, transcriptomics plays a crucial role in shaping global conservation policies and strategies. Transcriptomic data not only provide a detailed view of the molecular mechanisms enabling species to adapt to their environment but also offer essential insights for managing at-risk populations [44]. By identifying physiological thresholds and tipping points in adaptive responses, this technology helps predict species' recovery potential in the face of environmental disturbances. These insights have direct implications for critical conservation decisions, such as habitat restoration prioritization, species or population selection for translocation programs, and the definition of sustainable management strategies [45]. For example, transcriptomic studies have identified genes linked to physiological resilience in both fish and plants under environmental stress, enabling the design of more effective interventions tailored to specific ecosystem needs [36]. In the context of climate change, transcriptomics is indispensable for ensuring that genetic resources conserved in germplasm banks not only represent genetic diversity but also include variants with high adaptive potential.

In this way, transcriptomics has not only revolutionized our understanding of molecular processes in plants and other organisms but has also expanded the possibilities for conserving and sustainably utilizing genetic resources. When integrated with multi-omics approaches, this technology provides a powerful framework for identifying and characterizing key genes associated with resilience and adaptability. This combination of capabilities positions transcriptomics as an essential tool for addressing global challenges related to biodiversity, climate change, and food security, reinforcing the foundation of sustainable strategies that benefit both ecosystems and the human communities that depend on them [6, 12].

4 Proteomics

Proteomics, a key branch of omics sciences, enables a comprehensive study of the proteins expressed in an organism under specific conditions, offering a functional perspective that complements genomics and transcriptomics [46]. In the context of genetic resource conservation, its application is essential, as it reveals the molecular mechanisms underlying the physiological and adaptive responses of species to environmental stressors. Through advanced techniques such as mass spectrometry and protein separation platforms, it is possible to identify and quantify proteins involved in drought tolerance, disease resistance, and other critical factors for ecological resilience [47].

Proteomics plays a particularly vital role in the conservation of crop wild relatives, which serve as invaluable reservoirs of genetic diversity for food security and plant breeding. This genetic diversity is fundamental for developing crops more resilient to adverse environmental conditions such as drought, salinity, and disease. By analyzing protein expression profiles in response to different stressors, researchers can identify specific classes of proteins that are altered under these conditions, highlighting genetic traits with potential for conservation and genetic improvement [48, 49]. A concrete example of this application is the study of proteomic responses in plant roots subjected to abiotic stress, which has uncovered key molecular mechanisms regulating the adaptation of wild species to their environments. This knowledge is crucial for designing conservation strategies that prioritize genetic diversity and crop adaptability in the face of climate change [50]. Beyond conservation, proteomics also has direct applications in food security, contributing to improved crop yield and quality. Detailed proteomic analyses can identify proteins associated with beneficial traits such as enhanced nutritional value, disease resistance, and improved organoleptic properties. An example is the restoration of metabolites and flavor compounds in domesticated crops like tomatoes through gene introgression from their wild relatives, demonstrating proteomics' potential in optimizing crops for human consumption [43]. In terms of traceability and authenticity of agricultural products, proteomics enables the identification of chemical markers that differentiate closely related species, which is particularly relevant for plant-based products used in health and nutrition. Advanced techniques such as UPLC-IMS-QTOFMS (Ultra-Performance Liquid Chromatography Ion Mobility Spectrometry Quadrupole Time-of-Flight Mass Spectrometry) have been employed to identify specific biomarkers distinguishing between species of the Panax genus, which are often misidentified in the trade of natural products [43]. Beyond ensuring product authenticity, this analytical capability promotes biodiversity conservation by encouraging the sustainable use of a wider variety of species in agriculture. By identifying and selecting biomarkers, researchers can develop crop varieties with desirable traits such as higher yields or improved tolerance to abiotic stress [51]. Simultaneously, proteomics facilitates the exploration of protein-protein interactions and signaling networks that regulate essential processes for plant growth and adaptation, providing fundamental insights for breeding programs focused on enhancing crop resilience to climate change and ensuring agricultural production stability [52]. Finally, the integration of proteomics with other omics technologies amplifies its potential to develop innovative solutions for biodiversity conservation and management. Proteomic analyses have identified specific adaptations in plants and animals from extreme ecosystems, providing key insights for restoration and translocation strategies [53]. This ability to decipher adaptive responses at a functional level positions proteomics as an indispensable tool for addressing global challenges related to biodiversity and climate change, contributing to more effective and sustainable conservation strategies [52].

5 Metabolomics: Profiling the Full Set of Metabolites to Enhance Genetic Resource Conservation and Food Security

Metabolomics is a key field within omics sciences that enables the identification and quantification of metabolites present in cells, tissues, or entire organisms [54]. These metabolites, which generally have molecular weights below 2000 Da though in certain cases can range between 30 and 3000 Da constitute the metabolome, a collection of small molecules that reflect an organism's metabolic activity [55, 56]. The metabolome encompasses a wide array of compounds from primary metabolism, including peptides, amino acids, nucleic acids, carbohydrates, organic acids, and vitamins, as well as secondary metabolites such as flavonoids, polyphenols, alkaloids, terpenoids, hormones, and minerals, among others [57].

Metabolic diversity is a fundamental component of biodiversity, as it determines a species' ability to adapt to changing environmental conditions. In this context, metabolomics has become an essential tool for the conservation of genetic resources, enabling the identification of key compounds associated with the resilience of both wild species and domesticated varieties. Recent studies on agriculturally significant crops such as wheat, rice, tomato, and maize have revealed substantial variations in the accumulation of specialized metabolites, many of which are linked to resistance against biotic and abiotic stressors [58–60]. A concrete example of this application can be observed in wheat (*Triticum aestivum* L.) and pearl millet (*Cenchrus americanus* (L.) Morrone), where metabolomic analyses have identified compounds associated with drought tolerance and disease resistance, providing crucial insights for selecting and improving varieties adapted to extreme climates [61, 62]. Furthermore, studies on metabolic quantitative trait loci (mQTL) have facilitated the mapping of genomic regions responsible for the synthesis and regulation of secondary metabolites essential for survival in hostile environments. Integrating these data with transcriptomic information has enabled the identification of key genes involved in the biosynthesis of flavonoids, alkaloids, and other protective compounds in plants [63, 64]. Beyond its impact on biodiversity conservation, metabolomics plays a pivotal role in food security by enabling the identification and characterization of compounds that influence the quality, nutritional value, and traceability of agricultural products. One of the most significant aspects of this field is the detection of bioactive metabolites, which can profoundly affect the sensory and nutritional attributes of food. For instance, in strawberries and tomatoes, metabolomic analyses have identified flavonoids and other polyphenols responsible for antioxidant properties and sensory characteristics such as flavor and color, with direct implications for human health [64, 65]. In wheat and rice, variability in flavonoid glycoside accumulation has been observed, with these metabolites not only enhancing stress tolerance but also improving grain quality [58, 59]. Additionally, metabolomics has become a key tool in authenticity and traceability studies of agricultural products, enabling the detection of specific chemical markers that distinguish cultivated varieties from wild species. A notable example of this application is the use of advanced techniques such as UPLC-IMS-QTOFMS (Ultra-Performance Liquid Chromatography Ion Mobility Spectrometry Quadrupole Time-of-Flight Mass Spectrometry), which has been employed to chemically characterize species within the Panax genus, ensuring the authenticity of plant-derived products used in the food and pharmaceutical industries [61]. Beyond identifying specific metabolites, the potential of metabolomics is further amplified when integrated with other omics sciences, such as genomics and transcriptomics. This combination represents a powerful strategy for developing improved crops with enhanced resilience and greater nutritional value. The implementation of metabolomic studies in breeding programs will enable more precise selection of varieties with desirable traits, optimizing the use of available genetic resources and contributing to global food security [62, 66]. Moreover, artificial intelligence and machine learning are emerging as key tools for analyzing and predicting metabolomic data. The integration of these technologies will facilitate anticipating crop responses to different environmental conditions and enhancing the efficiency of breeding processes, accelerating the development of varieties adapted to future agricultural and climatic challenges [66]. In conclusion, metabolomics has not only revolutionized research in genetic resource conservation and food security but has also opened new opportunities for crop improvement and the design of sustainable agricultural strategies. With continuous advancements in analytical technologies and their integration with other omics disciplines, metabolomics will remain at the forefront of optimizing agricultural production and building more resilient and secure food systems for the future.

6 Metagenomics and Biotechnological Applications for the Conservation of Agricultural Biodiversity

6.1 Soil and the Importance of Plant Growth-Promoting Microorganisms

Soil is a highly complex and dynamic terrestrial environment, one of the main reservoirs of microbial diversity [67]. Soil contains more than 50,000 species in 1 g; they are usually very diverse because the environmental conditions of the soil are very heterogeneous [68]. The communities of microorganisms living in a specific environment with their structures and substances generated are known as the microbiome [69]. The microbiome is responsible for different processes in the soil; for example, it influences soil acidity, it is responsible for the cycling of various nutrients such as iron, sulfur, phosphorus, and nitrogen, it can affect the availability of water in the soil, sequestration of heavy metals, and degradation of xenobiotic compounds [68, 70]. Specifically, we can find microorganisms known as plant growth promoters within the microbiome, which can act through direct and indirect mechanisms [71]. Direct mechanisms include those that promote plant growth, such as the production of hormones, such as auxins, cytokinins, and gibberellins, as well as nitrogen fixation and phosphorus solubilization. Indirect mechanisms are related to inhibiting the function of one or more phytopathogenic organisms through the synthesis of antibiotics, enzymes, and siderophore production, which is related to the inhibition of pathogen quorum. They can also help contend with different types of biotic and abiotic stress [72–74].

6.2 Study of Microorganisms and Identification of Plant Growth Promoters with Metagenomics

Despite the fundamental role played by the microbiome in the soil, little has been studied, estimated at less than 1%. Most soil taxa (microorganisms grouped by similar characteristics) have not yet been described; therefore, the databases have no information [69, 75]. Most microbial studies are focused on the study of bacteria and less on fungi, so there is still much to learn about microorganisms in different systems. Several approaches to studying the microbiome can be classified as (1) culture-dependent and (2) culture-independent [76]. Culture-dependent approaches are based on isolating and identifying microorganisms through staining, morphology, and biochemical testing. Although this approach has been used for more than 200 years and has generated much information that has served to exploit microorganisms at a biotechnological level, it has been an austere approach to understanding all the microorganisms that may be present within a community [77]. In recent decades, the development of omics technologies, including genomics, transcriptomics, metagenomics, metatranscriptomics, metabolomics, and proteomics, has played a crucial role in detecting the composition and functions of nonculturable microorganisms, thus representing a good strategy for culture-independent studies [78, 79].

Metagenomics refers to the genetic material in a given community and the aggregate functions [80]. Metagenomics has been used to identify the diversity of microbial communities in different environments; it also allows us to verify the presence or absence of specific organisms or genes in a microbiome [81, 82]. Soil and plant microbial communities are critical agricultural players in determining plant health, soil characteristics, and biogeochemical features. The application of metagenomics to the study of these communities as well as to agriculture has not only allowed the profiling of plant growth-promoting rhizobacteria (PGPR) but has also described their functional characteristics [67]. This approach has proven appropriate for understanding the interactions in the soil environment, rhizosphere, and plant tissues [83]. For example, diazotrophs with genes related to nitrogen fixation (nifH) have been successfully identified [84]. Using metagenomics, described the differences in microbial functional genes in two corn crops under different management schemes (intensive and conservation), finding significant differences in genes such as *nifA*, *nifH*, *fixJ*, *pstA*, *kefA*, *kefB*, *ktrA*, and *ktrB* involved in the nutrient cycle, such as nitrogen, phosphorus, and potassium, which are involved in soil fertility [85]. Metagenomic sequencing allowed a comparison of genes in the microbiome of healthy and diseased tomato plants and associated soil. From this, it was possible to identify that more genes related to growth promotion and disease resistance are expressed in healthy plants compared to diseased plants (27 and 9 genes, respectively). Within the identified genes, genes involved in nitrogen fixation (*nifH*, *fixJ*, *iorA*), siderophore production (*iorB*, *tonB*), and Indole Acetic Acid synthesis (*iaaH*, *ipdC*) were found. Together, these results show that microorganisms present in healthy plants improve the health status of the plants and contribute to their growth [86]. On the other hand, it has been shown that when some bacterial genera capable of producing antimicrobial metabolites are enriched, for example, *Peanibacillus*, *Bacillus*, *Pseudomonas*, *Actinomycetes*, and *Burkholderia*, plant growth, seed germination, and crop yield are improved. These improvements have been associated with the enrichment of bacterial lineages that have members with plant growth-promoting activity, such as *Actinobacteria*, *Firmicutes*, *Cyanobacteria*, and *Actinobacteria* [87–89]. By inoculating some beneficial microorganisms such as *Bacillus velezensis* and *Pseudomonas psychrophila*, it was possible to modify the relative abundance of phyla (*Actinobacteria* > *Proteobacteria* > *Firmicutes*) and the genera *Bacillus* > *Arthrobacter* > *Pseudomonas*, as well as the abundance of some genes associated with Nitrogen fixation and nitrification, which had a positive impact on the growth and production of peanut (*Arachis hypogaea* L.) [90]. Enriching some growth-promoting microorganisms can modulate microbial communities and suppress present phytopathogens such as *Fusarium graminearum* and *Rhizoctonia solani* [91, 92]. The contribution that microorganisms have when introduced as potential biofertilizers favors sustainable agriculture.

6.3 Development of Agricultural Biofertilizers and the Importance of Preserving Agricultural Biodiversity

Due to the importance that microorganisms have had in agriculture, bioformulations have been developed over time for their application and use in agriculture [71]. A bioformulation has been defined as the preparation of beneficial microorganisms with a carrier that can stabilize and protect the strain during transport and storage [93]. Typically, these bioformulations are made from microorganisms isolated in the laboratory, so when introduced into the field, they go through a transition phase where they face acclimatization, survival, and interaction. Biotic factors such as predation, competition, and antagonism, as well as abiotic factors such as nutrients present in the soil, pH, temperature, humidity, light, and salinity, play an essential challenge in the successful development of the inoculant [94–96]. Therefore, preserving nonculturable microbiota and intact microbiome is a vital aspect to consider in microbiology. This could represent an approach to study the impact of ecological and environmental disturbance; this will allow the reconstruction and reestablishment of relative abundances and organization of the community, as well as the effect generated by microorganisms when inoculated to promote plant growth. By knowing the microbiomes, we can avoid losses or imbalances in the functioning of ecosystems and favor the functioning of ecosystems and sustainable development.

References

1. Lay Jr JO, Liyanage R, Borgmann S, Wilkins CL (2006) Problems with the "omics". TrAC, Trends Anal Chem 25(11):1046–1056
2. Wambugu PW, Ndjiondjop M-N, Henry RJ (2018) Role of genomics in promoting the utilization of plant genetic resources in genebanks. Briefings Funct Genomics 17(3):198–206
3. Vailati-Riboni M, Palombo V, Loor J (2017) What are omics sciences? Periparturient diseases of dairy cows. Springer, Berlin/Heidelberg, Germany
4. Kumar A, Anju T, Kumar S, Chhapekar SS, Sreedharan S, Singh S, Choi SR, Ramchiary N, Lim YP (2021) Integrating omics and gene editing tools for rapid improvement of traditional food plants for diversified and sustainable food security. Int J Mol Sci 22(15):8093
5. Chen Y, Zhang B, Annis GB (2024) Plant biotechnology and genetics for sustainable agriculture and global food security. Front Plant Sci 13:1–3. https://doi.org/10.3389/fpls.2024.1479632
6. Van Emon JM (2016) The omics revolution in agricultural research. J Agric Food Chem 64(1):36–44
7. Cooper M, Messina CD, Podlich D, Totir LR, Baumgarten A, Hausmann NJ, Wright D, Graham G (2014) Predicting the future of plant breeding: complementing empirical evaluation with genetic prediction. Crop Pasture Sci 65(4):311–336
8. Halewood M, Chiurugwi T, Sackville Hamilton R, Kurtz B, Marden E, Welch E, Michiels F, Mozafari J, Sabran M, Patron N (2018) Plant genetic resources for food and agriculture: opportunities and challenges emerging from the science and information technology revolution. New Phytol 217(4):1407–1419
9. Balotf S, Wilson R, Nichols DS, Tegg RS, Wilson CR (2022) Multi-omics reveals mechanisms of resistance to potato root infection by Spongospora subterranea. Sci Rep 12(1):10804.
10. Mahmood U, Li X, Fan Y, Chang W, Niu Y, Li J, Qu C, Lu K (2022) Multi-omics revolution to promote plant breeding efficiency. Front Plant Sci 13:1062952
11. Chen K, Wang Y, Zhang R, Zhang H, Gao C (2019) CRISPR/Cas genome editing and precision plant breeding in agriculture. Annu Rev Plant Biol 70(1):667–697
12. Naqvi RZ, Mahmood MA, Mansoor S, Amin I, Asif M (2024) Omics-driven exploration and mining of key functional genes for the improvement of food and fiber crops. Front Plant Sci 14:1273859
13. Aslam MQ, Naqvi RZ, Zaidi SS-e-A, Asif M, Akhter KP, Scheffler BE, Scheffler JA, Liu S-S, Amin I, Mansoor S (2022) Analysis of a tetraploid cotton line Mac7 transcriptome reveals mechanisms underlying resistance against the whitefly *Bemisia tabaci*. Gene 820:146200
14. Li B, Cai H, Liu K, An B, Wang R, Yang F, Zeng C, Jiao C, Xu Y (2023) DNA methylation alterations and their association with high temperature tolerance in rice anthesis. J Plant Growth Regul 42(2):780–794
15. Salgotra RK, Chauhan BS (2023) Genetic diversity, conservation, and utilization of plant genetic resources. Genes 14(1):174
16. Singh K, Kumar S, Kumar SR, Singh M, Gupta K (2019) Plant genetic resources management and pre-breeding in genomics era. Indian J Genet Plant Breed 79(Sup-01):117–130
17. Anilkumar C, Sunitha N, Harikrishna, Devate NB, Ramesh S (2022) Advances in integrated genomic selection for rapid genetic gain in crop improvement: a review. Planta 256(5):87
18. Volk GM, Byrne PF, Coyne CJ, Flint-Garcia S, Reeves PA, Richards C (2021) Integrating genomic and phenomic approaches to support plant genetic resources conservation and use. Plants 10(11):2260
19. Theissinger K, Fernandes C, Formenti G, Bista I, Berg PR, Bleidorn C, Bombarely A, Crottini

A, Gallo GR, Godoy JA (2023) How genomics can help biodiversity conservation. Trends Genet 39(7):545–559
20. Yanez JM, Barria A, Lopez ME, Moen T, Garcia BF, Yoshida GM, Xu P (2023) Genome-wide association and genomic selection in aquaculture. Rev Aquac 15(2):645–675
21. Bidyananda N, Jamir I, Nowakowska K, Varte V, Vendrame WA, Devi RS, Nongdam P (2024) Plant genetic diversity studies: insights from DNA marker analyses. Int J Plant Biol 15(3):607–640
22. Zolkin AL, Matvienko EV, Chistyakov MS, Suchkov DK, Ryabkova GV (2022) Biodiversity and ecosystem stability based on molecular and genetic technologies in plant breeding. In: IOP conference series: earth and environmental science, vol 1. IOP Publishing, p 012102
23. Khoury CK, Brush S, Costich DE, Curry HA, De Haan S, Engels JM, Guarino L, Hoban S, Mercer KL, Miller AJ (2022) Crop genetic erosion: understanding and responding to loss of crop diversity. New Phytol 233(1):84–118
24. Kumar A, Swapnil PS, Singh R, Singh D (2020) Prospect of molecular markers in precision plant breeding. In: Jha AK, Kumar U (Eds.), Recent advances in chemical sciences and biotechnology. New Delhi: New Delhi Publishers, p 131–142.
25. Tahir N, Lateef DD, Rasul KS, Rahim D, Mustafa KM, Sleman S, Mirza A, Aziz R (2023) Assessment of genetic variation and population structure in Iraqi barley accessions using ISSR, CDDP, and SCoT markers. Czech J Genet Plant Breed 59(2023):3
26. Amiteye S (2021) Basic concepts and methodologies of DNA marker systems in plant molecular breeding. Heliyon 7(10):e08153
27. Chen W, Hou L, Zhang Z, Pang X, Li Y (2017) Genetic diversity, population structure, and linkage disequilibrium of a core collection of Ziziphus jujuba assessed with genome-wide SNPs developed by genotyping-by-sequencing and SSR markers. Front Plant Sci 8:575
28. Kumar R, Das SP, Choudhury BU, Kumar A, Prakash NR, Verma R, Chakraborti M, Devi AG, Bhattacharjee B, Das R (2024) Advances in genomic tools for plant breeding: harnessing DNA molecular markers, genomic selection, and genome editing. Biol Res 57(1):80
29. Merga W, Getu A (2023) The application of genetic marker for diversity assessment and conserving of plant genetic resource. J Exp Mol Biol 24(3):195–206
30. Song L, Wang R, Yang X, Zhang A, Liu D (2023) Molecular markers and their applications in marker-assisted selection (MAS) in bread wheat (*Triticum aestivum* L.). Agriculture 13(3):642
31. Pathirana R, Carimi F (2022) Management and utilization of plant genetic resources for a sustainable agriculture. Plants 11(15):2038
32. Hasan N, Choudhary S, Naaz N, Sharma N, Laskar RA (2021) Recent advancements in molecular marker-assisted selection and applications in plant breeding programmes. J Genet Eng Biotechnol 19(1):128
33. De Mori G, Cipriani G (2023) Marker-assisted selection in breeding for fruit trait improvement: a review. Int J Mol Sci 24(10):8984
34. Liu Y, Cai L, Sun W (2024) Transcriptome data analysis provides insights into the conservation of *Michelia lacei*, a plant species with extremely small populations distributed in Yunnan province, China. BMC Plant Biol 24(1):200.
35. Abdelrahman M, Jogaiah S, Burritt DJ, Tran LSP (2018) Legume genetic resources and transcriptome dynamics under abiotic stress conditions. Plant Cell Environ 41(9):1972–1983
36. Connon RE, Jeffries KM, Komoroske LM, Todgham AE, Fangue NA (2018) The utility of transcriptomics in fish conservation. J Exp Biol 221(2):jeb148833
37. Lowe R, Shirley N, Bleackley M, Dolan S, Shafee T (2017) Transcriptomics technologies. PLoS Comp Biol 13(5):e1005457
38. Singh D, Singh CK, Taunk J, Tomar RSS, Chaturvedi AK, Gaikwad K, Pal M (2017) Transcriptome analysis of lentil (*Lens culinaris* Medikus) in response to seedling drought stress. BMC Genomics 18:1–20
39. Bashir K, Matsui A, Rasheed S, Seki M (2019) Recent advances in the characterization of plant transcriptomes in response to drought, salinity, heat, and cold stress. F1000Research 8:6.
40. Dries R, Chen J, Del Rossi N, Khan MM, Sistig A, Yuan G-C (2021) Advances in spatial transcriptomic data analysis. Genome Res 31(10):1706–1718
41. Diwan D, Rashid MM, Vaishnav A (2022) Current understanding of plant-microbe interaction through the lenses of multi-omics approaches and their benefits in sustainable agriculture. Microbiol Res 265:127180
42. Cao X, Hu Y, Song J, Feng H, Wang J, Chen L, Wang L, Diao X, Wan Y, Liu S (2022) Transcriptome sequencing and metabolome analysis reveals the molecular mechanism of drought stress in millet. Int J Mol Sci 23(18):10792
43. Wang N, Cao S, Liu Z, Xiao H, Hu J, Xu X, Chen P, Ma Z, Ye J, Chai L (2023) Genomic conservation of crop wild relatives: a case study of citrus. PLoS Genet 19(6):e1010811
44. D'Adamo GL, Widdop JT, Giles EM (2021) The future is now? Clinical and translational aspects of "Omics" technologies. Immunol Cell Biol 99(2):168–176

45. Gruszecka-Kosowska A, Ampatzoglou A, Aguilera M (2022) Integration of omics approaches enhances the impact of scientific research in environmental applications. Int J Environ Res Public Health 19(14):8758.
46. Noor Z, Ahn SB, Baker MS, Ranganathan S, Mohamedali A (2021) Mass spectrometry–based protein identification in proteomics—a review. Brief Bioinf 22(2):1620–1638
47. Tan BC, Lim YS, Lau S-E (2017) Proteomics in commercial crops: an overview. J Proteomics 169:176–188
48. Ghosh D, Xu J (2014) Abiotic stress responses in plant roots: a proteomics perspective. Front Plant Sci 5:6
49. Jha UC, Nayyar H, Parida SK, Deshmukh R, von Wettberg EJ, Siddique KH (2022) Ensuring global food security by improving protein content in major grain legumes using breeding and 'Omics' tools. Int J Mol Sci 23(14):7710
50. Pedrouso ML, Lorenzo JM, Ruiz DF (2022) Introduction Food proteomics: technological advances, current applications and future perspectives. In: Food Proteomics. Academic Press/Elsevier, London, pp 1–12.
51. Mustafa G, Komatsu S (2021) Plant proteomic research for improvement of food crops under stresses: a review. Mol Omics 17(6):860–880
52. Mathur M, Prajapat RK, Upadhyay TK, Lal D, Khatik N, Sharma D (2021) Advances in genomics and proteomics in agriculture. In: Crop improvement. CRC Press, Boca Raton, FL, USA, pp 23–35.
53. Jan N, Rather AM-U-D, John R, Chaturvedi P, Ghatak A, Weckwerth W, Zargar SM, Mir RA, Khan MA, Mir RR (2023) Proteomics for abiotic stresses in legumes: present status and future directions. Crit Rev Biotechnol 43(2):171–190
54. Tian H, Lam SM, Shui G (2016) Metabolomics, a powerful tool for agricultural research. Int J Mol Sci 17(11):1871
55. Wishart DS, Cheng LL, Copié V, Edison AS, Eghbalnia HR, Hoch JC, Gouveia GJ, Pathmasiri W, Powers R, Schock TB (2022) NMR and metabolomics—a roadmap for the future. Metabolites 12(8):678
56. Yan S, Bhawal R, Yin Z, Thannhauser TW, Zhang S (2022) Recent advances in proteomics and metabolomics in plants. Mol Hortic 2(1):17
57. Raza A (2022) Metabolomics: a systems biology approach for enhancing heat stress tolerance in plants. Plant Cell Rep 41(3):741–763
58. Chen J, Hu X, Shi T, Yin H, Sun D, Hao Y, Xia X, Luo J, Fernie AR, He Z (2020) Metabolite-based genome-wide association study enables dissection of the flavonoid decoration pathway of wheat kernels. Plant Biotechnol J 18(8):1722–1735
59. Li K, Wang D, Gong L, Lyu Y, Guo H, Chen W, Jin C, Liu X, Fang C, Luo J (2019) Comparative analysis of metabolome of rice seeds at three developmental stages using a recombinant inbred line population. PlJ 100(5):908–922
60. Zhou S, Kremling KA, Bandillo N, Richter A, Zhang YK, Ahern KR, Artyukhin AB, Hui JX, Younkin GC, Schroeder FC (2019) Metabolome-scale genome-wide association studies reveal chemical diversity and genetic control of maize specialized metabolites. Plant Cell 31(5):937–955
61. Wei W, Li S, Wang Y, Wang B, Fan G, Zeng Q, Zhao F, Xu C, Zhang X, Tang T (2021) Metabolome-based genome-wide association study provides genetic insights into the natural variation of foxtail millet. Front Plant Sci 12:665530
62. Yadav CB, Srivastava RK, Gangashetty PI, Yadav R, Mur LA, Yadav RS (2021) Metabolite diversity and metabolic genome-wide marker association studies (MGWAS) for health benefiting nutritional traits in pearl millet grains. Cells 10(11):3076
63. Kleessen S, Giavalisco P, Pleban T, Mueller-Roeber B, Zamir D, Nikoloski Z, Ferniea AR (2025) Identification and mode of inheritance of quantitative trait loci for secondary metabolite abundance in tomato. Plant Cell 27(3):485–512.
64. Labadie M, Vallin G, Petit A, Ring L, Hoffmann T, Gaston A, Potier A, Schwab W, Rothan C, Denoyes B (2020) Metabolite quantitative trait loci for flavonoids provide new insights into the genetic architecture of strawberry (Fragaria × ananassa) fruit quality. J Agric Food Chem 68(25):6927–6939
65. Zhu G, Wang S, Huang Z, Zhang S, Liao Q, Zhang C, Lin T, Qin M, Peng M, Yang C (2018) Rewiring of the fruit metabolome in tomato breeding. Cell 172(1):249–261
66. Shi T, Zhu A, Jia J, Hu X, Chen J, Liu W, Ren X, Sun D, Fernie AR, Cui F (2020) Metabolomics analysis and metabolite-agronomic trait associations using kernels of wheat (*Triticum aestivum*) recombinant inbred lines. PlJ 103(1):279–292
67. Nwachukwu BC, Babalola OO (2022) Metagenomics: a tool for exploring key microbiome with the potentials for improving sustainable agriculture. Front Sustain Food Syst 6:886987
68. Fierer N (2017) Embracing the unknown: disentangling the complexities of the soil microbiome. Nat Rev Microbiol 15(10):579–590

69. Nadeu E, Van Dijk R, Hiller N (2023) The soil microbiome: Its contribution to soil health and one health. Institute for European Environmental Policy. Available at: https://ieep.eu/wp-content/uploads/2023/12/The-Soil-Microbiome-ESAD-IEEP-2023.pdf
70. Gómez-Godínez LJ, Martínez-Romero E, Banuelos J, Arteaga-Garibay RI (2021) Tools and challenges to exploit microbial communities in agriculture. Curr Res Microb Sci 2:100062
71. Gómez-Godínez LJ, Aguirre-Noyola JL, Martínez-Romero E, Arteaga-Garibay RI, Ireta-Moreno J, Ruvalcaba-Gómez JM (2023) A look at plant-growth-promoting bacteria. Plants 12(8):1668
72. Beneduzi A, Ambrosini A, Passaglia LM (2012) Plant growth-promoting rhizobacteria (PGPR): their potential as antagonists and biocontrol agents. Genet Mol Biol 35:1044–1051
73. Etesami H, Adl SM (2020) Plant growth-promoting rhizobacteria (PGPR) and their action mechanisms in availability of nutrients to plants. In: Phyto-microbiome in stress regulation. Springer Nature, Singapore 147–203
74. Vocciante M, Grifoni M, Fusini D, Petruzzelli G, Franchi E (2022) The role of plant growth-promoting rhizobacteria (PGPR) in mitigating plant's environmental stresses. Appl Sci 12(3):1231
75. Ramirez KS, Leff JW, Barberán A, Bates ST, Betley J, Crowther TW, Kelly EF, Oldfield EE, Shaw EA, Steenbock C (2014) Biogeographic patterns in below-ground diversity in New York City's Central Park are similar to those observed globally. Proc R Soc B 281(1795):20141988
76. Altowayti WAH, Almoalemi H, Shahir S, Othman N (2020) Comparison of culture independent and dependent approaches for identification of native arsenic-resistant bacteria and their potential use for arsenic bioremediation. Ecotoxicol Environ Saf 205:111267
77. McIlwaine DB, Moore M, Corrigan A, Niemaseck B, Nicoletti D (2024) A comparison of the microbial populations in a culture-dependent and a culture-independent analysis of industrial water samples. ApMic 4(3):1079–1090
78. Dalal N, Jalandra R, Sharma M, Prakash H, Makharia GK, Solanki PR, Singh R, Kumar A (2020) Omics technologies for improved diagnosis and treatment of colorectal cancer: technical advancement and major perspectives. Biomed Pharmacother 131:110648
79. O'Donnell ST, Ross RP, Stanton C (2020) The progress of multi-omics technologies: determining function in lactic acid bacteria using a systems level approach. Front Microbiol 10:3084
80. Hugenholtz P, Tyson GW (2008) Metagenomics. Nature 455(7212):481–483
81. Bahram M, Netherway T, Frioux C, Ferretti P, Coelho LP, Geisen S, Bork P, Hildebrand F (2021) Metagenomic assessment of the global diversity and distribution of bacteria and fungi. Environ Microbiol 23(1):316–326
82. Wani AK, Roy P, Kumar V (2022) Metagenomics and artificial intelligence in the context of human health. Infect Genet Evol 100:105267
83. Gupta N, Vats S, Bhargava P (2018) Sustainable agriculture: role of metagenomics and metabolomics in exploring the soil microbiota. In: In Silico Approach for Sustainable Agriculture. Springer, Singapore, pp 183–199.
84. Sabale SN, Suryawanshi PP, Krishnaraj P (2019) Soil metagenomics: concepts and applications. In: Metagenomics: Basics, Methods and Applications. IntechOpen, London, UK.
85. Chukwuneme CF, Ayangbenro AS, Babalola OO (2021) Metagenomic analyses of plant growth-promoting and carbon-cycling genes in maize rhizosphere soils with distinct land-use and management histories. Genes 12(9):1431
86. Adedayo AA, Fadiji AE, Babalola OO (2023) Unraveling the functional genes present in rhizosphere microbiomes of *Solanum lycopersicum*. PeerJ 11:e15432
87. Kusstatscher P, Wicaksono WA, Thenappan DP, Adam E, Müller H, Berg G (2020) Microbiome management by biological and chemical treatments in maize is linked to plant health. Microorganisms 8(10):1506
88. Kaur S, Egidi E, Qiu Z, Macdonald CA, Verma JP, Trivedi P, Wang J, Liu H, Singh BK (2022) Synthetic community improves crop performance and alters rhizosphere microbial communities. J Sustainable Agric Environ 1(2):118–131
89. Wang J, Li Q, Xu S, Zhao W, Lei Y, Song C, Huang Z (2018) Traits-based integration of multi-species inoculants facilitates shifts of indigenous soil bacterial community. Front Microbiol 9:1692
90. Bigatton ED, Verdenelli RA, Haro RJ, Ayoub I, Barbero FM, Martín MP, Dubini LE, Jorrí n Novo JsV, Lucini EI, Castillejo MÁ (2024) Metagenomic analysis to assess the impact of plant growth-promoting rhizobacteria on peanut (*Arachis hypogaea* L.) crop production and soil enzymes and microbial diversity. J Agric Food Chem 72(40):22385–22397
91. Saravanakumar K, Li Y, Yu C, Wang Q-q, Wang M, Sun J, Gao J-x, Chen J (2017) Effect of *Trichoderma harzianum* on maize rhizosphere microbiome and biocontrol of Fusarium Stalk rot. Sci Rep 7(1):1771.

92. Passera A, Vacchini V, Cocetta G, Arpanahi AA, Casati P, Ferrante A, Piazza L (2020) Towards nutrition-sensitive agriculture: an evaluation of biocontrol effects, nutritional value, and ecological impact of bacterial inoculants. Sci Total Environ 724:138127
93. Chaudhary T, Dixit M, Gera R, Shukla AK, Prakash A, Gupta G, Shukla P (2020) Techniques for improving formulations of bioinoculants. 3 Biotech 10:1–9
94. Keswani C, Prakash O, Bharti N, Vílchez JI, Sansinenea E, Lally RD, Borriss R, Singh SP, Gupta VK, Fraceto LF (2019) Re-addressing the biosafety issues of plant growth promoting rhizobacteria. Sci Total Environ 690:841–852
95. Schütz L, Gattinger A, Meier M, Müller A, Boller T, Mäder P, Mathimaran N (2018) Improving crop yield and nutrient use efficiency via biofertilization—a global meta-analysis. Front Plant Sci 8:2204
96. Rai A, Saito K, Yamazaki M (2017) Integrated omics analysis of specialized metabolism in medicinal plants. Plant J 90(4):764–787. https://doi.org/10.1111/tpj.1348

Chapter 33

Perspectives of Gene Editing for the Conservation of Plant Genetic Resources

Alma Armenta-Medina, Javier Mora-Macias, Julio Armando Massange-Sánchez, Blanca Elvira López-Valenzuela and Cynthia Paola Rangel-Chavez

Abstract

The rapid development of genome editing tools, particularly CRISPR-Cas9 and virus-induced genome editing (VIGE), offers transformative opportunities for the conservation of plant genetic resources. These technologies enable precise modifications that can enhance genetic variation, restore endangered populations, and strengthen resilience against climate change and invasive species. This chapter reviews methodological frameworks, including guide design, delivery systems, regeneration protocols, and verification of off-target effects, as well as ecological and ethical considerations. Case studies are highlighted in both wild and cultivated species: American chestnut, ash, elm, Pacific yew, hemlock, and whitebark pine, alongside crops of global significance such as coffee, cacao, wild apple, and banana. Applications include resistance to blight, borers, Dutch elm disease, woolly adelgid, rust, rots, and endogenous viral elements. Finally, the chapter outlines future directions emphasizing biosafety, ecological validation, and equitable governance, underscoring the potential of CRISPR-based strategies to integrate conservation science with sustainable agricultural practices.

Keywords CRISPR-Cas9, Plant genetic resources conservation, Genome editing, Forest and crop resilience, Biodiversity preservation

1 Introduction

Forests and native high-value crops are indispensable for maintaining biodiversity, supporting livelihoods, and ensuring food security. Forests, for instance, serve as critical carbon sinks, helping to regulate global climate patterns while providing habitats for countless species. This makes them a cornerstone of ecological balance [1, 2]. Similarly, native high-value crops are not just sources of food but also reservoirs of genetic diversity. They hold cultural significance for many communities and contribute to economic stability, particularly in rural and indigenous areas [3]. Protecting these resources is essential for

Marco A. Ramírez-Mosqueda et al. (eds.), *Conservation of Plant Genetic Resources*, Methods in Molecular Biology, vol. 3011,
https://doi.org/10.1007/978-1-0716-5130-8_33,

long-term environmental sustainability and for building resilience against the impacts of climate change.

However, biodiversity is under unprecedented threat. Human activities such as deforestation, habitat destruction, and overexploitation of natural resources are driving species extinction and ecosystem degradation at alarming rates [4, 5]. Climate change, pollution, and the spread of invasive species further exacerbate these challenges, disrupting ecosystems and eroding genetic diversity [6]. Agricultural expansion and urbanization are also significant contributors, fragmenting habitats and displacing native species.

Addressing these challenges requires a multifaceted approach to conservation. Traditional methods, such as establishing protected areas, restoring degraded habitats, and promoting sustainable land management, remain vital [7]. At the same time, modern tools like biotechnology and seed banking offer innovative ways to preserve genetic diversity and restore ecosystems. Equally important are policies that encourage sustainable agriculture, responsible resource use, and climate change mitigation, all of which are critical for ensuring ecological stability in the long term.

The urgency of the global biodiversity crisis has spurred scientists and conservationists to explore cutting-edge solutions [8]. One such tool is CRISPR-Cas9, a groundbreaking technology originally discovered as part of the bacterial immune system. CRISPR allows for precise genetic modifications, offering exciting possibilities for conservation. For example, it could be used to enhance the resilience of crops to environmental stressors, restore endangered species, or even combat invasive species [9]. While its applications are still emerging, CRISPR holds significant promise for protecting forest resources and native high-value crops. This chapter delves into the potential of CRISPR-based strategies to safeguard plant genetic resources and strengthen ecosystem resilience.

2 Understanding CRISPR Technology

CRISPR-Cas9 technology enables precise and targeted modifications of DNA in living organisms. The system consists of two essential components: the Cas9 nuclease and the guide RNA (gRNA). The Cas9 nuclease is a DNA-cutting enzyme derived from *Streptococcus pyogenes* that introduces double-strand breaks (DSBs) at specific DNA sites. The guide RNA (gRNA) is a short RNA molecule that directs the Cas9 protein to the target DNA sequence through complementary base pairing [10].

The CRISPR-Cas9 mechanism involves three main steps. The first step is target recognition, where the guide RNA binds to the Cas9 protein and directs it to the target DNA sequence (Fig. 1). The gRNA spacer region, a 20-nucleotide sequence, must be complementary

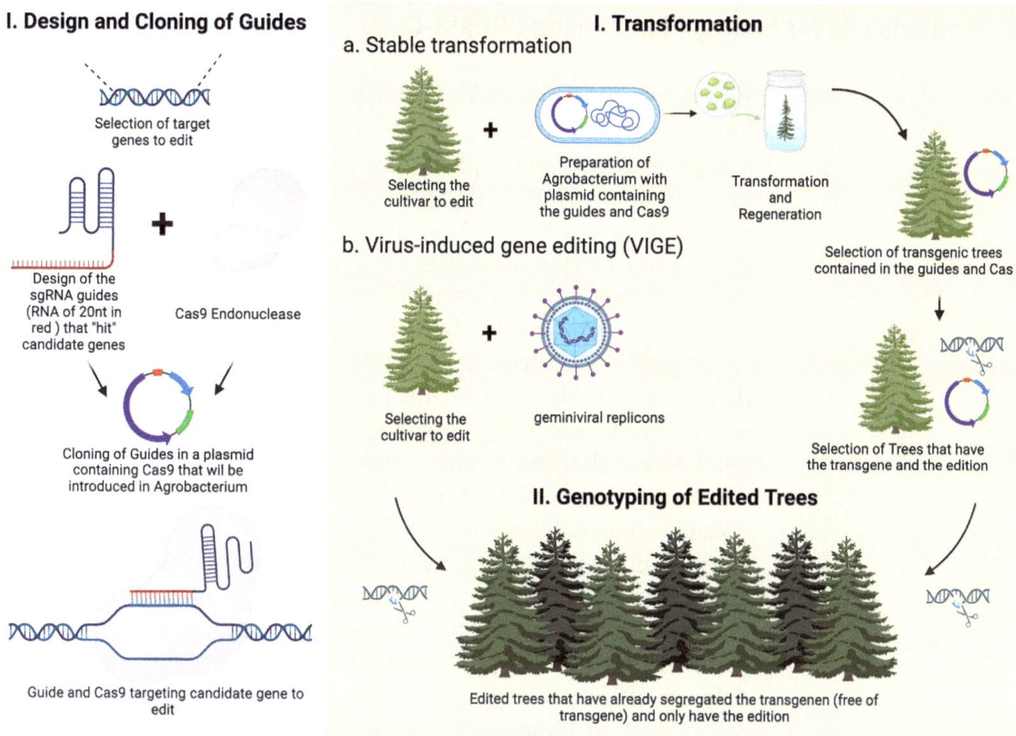

Fig. 1 CRISPR methodology divided in three phases: (1) design and cloning of guides in plasmids contain the Cas9 Endonuclease, (2) transformation of plants either (a) using stable transformation techniques that incorporate the guide and Cas9 to the genome, usually using Agrobacterium or (b) using virus induced gene editing, (3) genotyping of edited trees to select plants that has the edited sequence but segregated the guide and Cas9, free of transgene

to the target DNA and adjacent to a protospacer adjacent motif (PAM), typically "NGG." The PAM sequence is essential for Cas9 to bind and introduce a cut. The second step is double-strand break formation, in which Cas9 introduces a precise double-strand break at the target site. Finally, DNA repair occurs through two primary pathways: the nonhomologous end joining (NHEJ) pathway, which is error-prone and often results in small insertions or deletions (indels) that can knock out genes, or the homology-directed repair (HDR) pathway, a more precise mechanism that uses a donor DNA template for targeted insertions or replacements [10].

By utilizing these mechanisms, the CRISPR-Cas9 system allows precise genome edits, including gene knockouts, insertions, and replacements, making it a versatile tool for applications such as biodiversity conservation [11] (Fig. 1).

3 Methodology for Editing Plants Using CRISPR-Cas9

Editing plant genomes with CRISPR-Cas9 involves a series of well-defined steps. The process begins with the design of the gRNA, where a target sequence in the plant genome is identified. A 20-nucleotide spacer sequence is then designed to be complementary to the target site and adjacent to a PAM sequence (NGG) [10].

The next step is the construct preparation, during which the Cas9 gene and the guide RNA are cloned into a suitable vector. Typically, vectors compatible with *Agrobacterium tumefaciens* are used.

Once the construct is ready, the CRISPR-Cas9 system must be delivered into plant cells. This can be accomplished through various methods, including Agrobacterium-mediated transformation, particle bombardment (biolistics), or protoplast transformation. Agrobacterium-mediated transformation is typically used for dicot species, whereas biolistic approaches are more common for monocots. Another emerging technique involves the use of viral vectors, which are increasingly recognized as efficient carriers for delivering CRISPR components into plant cells. This strategy, known as virus-induced gene editing (VIGE) [12], offers notable advantages such as high delivery efficiency, a simplified process for producing gene-edited plants, and, in some cases, the possibility of in situ transformation. The latter could be particularly valuable in plant genetic conservation, enabling precise genetic modifications directly within natural or field populations.

After delivery, the regeneration of edited plants follows, edited plant cells are cultured under controlled conditions to regenerate whole plants using tissue culture techniques. Selectable markers, such as antibiotic resistance genes, are used to identify successfully transformed plants.

The next stage is the validation of edits, where the accuracy of the genome edits is confirmed. Techniques such as polymerase chain reaction (PCR), Sanger sequencing, next-generation sequencing (NGS), and bioinformatics tools are used to verify the presence and precision of the desired edits.

Finally, off-target analysis is conducted to ensure the specificity of the CRISPR-Cas9 system. Computational tools and sequencing methods are applied to identify any unintended modifications in the genome. This step is crucial for confirming the precision of the edits and ensuring minimal off-target effects.

Through these steps, CRISPR-Cas9 provides a robust and efficient methodology for plant genome editing, enabling the development of improved plant varieties for applications such as agriculture, conservation, and biotechnology (Fig. 1).

4 Potential Use of CRISPR for Plant Conservation

4.1 Generate Genetic Variation

Endangered species often suffer from genetic bottlenecks and inbreeding depression, which reduce their genetic diversity and adaptability. CRISPR-Cas9 can help by introducing genetic variation into small populations. For example, scientists have proposed using CRISPR to revive genetic diversity in the endangered chestnut tree by editing genes associated with chestnut blight resistance and reproductive success [13].

4.2 Combatting Invasive Species

Invasive species pose significant threats to native biodiversity. CRISPR-based gene drives, which bias the inheritance of specific genes, can be used to control or eradicate invasive populations. This approach has been explored for controlling the spread of plage (Fusarium, etc.), which are notorious for devastating native populations [13].

4.3 Enhancing Climate Resilience

Climate change is altering habitats and threatening species unable to adapt quickly enough. CRISPR can be employed to enhance the climate resilience of key species by editing genes related to stress tolerance. For instance, researchers are investigating ways to make more resilient trees by modifying genes involved in heat or drought tolerance, such as those controlling stomatal function, which can be edited to improve water-use efficiency during critical reproductive phases [13].

5 Cases for Applications of CRISPR in Conservation

5.1 Conservation in Forestry Using CRISPR

Earth is home to ~73,000 tree species and the latest assessment indicates that at least 31% of all tree species are threatened with extinction globally. The challenges of forest management are significantly compounded by the rapid changes in climate. As climatic conditions change, the potential distribution of plant species will also shift, leading to changes in species composition and a potential decline in provided ecosystem services [14].

Methods that promote adaptation contrast with the typical paradigm of conservation breeding programs which seek to minimize inbreeding and maintain genetic diversity [15].

Several projects are underway to test the efficacy of synthetic biology for conservation. Most are in the early stages, but the results are promising. Currently, no genetically altered organisms have been approved for release for conservation purposes, although approval is pending for transgenic American chestnut trees (*Castanea dentata*) [16]. After near extirpation from chestnut blight in the early 1900s, research has aimed to understand and increase the mechanisms of blight resistance [16]. Transgenic methods have generated a blight-tolerant strain of trees which, pending regulatory approval, may soon be planted to restore this species to American forests [16, 17].

Regulatory requirements for genetically altered organisms vary considerably by country but, in agriculture, organisms modified by selective breeding or gene editing are better tolerated than those modified by transgenic approaches, suggesting that the same may apply for conservation. The American chestnut breeding program involves both transgenic and selective breeding approaches in case the release of transgenic trees is not approved, or one method proves to be more effective.

CRISPR-mediated genome editing in forest trees has been mainly achieved in poplar, and, for the last years, in the tropical tree parasponia (*Parasponia andersonii*), eucalypts (*Eucalyptus grandis* × *urophylla* hybrid), rubber tree (*Hevea brasiliensis*), monterey pine (*Pino insigne*), and European chestnut (*Castanea sativa*) [18].

The conservation of genetic resources is crucial for maintaining biodiversity and ensuring ecosystem resilience against threats such as climate change and habitat loss. In forestry, the application of CRISPR technology offers innovative opportunities to enhance conservation efforts. According to the International Union for Conservation of Nature (IUCN), it is essential to implement strategies that promote the sustainable use and equitable sharing of benefits derived from these resources, in line with international agreements like the Nagoya Protocol. The use of CRISPR in genetic editing of tree species can aid in developing varieties more resistant to diseases and adaptable to changing environmental conditions, aligning with the conservation and sustainable use objectives advocated by the IUCN [13, 19, 20] (Table 1). Additionally, the IUCN emphasizes the need to strengthen local and regional capacities to manage and protect genetic diversity, recognizing the vital role of indigenous and local communities in this process. Integrating advanced technologies like CRISPR should be conducted inclusively and participatively, ensuring that benefits are shared equitably and respecting traditional knowledge and the rights of local communities.

American Chestnut (*Castanea dentata*): The American Chestnut has been heavily impacted by chestnut blight, a fungal disease that has devastated native populations. CRISPR could help by editing genes to confer resistance, building on recent transgenic methods aimed at producing blight-resistant chestnut varieties. Some foundational research has paved the way for these advancements. For instance, researchers have looked at how genetically modified chestnut trees might impact insect populations [21], underlining the importance of considering broader ecological effects. A couple of years later, Zhang et al. [22], focused on the role of oxalate oxidase transgene

Table 1
Examples of conservation projects using CRISPR for genetic modification in forestry and Native species of commercial important crops

Species	Trait to modify by CRISPR(s)	IUCN status	Susceptible to transformation	Proof of transformation
American chestnut trees (*Castanea dentata*)	Chestnut blight resistance	Critically Endangered (CE)	Yes	[16, 17]
American elm trees (*Ulmus americana*) and European Elm (*Ulmus* spp.)	Dutch elm resistance	Critically Endangered (CE)	Yes	[34]
Ash Trees (*Fraxinus* spp.)	Emerald ash borer	Critically Endangered (CE)	Yes	[66–68]
Pacific Yew (*Taxus brevifolia*)	Improve the growth rate	Near Threatened (NT)	Yes	[69]
Hemlock (*Tsuga* spp.)	hemlock woolly adelgid resistance	Near Threatened (NT)	No data found	No data found
Whitebark Pine (*Pinus albicaulis*)	Pine blister rust	Endangered (EN)	No data found	No data found
NATIVE SPECIES NOT ENDANGERED BUT OF COMMERCIAL INTEREST				
Species	**Trait to modify by CRISPR(s)**	**IUCN status**	**Susceptible to transformation**	**Proof of transformation**
Apple (*Malus sieversii*)	Stress-resistant plants	Vulnerable (VU)	Yes	[56, 70]
Cacao (*Theobroma cacao*)	Pod disease and witches' broom disease resistance AND *Phytophthora tropicalis*	Not Endangered	Yes	[71]
Coffee (*Coffea arabica*)	Leaf Rust resistance	Least Concern (LC)	Yes	[51, 72]
Banana (*Musa* spp.)	*Fusarium oxysporum* resistance	Least Concern (LC)	Yes	[63, 73, 74]

expression in minimizing blight damage, marking a significant step toward resilience. Around the same time, D'Amico et al. [23], compared ectomycorrhizal fungal interactions between transgenic and wild-type chestnuts, adding valuable insights on maintaining ecosystem balance. More recently, other researchers [24] evaluated the consistency of blight resistance in transgenic *Castanea dentata* lines, highlighting the potential for reliable resistance traits. Most recently, Sandercock et al. [25] proposed a new approach to boost climate resilience in restoration efforts, underscoring adaptability as a key factor for long-term success. Together, these studies reflect a comprehensive approach to combatting chestnut blight and ensuring the survival of this important species.

Ash Trees (*Fraxinus* spp.): Ash trees face a serious threat from the emerald ash borer, an invasive insect that has devastated populations across North America. Genome editing offers a promising way to boost the natural defenses of these trees. Research has already provided valuable insights into the defensive capabilities of ash trees. For example, Whitehill et al. [26] found that treating ash trees with methyl jasmonate (MeJA) prior to an emerald ash borer attack can enhance resistance by inducing general defense traits and the production of the toxic compound verbascoside. Meanwhile, Peterson and Cipollini [27] observed that non-native hosts like North American ash trees can often support better larval performance for the emerald ash borer compared to their original hosts, making these trees particularly vulnerable. In a different line of research, Cao et al. [28] examined the gut microbiota of wood borers and found that certain microbes help break down defensive compounds in ash, suggesting another angle to consider in bolstering tree resistance. Liu et al. [29] added that increased lignan production in velvet ash (*Fraxinus velutina*) significantly enhances resistance to the pest. Environmental factors also play a role; for instance, Chen and Poland [30] demonstrated how light intensity and leaf age influence the production of defensive compounds in green ash. Together, these findings point to genome editing as a way to strengthen ash trees—potentially by increasing lignan production or activating jasmonate pathways to better resist emerald ash borer infestations.

European Elm (*Ulmus* spp.): European elm trees are highly susceptible to Dutch elm disease (DED), a fungal pathogen that has devastated elm populations worldwide. CRISPR gene editing could offer a way to enhance resistance in these trees. Research into the factors behind DED tolerance provides a strong foundation for these efforts. For example, Kännaste et al. [31] studied how DED fungi affect photosynthesis and volatile compounds in elm leaves, with a focus on how global warming might influence these responses. Additionally, Martín et al. [32] explored the relationship between DED resistance and xylem endophytic fungi, suggesting that breeding for resistance could unintentionally impact the diversity of these beneficial microbes and related defense compounds. Looking at tissue-level responses, Sherif et al. [33] found that American elms with faster activation of disease-response genes in bark and wood tissues showed higher tolerance. In a similar line of inquiry, Newhouse et al. [34] described a genetically modified American elm that exhibited fewer DED symptoms and maintained healthy mycorrhizal relationships, underlining the historical toll DED has taken on these trees. Moreover, Sherif et al. demonstrated that the activation of both jasmonic acid and disease-response genes could be a key marker of tolerance in American elms. Pouzoulet et al. [35] offered further insight by suggesting that vessel size in elm trees might contribute to their resistance to vascular wilt diseases like DED. Altogether, these studies suggest that CRISPR could potentially target pathways related to defense compounds, vessel size, and hormone signaling to build more resilient elm trees in the fight against Dutch elm disease.

Pacific Yew (*Taxus brevifolia*): Pacific yew is a valuable species used in producing cancer-treatment drugs, notably paclitaxel (Taxol). Due to overharvesting and its slow growth rate, this tree faces conservation challenges. Genome editing could offer a solution by potentially enhancing growth rates or increasing yields of desired compounds. Research into Taxus species and their applications in cancer treatment has laid the groundwork for such advancements. For instance, Gao et al. [36] provided an in-depth review of Taxus extracts and their applications in modern cancer research, while Sati et al. [37] examined paclitaxel and its semi-synthetic derivatives, discussing their structures, mechanisms, and limitations in chemotherapy. Zhang et al. [38] further highlighted the unique bioactivities of *Taxus chinensis*, *Taxus cuspidata*, and *Taxus media*, underlining the challenges posed by their diversity and endangered status. Historically, paclitaxel's development from the western yew tree for breast cancer treatment, as discussed by Arbuck et al. [39], paved the way for its widespread therapeutic use. Additionally, studies have focused on alternatives and enhancements for paclitaxel production; Kovacs et al. [40] investigated the redirection of carotenoid metabolism in transgenic tomatoes to produce taxadiene, an approach that could address cost and resistance issues associated with taxol. Together, these studies suggest that genome editing in *Taxus brevifolia* could play a key role in sustainably meeting the demand for this critical anticancer compound.

Hemlock (*Tsuga* spp.): Hemlock trees are under significant threat from the invasive hemlock woolly adelgid, an insect that feeds on sap and causes extensive damage to hemlock forests. Genetic editing to enhance resistance to this pest could be an effective strategy for conservation efforts. Research into hemlock defenses has provided valuable insights that could inform such efforts. For example, Ayayee et al. [41] examined the micromechanical properties of hemlock needles, finding that these characteristics can impact herbivore access, especially for piercing-sucking insects like the adelgid. McKenzie et al. [42] found that resistant eastern hemlocks contain higher concentrations of terpenoids, suggesting that these compounds play a role in natural resistance. Additionally, Jetton et al. [43] developed a technique to simulate adelgid infestation in hemlocks, aiding in resistance studies. Wilson et al. [44] explored how adelgid infestation affects native forest defoliators, observing that infested foliage altered looper performance and preference. Montgomery et al. [45] also identified promising hemlock hybrids with natural resistance to adelgid, offering potential landscaping options. Other research, such as that by Crandall et al. [46], compared predator-prey dynamics in the adelgid's native and invasive ranges, providing ecological context to the pest's abundance. Furthermore, Schaeffer et al. [47] investigated the role of phytohormone pathways in response to adelgid and another invasive pest, showing how hemlocks signal defensive responses. Together, these studies highlight key defensive traits and ecological dynamics that could guide genome editing strategies to bolster hemlock resilience against the hemlock woolly adelgid.

Whitebark Pine (*Pinus albicaulis*): Whitebark pine is highly vulnerable to diseases that threaten its population, particularly the white pine blister rust caused by the invasive fungus *Cronartium ribicola*. This fungal infection has severe consequences, not only for the trees themselves but also for the high-elevation ecosystems they support. For instance, McKinney et al. [48] highlighted how this pathogen disrupts the mutualistic relationship between whitebark pine and birds by reducing cone production, which in turn affects seed dispersal. Mortality rates in some areas have reached 90% or higher, indicating the scale of this threat. In terms of early defense mechanisms, Figueroa-Corona et al. [49] investigated the transcriptional responses of *Pinus albicaulis* upon infection, focusing on the tree's pathogen perception and activation of signal transduction cascades to initiate defense. Given these insights, CRISPR technology offers a promising path forward by allowing researchers to target and modify specific genes involved in these defense responses, potentially enhancing resistance and contributing to the conservation of this keystone species.

5.2 Conservation of Native Species of Commercial Interest Using CRISPR (Native High-Value Crops)

Conserving native species of commercial importance is essential for preserving genetic diversity, which serves as a critical resource for future crop improvement and the development of resistance to biotic stresses (such as pests and diseases) and abiotic stresses (including drought, salinity, and extreme temperatures). While some species are not currently endangered due to their widespread cultivation, their native populations may be diminishing due to factors such as habitat loss, pest outbreaks, or climate change (Table 1). The following examples highlight such species and their significance.

Coffee (*Coffea arabica*): Coffee species face significant challenges, not only from habitat loss but also from pests and diseases that threaten their productivity and survival. Agrobacterium-mediated transformation has been employed to enhance disease resistance, especially against coffee leaf rust. Additional research has focused on bolstering resistance to other threats. For example, Hendre et al. [50] isolated resistance gene analogues from wild coffee relatives in India to improve disease resistance in cultivated coffee varieties. Valencia-Lozano et al. [51] investigated the expression of the *Bacillus thuringiensis* Cry10Aa protein in *Coffea arabica* as a defense against the Coffee Berry Borer (CBB), one of the major pests affecting coffee crops. The impact of bacterial pathogens is also significant; Monteiro et al. [52] explored the stable transformation of *Xylella fastidiosa*—a bacterium impacting coffee and other crops—providing insights into managing bacterial diseases. Furthermore, Armesto et al. [53] used genetic transformation to study *Colletotrichum gloeosporioides*, a fungus causing blister spot disease in coffee, which has become a critical issue in Brazilian coffee production. Together, these studies highlight the need for comprehensive genetic strategies to protect coffee species from both environmental and biological threats.

Apple (*Malus sieversii* Ledeb.): Wild apple is native to the mountains of Central Asia and is recognized as a vulnerable species on the IUCN Red List. As the progenitor of modern domesticated apples, *Malus sieversii* holds substantial genetic importance, yet it faces multiple threats from habitat loss, deforestation, and crossbreeding with cultivated varieties [54]. These pressures are further compounded by human activities, including agricultural expansion and urban development, which continue to diminish its natural populations. Research has also shown the impact of environmental conditions on this species; for example, Jiao et al. [55] examined the rhizosphere microbiome of *Malus sieversii* in Xinjiang, China, revealing how environmental gradients affect plant health and microbial interactions. Furthermore, Liu et al. [56] demonstrated that genetic modifications, such as the overexpression of *MdBAK1*, can enhance resistance in apple species against diseases like replant disease caused by *Fusarium oxysporum*. Studies on apple orchard biodiversity also emphasize the importance of ecosystem balance for pest control, as Judt et al. [57] highlighted the role of floral resources and ground covers in promoting natural enemies of pests. Together, these insights reinforce the need to combine conservation efforts for *Malus sieversii* with advanced biotechnological approaches like CRISPR/Cas9 gene editing to help preserve and utilize its unique genetic heritage in modern agriculture.

Cacao (*Theobroma cacao*): While not endangered, *Theobroma cacao* faces substantial challenges due to diseases like black pod disease and witches' broom, which severely impact yields and threaten the livelihoods of farmers. Efforts to conserve cacao diversity and ensure sustainable cultivation are ongoing. For example, Osorio-Guarín et al. [58] highlighted the impact of diseases like frosty pod rot and witches' broom on cacao yields, while McElroy et al. [59] noted that genome-wide association studies (GWAS) markers have not been predictive enough for effective marker-assisted selection (MAS) across diverse cacao germplasm. This has led to an emphasis on other breeding methods, including CRISPR, to introduce disease-resistant traits directly into high-value cultivars. In line with this, Tscharntke et al. [60] discussed current selection programs focused on developing high-yielding, disease-resistant, fine-flavor genotypes to stabilize market outcomes for farmers. Furthermore, Winters et al. [61] emphasized the importance of utilizing the genetic diversity found in wild cacao relatives to bolster resistance traits in cultivated cacao, an approach that could enhance sustainability. Additionally, Camillo et al. [62] found that APX isoenzymes play a crucial role in eliminating intracellular H_2O_2 during infection by *Moniliophthora perniciosa*, a common cacao pathogen, highlighting a potential target for CRISPR to improve stress responses. Together, these studies underscore the importance of genetic conservation and innovative breeding approaches, like CRISPR, to safeguard cacao against disease and support sustainable cultivation.

Bananas (*Musa* spp.): A staple food crop and major agricultural product, face severe challenges from pests, diseases like Fusarium wilt and Banana Xanthomonas Wilt (BXW), and abiotic stresses such as drought and salinity. BXW, caused by *Xanthomonas campestris* pv. *musacearum*, is among the most destructive diseases in East and Central Africa, with all cultivated varieties being susceptible. However, the wild progenitor *Musa balbisiana* shows natural resistance to BXW, highlighting the importance of native species for genetic improvement [63]. Advancements in CRISPR technology have opened new avenues for introducing beneficial traits from wild relatives, such as *Musa acuminata* and *Musa balbisiana*, into cultivated bananas. These wild species harbor valuable genetic variations for disease resistance, stress tolerance, and nutrient-use efficiency. For example, *Musadmr6* mutants, created by editing the susceptibility gene *DMR6* encoding 2-oxoglutarate Fe(II)-dependent oxygenase (2OGO), exhibited enhanced BXW resistance without detrimental effects on plant growth [63]. Furthermore, CRISPR has been employed to inactivate the endogenous banana streak virus (eBSV) found in the B genome of plantains, which activates under stress and poses a significant challenge for breeding programs. Genome-edited lines, such as Gonja Manjaya, demonstrated effective suppression of eBSV activation and remained asymptomatic under water stress, enabling their potential use in hybrid development [64]. By leveraging CRISPR to precisely edit the banana genome, it is possible to enhance resistance to pathogens and environmental stressors, ensuring the sustainability and productivity of banana cultivation while safeguarding global food security.

6 Ethical and Ecological Considerations

While CRISPR offers powerful tools for conservation, its application raises ethical and ecological concerns. Potential risks include off-target effects, unintended consequences for ecosystems, and ethical dilemmas regarding genetic modifications of wild organisms. It is crucial to conduct thorough risk assessments and engage in public discourse to navigate these challenges responsibly [65].

7 Future Directions

The integration of CRISPR technology into conservation biology is still in its infancy, but its potential is immense. Future research should focus on refining gene-editing techniques to ensure precision and safety. Additionally, interdisciplinary collaborations between geneticists, ecologists, ethicists, and policymakers are essential to develop comprehensive strategies for the responsible use of CRISPR in conservation.

8 Conclusion

CRISPR represents a groundbreaking tool with the potential to make a substantial contribution to the conservation of genetic resources. In this review, we have outlined strategies that could aid in the restoration of ecologically and economically important tree species—such as American chestnut, ash, European elm, Pacific yew, hemlock, and whitebark pine—as well as native species of major commercial crops including coffee, cacao, apple, and banana. By enabling precise genetic modifications, CRISPR offers promising solutions for enhancing biodiversity, combating invasive species, and increasing the resilience of ecosystems to climate change. Nonetheless, its successful integration into conservation programs demands careful consideration of ethical, ecological, and regulatory factors. When applied thoughtfully and with rigorous oversight, CRISPR holds the potential to become a powerful ally in safeguarding the planet's genetic heritage.

References

1. Pan Y, Birdsey RA, Fang J, Houghton R, Kauppi PE, Kurz WA, Hayes D (2011) A large and persistent carbon sink in the world's forests. Science 333(6045):988–993. https://doi.org/10.1126/science.1201609
2. Watson JE, Evans T, Venter O, Williams B, Tulloch A, Stewart C, Lindenmayer D (2018) The exceptional value of intact forest ecosystems. Nat Ecol Evol 2(4):599–610. https://doi.org/10.1038/s41559-018-0490-x
3. Jarvis DI, Brown AH, Cuong PH, Collado-Panduro L, Latournerie-Moreno L, Gyawali S, Hodgkin T (2008) A global perspective of the richness and evenness of traditional crop-variety diversity maintained by farming communities. Proc Natl Acad Sci 105(14):5326–5331. https://doi.org/10.1073/pnas.0800607105
4. Foley JA, DeFries R, Asner GP, Barford C, Bonan G, Carpenter SR, Snyder PK (2005) Global consequences of land use. Science 309(5734):570–574. https://doi.org/10.1126/science.1111772
5. IPBES (2019) Global assessment report on biodiversity and ecosystem services of the Intergovernmental Science-Policy Platform on Biodiversity and Ecosystem Services. In: Brondizio ES, Settele J, Díaz S, HT Ngo HT (eds). IPBES Secretariat, Bonn, Germany. p 1148. https://doi.org/10.5281/zenodo.3831673
6. Díaz S, Settele J, Brondízio ES, Ngo HT, Agard J, Arneth A, Zayas CN (2019) Pervasive human-driven decline of life on Earth points to the need for transformative change. Science 366(6471):eaax3100. https://doi.org/10.1126/science.aax3100
7. Watson JE, Dudley N, Segan DB, Hockings M (2014) The performance and potential of protected areas. Nature 515(7525):67–73. https://doi.org/10.1038/nature13947
8. Ceballos G, Ehrlich PR, Dirzo R (2017) Biological annihilation via the ongoing sixth mass extinction signaled by vertebrate population losses and declines. Proc Natl Acad Sci 114(30):E6089–E6096. https://doi.org/10.1073/pnas.1704949114
9. Esvelt KM, Gemmell NJ (2017) Conservation demands safe gene drive. PLoS Biol 15(11):e2003850. https://doi.org/10.1371/journal.pbio.2003850
10. Jinek M, Chylinski K, Fonfara I, Hauer M, Doudna JA, Charpentier E (2012) A programmable dual-RNA–guided DNA endonuclease in adaptive bacterial immunity. Science 337(6096):816–821. https://doi.org/10.1126/science.1225829
11. Cardi T, Murovec J, Bakhsh A, Boniecka J, Bruegmann T, Bull SE, van Laere K (2023) CRISPR/Cas-mediated plant genome editing: outstanding challenges a decade after implementation. Trends Plant Sci 28(10):1144–1165. https://doi.org/10.1016/j.tplants.2023.05.012
12. Yin K, Han T, Liu G, Chen T, Wang Y, Yu AYL, Liu Y (2015) A geminivirus-based guide RNA delivery system for CRISPR/Cas9 mediated plant genome editing. Sci Rep 5(1):14926. https://doi.org/10.1038/srep14926
13. Heidari R, Shaw DM, Elger BS (2017) CRISPR and the rebirth of synthetic biology. Sci Eng Ethics 23(2):351–363. https://doi.org/10.1007/s11948-016-9768-z
14. Van Tiel N, Fopp F, Brun P, van den Hoogen J, Karger DN, Casadei CM, Lyu L, Tuia D, Zimmermann NE, Crowther TW TW, Pellissier L (2024) Regional uniqueness of tree species composition and response to forest loss and climate change. Nat Commun 15(1):4375. https://doi.org/10.1038/s41467-024-48276-3
15. Kosch TA, Waddle AW, Cooper CA, Zenger KR, Garrick DJ, Berger L, Skerratt LF (2022) Genetic approaches for increasing fitness in endangered species. Trends Ecol Evol 37(4):332–345. https://doi.org/10.1016/j.tree.2021.12.003
16. Powell WA, Newhouse AE, Coffey V (2019) Developing blight-tolerant American chestnut trees. Cold Spring Harbor Perspect Biol 11(7):a034587. https://doi.org/10.1101/cshperspect.a034587
17. Newhouse AE, Powell WA (2021) Intentional introgression of a blight tolerance transgene to rescue the remnant population of American chestnut. Conserv Sci Pract 3(4)e348. https://doi.org/10.1111/csp2.348
18. Cao HX, Vu GTH, Gailing O (2022) From genome sequencing to CRISPR-based genome editing for climate-resilient forest trees. Int J Mol Sci 23(2):966. https://doi.org/10.3390/ijms23020966
19. Phelps MP, Seeb LW, Seeb JE (2020) Transforming ecology and conservation biology through genome editing. Conserv Biol 34(1):54–65. https://doi.org/10.1111/cobi.13292
20. Redford KH, Adams W, Carlson R, Mace GM, Woodruff DS (2019) Genetic frontiers for conservation: an assessment of synthetic

biology and biodiversity conservation. Technical Assessment. IUCN
21. Post KH, Parry D (2011) Non-target effects of transgenic blight-resistant American chestnut (*Fagales: Fagaceae*) on insect herbivores. Environ Entomol 40(4):955–963. https://doi.org/10.1603/EN10063
22. Zhang B, Oakes AD, Newhouse AE, Baier KM, Maynard CA, Powell WA (2013) A threshold level of oxalate oxidase transgene expression reduces *Cryphonectria parasitica*-induced necrosis in a transgenic American chestnut (*Castanea dentata*) leaf bioassay. Transgenic Res 22(5):973–982. https://doi.org/10.1007/s11248-013-9708-5
23. D'Amico KM, Horton TR, Maynard CA, Stehman SV, Oakes AD, Powell WA (2015) Comparisons of ectomycorrhizal colonization of transgenic American chestnut with those of the wild type, a conventionally bred hybrid, and related fagaceae species. Appl Environ Microbiol 81(1):100–108. https://doi.org/10.1128/AEM.02169-14
24. Carlson E, Stewart K, Baier K, McGuigan L, Culpepper T, Powell W (2022) Pathogen-induced expression of a blight tolerance transgene in American chestnut. Mol Plant Pathol 23(3):370–382. https://doi.org/10.1111/mpp.13165
25. Sandercock AM, Westbrook JW, Zhang Q, Holliday JA (2024) A genome-guided strategy for climate resilience in American chestnut restoration populations. Proc Natl Acad Sci 121(30):e2403505121. https://doi.org/10.1073/pnas.2403505121
26. Whitehill JG, Rigsby C, Cipollini D, Herms DA, Bonello P (2014) Decreased emergence of emerald ash borer from ash treated with methyl jasmonate is associated with induction of general defense traits and the toxic phenolic compound verbascoside. Oecologia 176(4):1047–1059. https://doi.org/10.1007/s00442-014-3082-8
27. Peterson DL, Cipollini D (2020) Larval performance of a major forest pest on novel hosts and the effect of stressors. Environ Entomol 49(2):482–488. https://doi.org/10.1093/ee/nvz160
28. Cao Q, Wang C, Koski TM, Bu Z, Li H, Sun J (2023) Gut bacteria of wood-boring *Streltzoviella insularis* and their involvement in pinoresinol degradation of ash tree (*Fraxinus* sp.) host. J Agricult Food Chem 71(23):8941–8951. https://doi.org/10.1021/acs.jafc.3c01307
29. Liu J, Feng R, Fu X, Zhao J, Zhang S, Wang J, Wei J (2022) Lignans dramatically enhance the resistance of Fraxinus velutina Torr. by adjusting the dominant bacterium group of *Agrilus planipennis* Fairmaire. Pest Manag Sci 78(4):1386–1397. https://doi.org/10.1002/ps.6755
30. Chen Y, Poland TM (2009) Interactive influence of leaf age, light intensity, and girdling on green ash foliar chemistry and emerald ash borer development. J Chem Ecol 35(7):806–815. https://doi.org/10.1007/s10886-009-9661-1
31. Kännaste A, Jürisoo L, Runno-Paurson E, Kask K, Talts E, Pärlist P, Drenkhan R, Niinemets Ü (2023) Impacts of Dutch elm disease-causing fungi on foliage photosynthetic characteristics and volatiles in *Ulmus* species with different pathogen resistance. Tree Physiol 43(1):57–74. https://doi.org/10.1093/treephys/tpac108
32. Martín JA, Witzell J, Blumenstein K, Rozpedowska E, Helander M, Sieber TN, Gil L (2013) Resistance to Dutch elm disease reduces presence of xylem endophytic fungi in Elms (*Ulmus* spp.). PLoS One 8(2):e56987. https://doi.org/10.1371/journal.pone.0056987
33. Sherif SM, Erland LA, Shukla MR, Saxena PK (2017) Bark and wood tissues of American elm exhibit distinct responses to Dutch elm disease. Sci Rep 7(1):7114. https://doi.org/10.1038/s41598-017-07779-4
34. Newhouse AE, Schrodt F, Liang H, Maynard CA, Powell WA (2007) Transgenic American elm shows reduced Dutch elm disease symptoms and normal mycorrhizal colonization. Plant Cell Rep 26(7):977–987. https://doi.org/10.1007/s00299-007-0313-z
35. Pouzoulet J, Pivovaroff AL, Santiago LS, Rolshausen PE (2014) Can vessel dimension explain tolerance toward fungal vascular wilt diseases in woody plants? Lessons from Dutch elm disease and esca disease in grapevine. Front Plant Sci 5:253. https://doi.org/10.3389/fpls.2014.00253
36. Gao X, Chen K, Xie W (2024) Research progress on taxus extraction and formulation preparation technologies. Molecules 29(10):2291. https://doi.org/10.3390/molecules29102291
37. Sati Priyanka, Sharma Eshita, Dhyani Praveen, Attri Dharam, Rana Rohit, Kiyekbayeva Lashyn, Büsselberg Dietrich, Samuel Samson, Sharif-Rad Javad (2024) Paclitaxel and its semi-synthetic derivatives: comprehensive insights into chemical structure, mechanisms of action, and anticancer properties. Eur J Med Res 29:1–26. https://doi.org/10.1186/s40001-024-01657-2
38. Zhang S, Lu X, Zheng T, Guo X, Chen Q, Tang Z (2021) Investigation of bioactivities of *Taxus chinensis*, *Taxus cuspidata*, and *Taxus* × *media* by gas chromatography-mass spectrometry. Open Life Sci 16(1):287–296. https://doi.org/10.1515/biol-2021-0032
39. Arbuck SG, Dorr A, Friedman MA (1994) Paclitaxel (Taxol) in breast cancer. Hematol

Oncol Clin North Am 8(1):121–140. https://doi.org/10.1016/S0889-8588(18)30191-6
40. Kovacs K, Zhang L, Linforth RS, Whittaker B, Hayes CJ, Fray RG (2007) Redirection of carotenoid metabolism for the efficient production of taxadiene [taxa-4(5),11(12)-diene] in transgenic tomato fruit. Transgenic Res 16(1):121–126. https://doi.org/10.1007/s11248-006-9039-x
41. Ayayee P, Yang F, Rieske LK (2014) Biomechanical properties of hemlocks: a novel approach to evaluating physical barriers of the plant-insect interface and resistance to a phloem-feeding herbivore. Insects 5(2):364–376. https://doi.org/10.3390/insects5020364
42. McKenzie EA, Elkinton JS, Casagrande RA, Preisser EL, Mayer M (2014) Terpene chemistry of eastern hemlocks resistant to hemlock woolly adelgid. J Chem Ecol 40(9):1003–1012. https://doi.org/10.1007/s10886-014-0495-0
43. Jetton RM, Mayfield AE, Powers ZL (2014) Development of a rain down technique to artificially infest hemlocks with the hemlock woolly adelgid, *Adelges tsugae*. J Insect Sci 14:106. https://doi.org/10.1673/031.014.106
44. Wilson CM, Vendettuoli JF, Orwig DA, Preisser EL (2016) Impact of an invasive insect and plant defense on a native forest defoliator. Insects 7(3):45. https://doi.org/10.3390/insects7030045
45. Montgomery ME, Bentz SE, Olsen RT (2009) Evaluation of hemlock (Tsuga) species and hybrids for resistance to *Adelges tsugae* (Hemiptera: Adelgidae) using artificial infestation. J Econ Entomol 102(3):1247–1254. https://doi.org/10.1603/029.102.0351
46. Crandall RS, Lombardo JA, Elkinton JS (2022) Top-down regulation of hemlock woolly adelgid (*Adelges tsugae*) in its native range in the Pacific Northwest of North America. Oecologia 199(3):599–609. https://doi.org/10.1007/s00442-022-05214-8
47. Schaeffer RN, Wang Z, Thornber CS, Preisser EL, Orians CM (2018) Two invasive herbivores on a shared host: patterns and consequences of phytohormone induction. Oecologia 186(4):973–982. https://doi.org/10.1007/s00442-018-4063-0
48. McKinney ST, Fiedler CE, Tomback DF (2009) Invasive pathogen threatens bird-pine mutualism: implications for sustaining a high-elevation ecosystem. Ecol Appl 19(3):597–607. https://doi.org/10.1890/08-0151.1
49. Figueroa-Corona L, Baesen K, Bhattarai A, Kegley A, Sniezko RA, Wegrzyn J, De La Torre AR (2024) Transcriptional profiling of early defense response to white pine blister rust infection in *Pinus albicaulis* (Whitebark Pine). Genes 15(5):602. https://doi.org/10.3390/genes15050602
50. Hendre PS, Bhat PR, Krishnakumar V, Aggarwal RK (2011) Isolation and characterization of resistance gene analogues from *Psilanthus* species that represent wild relatives of cultivated coffee endemic to India. Genome 54(5):377–390. https://doi.org/10.1139/g11-004
51. Valencia-Lozano E, Cabrera-Ponce JL, Noa-Carrazana JC, Ibarra JE (2021) *Coffea arabica* L. resistant to coffee berry borer (*Hypothenemus hampei*) mediated by expression of the *Bacillus thuringiensis* Cry10Aa protein. Front Plant Sci 12:765292. https://doi.org/10.3389/fpls.2021.765292
52. Monteiro PB, Teixeira DC, Palma RR, Garnier M, Bové JM, Renaudin J (2001) Stable transformation of the *Xylella fastidiosa* citrus variegated chlorosis strain with *oriC* plasmids. Appl Environ Microbiol 67(5):2263–2269. https://doi.org/10.1128/AEM.67.5.2263-2269.2001
53. Armesto C, Maia FG, de Abreu MS, Figueira A, dos R, da Silva BM, Monteiro FP (2012) Genetic transformation with the gfp gene of *Colletotrichum gloeosporioides* isolates from coffee with blister spot. Braz J Microbiol 43(3):1222–1229. https://doi.org/10.1590/S1517-838220120003000050
54. Tian Z, Song H, Wang Y, Li J, Maimaiti M, Liu Z, Zhang H, Zhang J (2022) Wild apples are not that wild: conservation status and potential threats of *Malus sieversii* in the mountains of Central Asia biodiversity hotspot. Diversity 14(6):489. https://doi.org/10.3390/d14060489
55. Jiao H, Liu L, Wang R, Qin W, Zhang B (2023) The rhizosphere microbiome of *Malus sieversii* (Ldb.) Roem. in the geographic and environmental gradients of China's Xinjiang. BMC Microbiol 23(1):26. https://doi.org/10.1186/s12866-023-02763-7
56. Liu X, Xu S, Wang X, Xin L, Wang L, Mao Z, Wu S (2022) MdBAK1 overexpression in apple enhanced resistance to replant disease as well as to the causative pathogen *Fusarium oxysporum*. Plant Physiol Biochem 179:144–157. https://doi.org/10.1016/j.plaphy.2022.03.014
57. Judt C, Korányi D, Zaller JG, Batáry P (2023) Floral resources and ground covers promote natural enemies but not pest insects in apple orchards: A global meta-analysis. Sci Total Environ 903:166139. https://doi.org/10.1016/j.scitotenv.2023.166139
58. Osorio-Guarín JA, Berdugo-Cely JA, Coronado-Silva RA, Baez E, Jaimes Y, Yockteng R (2020) Genome-wide association study reveals novel candidate genes associated with productivity and disease resistance to *Moniliophthora* spp. in Cacao (*Theobroma cacao* L.). G3 (Bethesda, Md.) 10(5):1713–1725. https://doi.org/10.1534/g3.120.401153

59. McElroy MS, Navarro AJR, Mustiga G, Stack C, Gezan S, Peña G, Sarabia W, Saquicela D, Sotomayor I, Douglas GM, Migicovsky Z, Amores F, Tarqui O, Myles S, Motamayor JC (2018) Prediction of *Theobroma cacao* resistance to *Moniliophthora* spp. diseases via genome-wide association analysis and genomic selection. Front Plant Sci 9:343. https://doi.org/10.3389/fpls.2018.00343
60. Tscharntke T, Ocampo-Ariza C, Vansynghel J, Ivañez-Ballesteros B, Aycart P, Rodriguez L, Ramirez M, Steffan-Dewenter I, Maas B, Thomas E (2023) Socio-ecological benefits of fine-flavor cacao in its center of origin. Conserv Lett 16(1):e12936. https://doi.org/10.1111/conl.12936
61. Winters NP, Wafula EK, Knollenberg BJ, Hämälä T, Timilsena PR, Perryman M, Zhang D, Sheaffer LL, Praul CA, Ralph PE, Prewitt S, Leandro-Muñoz ME, Delgadillo-Duran DA, Altman NS, Tiffin P, Maximova SN, dePamphilis CW, Marden JH, Guiltinan MJ (2024) A combination of conserved and diverged responses underlies *Theobroma cacao*'s defense response to *Phytophthora palmivora*. BMC Biol 22(1), 38. https://doi.org/10.1186/s12915-024-01831-2
62. Camillo LR, Filadelfo CR, Monzani PS, Corrêa RX, Gramacho KP, Micheli F, Pirovani CP (2013) Tc-cAPX, a cytosolic ascorbate peroxidase of *Theobroma cacao* L. engaged in the interaction with *Moniliophthora perniciosa*, the causing agent of witches' broom disease. Plant Physiol Biochem 73:254–265. https://doi.org/10.1016/j.plaphy.2013.10.009
63. Tripathi JN, Ntui VO, Shah T, Tripathi L (2021) CRISPR/Cas9-mediated editing of DMR6 orthologue in banana (Musa spp.) confers enhanced resistance to bacterial disease. Plant Biotechnol J 19(7):1291–1293. https://doi.org/10.1111/pbi.13614
64. Tripathi JN, Ntui VO, Ron M et al (2019) CRISPR/Cas9 editing of endogenous *banana streak virus* in the B genome of *Musa* spp. overcomes a major challenge in banana breeding. Commun Biol 2:46. https://doi.org/10.1038/s42003-019-0288-7
65. Sandler R (2019) The ethics of genetic engineering and gene drives in conservation. Conserv Biol 34(2):378–385
66. Palla KJ, Pijut PM (2015) *Agrobacterium*-mediated genetic transformation of *Fraxinus americana* hypocotyls. Plant Cell Tissue Organ Cult 120(3):631–641. https://doi.org/10.1007/s11240-014-0630-1
67. Du N, Pijut PM (2009) *Agrobacterium*-mediated transformation of *Fraxinus pennsylvanica* hypocotyls and plant regeneration. Plant Cell Rep 28(6):915–923. https://doi.org/10.1007/s00299-009-0697-z
68. Stevens ME, Pijut PM (2014) *Agrobacterium*-mediated genetic transformation and plant regeneration of the hardwood tree species *Fraxinus profunda*. Plant Cell Rep 33(6):861–870. https://doi.org/10.1007/s00299-014-1562-2
69. Luan EC, Allen SE, Bolyard MG (1996) Transient gus expression in zygotic embryos of *Taxus Brevifolia*. In Vitro Cell Dev Biol - Plant 32(2):81–85. http://www.jstor.org/stable/20064882
70. Zhang Y, Zhou P, Bozorov TA, Zhang D (2021) Application of CRISPR/Cas9 technology in wild apple (*Malus sieversii*) for paired sites gene editing. Plant Methods 17:79. https://doi.org/10.1186/s13007-021-00769-8
71. Fister AS, Landherr L, Maximova SN, Guiltinan MJ (2018) Transient expression of CRISPR/Cas9 machinery targeting *TcNPR3* enhances defense response in *Theobroma cacao*. Front Plant Sci 9:268. https://doi.org/10.3389/fpls.2018.00268
72. Ribas AF, Dechamp E, Champion A, Bertrand B, Combes MC, Verdeil JL, Lapeyre F, Lashermes P, Etienne H (2011) *Agrobacterium*-mediated genetic transformation of *Coffea arabica* (L.) is greatly enhanced by using established embryogenic callus cultures. BMC Plant Biol 11:92. https://doi.org/10.1186/1471-2229-11-92
73. Ganapathi TR, Negi S, Tak H, Bapat VA (2021) Transgenic banana: current status, opportunities, and challenges. In: Kavi Kishor, P.B., Rajam, M.V., Pullaiah, T. (eds) Genetically Modified Crops. Springer, Singapore. https://doi.org/10.1007/978-981-15-5932-7_5
74. Ntui VO, Uyoh EA, Ita EE, Markson AA, Tripathi JN, Okon NI et al (2021) Strategies to combat the problem of yam anthracnose disease: status and prospects. Mol Plant Pathol 22(10):1302–1314. https://doi.org/10.1111/mpp.13107

Chapter 34

Perspectives on Plant Germplasm Conservation and Biobanking

Marco A. Ramírez-Mosqueda

Abstract

Safeguarding plant species used for the benefit of humanity is a priority action to be carried out by various social sectors, including the scientific community. The implementation of innovative conservation strategies is a tool that science provides to safeguard our plant genetic resources for the long term and allow their use. In this sense, this book compiles different strategies for the conservation of plant germplasm, covering short-, medium-, and long-term ex situ conservation as well as in situ conservation. Each of the protocols presented stands out for its effectiveness and ease of implementation. Likewise, the literature reviews expand knowledge on conservation issues. Furthermore, the conservation of plant genetic resources in biobanks is a current trend aimed at ensuring food security for future generations. All of this aims to achieve food and environmental sustainability in the near future.

Keywords Sustainability, In situ conservation, Ex situ conservation, Plant genetic resources

1 Introduction

Plant germplasm is the set of genes transmitted to progeny found in any part of a plant and capable of generating a complete individual [1]. The safeguarding of this genetic material must be aimed at conserving the greatest possible genetic diversity of each species [2]. It should be understood as the variety of genetic characteristics present in each species [3]. This diversity of genetic characteristics allows species to adapt to potential environmental changes and is responsible for their genotypic and phenotypic plasticity [4].

The conservation of plant germplasm of different genetic resources will ensure food security at the global level and enable its use in the future. Thus, the safeguarding of this plant material in germplasm banks will encourage the conservation of the greatest genetic diversity of agricultural species [5, 6]. However, effective conservation strategies are needed to reduce the risk of material loss.

Marco A. Ramírez-Mosqueda et al. (eds.), *Conservation of Plant Genetic Resources*, Methods in Molecular Biology, vol. 3011,
https://doi.org/10.1007/978-1-0716-5130-8_34,

For this reason, different techniques for in situ and ex situ conservation of diverse plant genetic resources are compiled.

2 Conclusions

In this book different plant germplasm conservation techniques were described, ranging from short-term strategies such as seed banking and plant tissue culture, medium-term strategies such as in vitro conservation, and long-term strategies such as cryopreservation. In situ conservation strategies are also described, emphasizing the importance of social organization in achieving this goal. Each of these strategies provides different types of advantages and is applied in the preservation of germplasm of various plant species.

The reviews presented in this book expand knowledge on the conservation of plant genetic resources, including the description and application of innovative and cutting-edge approaches. It also compiles the factors that influence the success of conservation techniques and the proper use of plant germplasm to reduce the risk of loss of this valuable material. The appropriate choice of the conservation period for plant material (short, medium, or long term) is fundamental and will depend on the type of plant genetic resource to be protected. However, it is advisable to look for the implementation of integral conservation techniques, for which it is necessary to establish short-term strategies, then seek the conditions for medium-term conservation, and, finally, establish long-term conservation, with the aim of having plant material available for use and protected core collection. Both conservation strategies seek the sustainable use of plant genetic resources to ensure food security in the event that it is jeopardized.

3 Perspectives

The protection of plant material, especially that used for human consumption, should be a guaranteeing activity that involves different social sectors. However, from the scientific point of view, the development of innovative strategies for the conservation of plant genetic resources should be a priority. It should also be a recurring topic of scientific research and government legislation. Each territory has a list of plant species in critical danger of extinction, to which it is essential to apply conservation strategies such as those described in this summary. Finally, plant germplasm conservation alone will not guarantee global food security; it is always necessary to establish techniques for the sustainable use of these genetic resources. These two actions, conservation and sustainable use, will guarantee our food sovereignty in the near future.

References

1. Popova E, Kulichenko I, Kim HH (2023) Critical role of regrowth conditions in post-cryopreservation of in vitro plant germplasm. Biology 12(4):542
2. Salgotra RK, Chauhan BS (2023) Genetic diversity, conservation, and utilization of plant genetic resources. Genes 14(1):174
3. Khoury CK, Brush S, Costich DE, Curry HA, De Haan S, Engels JM, Thormann I (2022) Crop genetic erosion: understanding and responding to loss of crop diversity. New Phytol 233(1):84–118
4. Ali Z, Maryam H, Saddique MAB, Ikram RM (2023) Exploiting genetic diversity in enhancing phenotypic plasticity to develop climate-resilient cotton. Genet Resour Crop Evol 70(5):1305–1320
5. Priyanka V, Kumar R, Dhaliwal I, Kaushik P (2021) Germplasm conservation: instrumental in agricultural biodiversity—a review. Sustainability 13(12):6743
6. Quazi S, Golani T, Capuzzo AM (2021). Germplasm conservation. In: Kumar S (ed) Endangered plants. IntechOpen, London, UK, pp 123-150.

Index

Marco A. Ramírez-Mosqueda et al. (eds.), *Conservation of Plant Genetic Resources*, Methods in Molecular Biology, vol. 3011,
https://doi.org/10.1007/978-1-0716-5130-8

V

W

MIX
Papier aus verantwortungsvollen Quellen
Paper from responsible sources
FSC® C105338

If you have any concerns about our products,
you can contact us on
ProductSafety@springernature.com

In case Publisher is established outside the EU,
the EU authorized representative is:
Springer Nature Customer Service Center GmbH
Europaplatz 3, 69115 Heidelberg, Germany

Printed by Libri Plureos GmbH
in Hamburg, Germany